# NUCLEAR MATERIALS

# PHYSICS RESEARCH AND TECHNOLOGY

Additional books in this series can be found on Nova's website
under the Series tab.

Additional E-books in this series can be found on Nova's website
under the E-books tab.

# MATERIALS SCIENCE AND TECHNOLOGIES

Additional books in this series can be found on Nova's website
under the Series tab.

Additional E-books in this series can be found on Nova's website
under the E-books tab.

PHYSICS RESEARCH AND TECHNOLOGY

# NUCLEAR MATERIALS

## MICHAEL P. HEMSWORTH
### EDITOR

Nova Science Publishers, Inc.

*New York*

**Library of Congress Cataloging-in-Publication Data**

Nuclear materials / editor, Michael P. Hemsworth.
    p. cm.
  Includes index.
  ISBN 978-1-61324-010-6 (hardcover)
  1. Nuclear reactors--Materials. 2. Nuclear chemistry. 3. Nuclear waste. I. Hemsworth, Michael P.
  TK9185.N75 2011
  621.48'33--dc22

Published by Nova Science Publishers, Inc. † New York

# CONTENTS

# PREFACE

In this new book, the authors gather topical research in the study of nuclear materials. Topics discussed include experimental studies in nuclear fuel alloys for research reactors; the removal of arsenic from ground and surface waters using lanthanides; microstructural characterization of zirconium based alloys and current trends in the mathematical modeling and simulation of fission product transport from fuel to primary coolant of PWRs.

Chapter 1 – Arsenic is a toxic element for animals and the majority of plants. The long-term intake of small doses of arsenic has a carcinogenic effect. Concentrations of arsenic in soils and waters can become elevated as a result of application of arsenical pesticides, disposal of fly ash, mineral dissolution, mine drainage, and geothermal discharge. In recognition of the hazards arsenic (As) poses to the welfare of humans and domestic animals, the U.S. Environmental Protection Agency recently lowered the drinking water standard of As from 50 ppb to 10 ppb effective January 2006. Various methods have been proposed for removal of arsenic from water, such as chemical precipitation-coagulation, adsorption, reverse osmosis, electrodialysis, and ion exchange. Chemical precipitation-coagulation is a simple, economical method hence has been widely used for removing arsenic ions from large quantities of wastewater. Chemicals are added to water to form arsenic-containing precipitates of low solubility that are removed by subsequent sedimentation processes. Iron (III), calcium salts, magnesium salts, and aluminum salts are being used as As precipitants or coagulants in water and waste water systems. However, water purification and waste treatment techniques based on precipitation of calcium, magnesium, iron (III) and aluminum arsenates, have been found less effective or unlikely to produce aqueous solutions with arsenic concentrations below the guideline values proposed for arsenic dissolved in potable water and treated sewage effluents. On the other hand, surface precipitation/co-precipitation of radionuclide with secondary solids is currently thought to be important process limiting radionuclide solution concentrations. Lanthanides are radioactive, long-lived, and highly toxic. Lanthanides also have high selectivity for arsenic. Arsenic (V) and arsenic (III) of lanthanum salts possess extremely lower solubility compared to the iron (III), aluminum, calcium and magnesium salts of arsenic (V) and arsenic (III). Salts of lanthanides are also cheap and found in abundance in nuclear waste piles. This method also effectively removes the radioactive lanthanides from aqueous solutions.

In this chapter an attempt was made to use lanthanides as a new precipitant for removing arsenic from water and waste water; such compounds have not been used efficiently by the industry in spite of their abundance. The reactions of lanthanide ions with arsenic (V) ions were studied in detail. Interactions of arsenic (V) ions with lanthanides were studied with the aim of developing a new precipitation method for the removal of arsenic as well as the lanthanides. Ability of lanthanides to remove arsenic was evaluated and compared, under the same conditions, with those of iron (III), aluminum, calcium and magnesium salts. The solubility products of 15 lanthanide arsenates in aqueous solution have been determined at $25 \pm 1°C$. The most soluble compound is lanthanum arsenate (pK, = 21.45). The least soluble is scandium arsenate (pK, = 26.72). A predictive model was developed to study the effect of soil parameters on adsorption of arsenic from ground water. A co-precipitation model for removal of arsenic (V) using lanthanides showed that surface precipitation of arsenates of lanthanides on hydroxides of lanthanides can immobilize both arsenic and lanthanides. Lanthanides are shown as effective precipitants of arsenic (V) from aqueous solutions.

Chapter 2 – We present here a review of our recent years work and some new results in phase stability determination in uranium alloys, as a contribution to progress in nuclear fuel alloys for research reactors development. In the frame of RERTR Program (Reduced Enrichment for Research and Test Reactors) it is being developed a high density uranium based fuel that could remain stable in the body cubic centered (bcc) phase during fabrication and irradiation in the reactor. Research is focused in an U-Mo alloy dispersed fuel in aluminum matrix. The main problem focuses in an undesirable growth of the interface between fuel and Al matrix. This problem could be reduced with the addition of Si or/and Zr. Our efforts have been devoted to enlighten phase stability knowledge. U-Mo, U-Al and pseudo-binary $UAl_3$-$USi_3$ systems have been studied with DFT first principles methods to establish their ground states. U-Mo finite temperature bcc phases equilibria have been determined from first principles calculations applying the cluster variation method, while MonteCarlo simulations were applied for the finite temperature study of $UAl_3$-$USi_3$ system. U-Al system was also interest from a kinetic point of view since compounds formation and growth are crucial for nuclear fuel performance. Thus, phases equilibria have been described with a semiempirical CALPHAD method capable of being used as input for kinetic calculations and diffusion parameters have also been evaluated so as to use DICTRA package to simulate interphase growth in a $UAl_3$-Al diffusion couple. Finally, the effect of the addition of Zr and Si has been analyzed from an experimental approach. Overall results have succeeded in reproducing known data in cases where it was available and to provide with novel insights of bonding interactions in the alloys formation as well as new valuable experimental data.

Chapter 3 – A majority of radiation effects studies are connected with creation of radiation-induced defects in the crystal bulk, which causes the observed degradation of material properties. The main objective of this chapter is to describe the mechanisms of recovery from radiation damage, which operate during irradiation but are usually obscured by the concurrent process of defect creation. Accordingly, the conventional rate theory is modified with account of radiation-induced Schottky defect formation at extended defects, which often acts against the mechanisms based on the Frenkel pair production in the crystal bulk. The theory is applied for the description of technologically and fundamentally important phenomena such as irradiation creep, radiation-induced void "annealing" and the void ordering.

Chapter 4 – The fuel irradiation test (B14) was carried out in the experimental fast reactor Joyo from May 8 to 11, 2007, to evaluate the thermal behaviors of low-density uranium and plutonium mixed oxide fuels containing several percent of americium (Am-MOX fuels). The test objective is to research early thermal behaviors of Am-MOX fuels such as fuel restructuring and radial redistribution of actinides. The test results are expected to reduce the margin of the thermal design for minor actinide (MA)-MOX fuels.

Pellet-cladding gap width and the oxygen-to-metal (O/M) molar ratio of oxide fuels were specified as experimental parameters. The contents of Pu and Am in the fuel pellets were 31 wt% and 2.4 wt%, respectively. Four fuel pins were irradiated step-by-step in consideration of fuel restructuring during 48 h as a preconditioning before full power reactor operation. The irradiation history, i.e. linear power, was simulated using the conventional FBR oxide fuel pins, and to simulate the transient condition, the linear power was rapidly increased to 470 W $cm^{-1}$ for 10 min.

After the irradiation, non-destructive and destructive post-irradiation examinations were conducted. The fuel restructuring (microstructure change) was examined with an optical microscope and a scanning electron microscope, and the radial redistribution of actinides was measured with an electron probe micro analyzer. Ceramography specimens were taken from the axial position of each fuel pin where the fuel centerline temperature reached the maximum during irradiation.

The influences of both pellet-cladding gap width and O/M molar ratio on the fuel restructuring were observed, but no fuel melting was seen. Fuel restructuring and radial distributions of actinides were investigated relative to those of other irradiated fuels. It seemed that fuel restructuring before the transient condition due to preconditioning was enough to prevent fuel melting. Under the transient condition, additional radial migration of lenticular voids toward the fuel center resulted in an insignificant increase of the central void diameter. The degree of fuel restructuring was found to be more dependent on the linear heating rate than on the small addition of americium and on the fuel O/M molar ratio. In the fuel pin with a large as-fabricated gap width, significant fuel relocation and off-centered fuel restructuring were observed, but no enhancement in the maximum fuel temperature was identified.

The radial profiles of Am and Pu contents indicated that these elements had similar redistribute tendencies in the columnar grain region. The extent of increase in Am and Pu contents around the central void showed a significant dependence on the O/M molar ratio. The fuel having a higher O/M molar ratio had a larger redistribution of Am and Pu towards the central void. In addition the difference in the Am and Pu contents was larger for the fuel having a higher O/M molar ratio.

Chapter 5 – Zirconium based alloys are extensively used in Pressurized Heavy Water Reactor (PHWR) as structural materials due to their low neutron absorption cross-section, acceptable corrosion resistance in hot pressurized water, adequate mechanical properties at elevated temperature and irradiation induced creep and growth resistance properties. Among them, Zircaloy-2, Zircaloy-4 and Zr-2.5% Nb are the main alloys of zirconium used in Pressurized Heavy Water Reactor (PHWR). These alloys are used for fabricating the structural materials like fuel tube, coolant tube, calandria tube, spacers, garter spring, end caps and end plates etc. Several advanced alloys like Zr-1%Nb, Zr-1%Nb-1%Sn-0.1Fe has been designed as candidate materials for fuel cladding tube to withstand higher burn up of the fuel and are found to possess desired combination of strength, higher creep and growth resistance, corrosion resistance and fabricability. The initial microstructures of the fuel cladding tubes and pressure tubes are of great concern from the point of view of their structural integrity as well as irradiation induced creep and irradiation induced growth resistance properties inside the reactor. The pre-irradiation dislocation network is thought to contribute to irradiation growth at high neutron fluencies. This growth rate increases as the degree of cold work increases and has been reported to vary with the dislocation density. The mechanical behavior of these alloys also gets altered by the irradiation induced defects like vacancy, vacancy clusters, dislocation loops etc. As a result, irradiation hardening occurs, yield strength and tensile strength of the materials increase and ductility drastically reduces. All these structure-sensitive properties are dependent on the microstructure of the materials, mobility of vacancies, the interaction of dislocations with point defects, formation of dislocation substructure or cell etc. Therefore it is of utmost importance to characterize the microstructure of these alloys, their defect states, dislocation density, stacking and growth faults. The present article mainly deals with the characterization of several microstructural imperfections introduced by cold deformation and irradiation using light ion (proton) and heavy ions ($O^{5+}$, $Ne^{6+}$) in zirconium based alloys. The microstructural parameters like domain size, microstrain within the domains, density of dislocations, stacking faults and twin or growth fault probabilities have been characterized using X-Ray Diffraction Line Profile Analysis (XRDLPA).

Chapter 6 – The primary coolant activity has been of great concern for many decades mainly due to the persisting high post-shutdown radiation levels resulting in prolongation of reactor down time entailing substantial economic repercussions. Along with the activation products, the fission products released from defective fuel pins contribute towards these radiation fields. Both static and kinetic models have been prepared for the estimation of fission product activity in the primary coolant in steady-state as well as for power and flow rate transients. These models are based on two-stage processes: fuel-to-gap, and gap-to-coolant releases. Here, we present a review of multi-step models based on mass transport equations and models with details on the fission product activity transport to coolant using the FPCART code. In these simulations, the primary circuit has been treated using one-dimensional nodal scheme. The resulting set of governing kinetics equations have been implemented in the FPCART code which uses the LEOPARD and the ODMUG codes as subroutines.

For steady-state operation, the predictions of the FPCART code have been found in good agreement with the results obtained using the ODIGEN-2.0 code with maximum difference in the corresponding values below 0.36% for over 39 dominant fission products studied. Also, the FPCART predictions agree within 2.4% with the corresponding data from the ANS-18.1 standard as well as with the experimentally measured power plant data.

According to FPCART simulations, the fission product activity in primary coolant has strong dependence on flow rate. In the case of slow flow rate coast-downs ($\tau = 2000$ h), up to 8.6% increase in the fission product total specific activity has been observed. The power transients result in substantial changes in the fission product activity. The FPCART predicted values of $^{131}$I activity have been found in good agreement with the corresponding measured data from the Beznau and the Surry PWRs.

Due to inherent randomness of the fuel failure process, the stochastic simulation technique has also been found suitable for predicting the fission product activity in primary coolants of PWRs. For this purpose, three-step based model has been developed and implemented in the FPCART-ST code which simulates the burst release of fission products from random fuel failure. The predicted values of specific activities using stochastic simulations have been found in good agreement with the corresponding measured data of the ANGRA-I nuclear power plant as well as that of EDITHMOX-1 experiments.

Chapter 7 – Environmental sustainability and higher burn-up of nuclear fuel rely on new cladding materials capable of withstanding harsh environmental influence without exhibiting significant degradation of service properties. In order to optimize these materials for safe operating conditions it is important to quantify possible detrimental effects. The experimental work necessary to reach this objective is often time-consuming and costly. At the same time, due to their linear and/or temporal scale not all phenomena are amenable to experimental investigation. It is therefore essential to complement experiments by using mathematical modelling and scientific computing.

In our research we have focused on systematic study of radiation effects in zirconium that is widely used as a cladding material in modern commercial light water nuclear reactors. Large-scale molecular dynamics (MD) modelling is conducted for an investigation of primary damage creation, self-interstitial and vacancy clusters formation, and their stability in high energy displacement cascades. The simulations are carried out for a wide range of temperatures ($100 \text{ K} \leq T \leq 600 \text{ K}$) and primary knock-on atom (PKA) energies $5 \text{ keV} \leq E_{pka} \leq 25 \text{ keV}$. This study of over 300 cascades is the largest yet reported for this metal. At least 25 cascades for each ($E_{pka}$, T) pair are simulated in order to ensure statistical reliability of the results. The high number of simulations for each condition of temperature and energy has revealed the wide variety of defect clusters that can be created in cascades. Mobile or sessile, two-dimensional or three-dimensional clusters of both vacancy and interstitial type can be formed. The number of Frenkel pairs, population statistics of clusters of each type, the fraction of vacancies and self-interstitial atoms (SIA) in point defect clusters, cluster per cascade yield etc. are obtained and their dependence on the temperature and PKA energy is investigated. Strong spatial and size correlations of SIA and vacancy clusters formed in displacement cascades are observed. Both vacancy and SIA clusters can be mobile. However, depending on their type self-interstitial clusters exhibit one-dimensional, planar or three dimensional motions, whereas vacancy clusters of only one type can glide and in one dimension only. Separate MD simulations of some SIA and vacancy clusters are performed to study their thermal stability and possible transformations.

Typical clusters of point defects found in displacement cascades are extracted for investigation of radiation hardening of zirconium. Two vacancy loops (in the basal plane and prism plane) and three SIA clusters (a SIA dislocation loop in prism plane, a small triangular extrinsic fault in the basal plane and a disordered three dimensional SIA cluster) are considered. Atomic-scale details of the interaction of typical vacancy and SIA clusters placed in the gliding planes of two edge dislocations, $1/3[11\bar{2}0](0001)$ and $1/3[11\bar{2}0]\{1\bar{1}00\}$, are studied by large-scale MD modelling. Interaction mechanisms ranging from cluster dragging to partial or complete absorption of point defect clusters accompanied by dislocation climbing up or down are revealed. Stress–strain curves and the critical stress for a dislocation breakaway from a cluster are obtained for all the configurations studied.

In: Nuclear Materials
Editor: Michael P. Hemsworth, pp. 1-46

ISBN 978-1-61324-010-6
© 2011 Nova Science Publishers, Inc.

*Chapter 1*

# CO-PRECIPITATION MODEL COUPLED WITH PREDICTION MODEL FOR THE REMOVAL OF ARSENIC FROM GROUND AND SURFACE WATERS USING LANTHANIDES

*Anpalaki J. Ragavan[1,2] and Dean V. Adams[2]*
Departments of Electrical and Biomedical Engineering[1],
Civil and Environmental Engineering[2],
University of Nevada, Reno, Nevada, USA

## ABSTRACT

Arsenic is a toxic element for animals and the majority of plants. The long-term intake of small doses of arsenic has a carcinogenic effect. Concentrations of arsenic in soils and waters can become elevated as a result of application of arsenical pesticides, disposal of fly ash, mineral dissolution, mine drainage, and geothermal discharge. In recognition of the hazards arsenic (As) poses to the welfare of humans and domestic animals, the U.S. Environmental Protection Agency recently lowered the drinking water standard of As from 50 ppb to 10 ppb effective January 2006. Various methods have been proposed for removal of arsenic from water, such as chemical precipitation-coagulation, adsorption, reverse osmosis, electrodialysis, and ion exchange. Chemical precipitation-coagulation is a simple, economical method hence has been widely used for removing arsenic ions from large quantities of wastewater. Chemicals are added to water to form arsenic-containing precipitates of low solubility that are removed by subsequent sedimentation processes. Iron (III), calcium salts, magnesium salts, and aluminum salts are being used as As precipitants or coagulants in water and waste water systems. However, water purification and waste treatment techniques based on precipitation of calcium, magnesium, iron (III) and aluminum arsenates, have been found less effective or unlikely to produce aqueous solutions with arsenic concentrations below the guideline values proposed for arsenic dissolved in potable water and treated sewage effluents. On the other hand, surface precipitation/co-precipitation of radionuclide with secondary solids is currently thought to be important process limiting radionuclide solution concentrations. Lanthanides are radioactive, long-lived, and highly toxic. Lanthanides also have high selectivity for arsenic. Arsenic (V) and arsenic (III) of lanthanum salts possess extremely lower solubility compared to the iron (III), aluminum, calcium and magnesium salts of arsenic (V) and arsenic (III). Salts of lanthanides are also cheap and found in abundance in nuclear waste piles. This method also effectively removes the radioactive lanthanides from aqueous solutions.

In this chapter an attempt was made to use lanthanides as a new precipitant for removing arsenic from water and waste water; such compounds have not been used efficiently by the industry in spite of their abundance. The reactions of lanthanide ions with arsenic (V) ions were studied in detail. Interactions of arsenic (V) ions with lanthanides were studied with the aim of developing a new precipitation method for the removal of arsenic as well as the lanthanides. Ability of lanthanides to remove arsenic was evaluated and compared, under the same conditions, with those of iron (III), aluminum, calcium and magnesium salts. The solubility products of 15 lanthanide arsenates in aqueous solution have been determined at 25±1°C. The most soluble compound is lanthanum arsenate (pK, = 21.45). The least soluble is scandium arsenate (pK, = 26.72). A predictive model was developed to study the effect of soil parameters on adsorption of arsenic from ground water. A co-precipitation model for removal of arsenic (V) using lanthanides showed that surface precipitation of arsenates of lanthanides on hydroxides of lanthanides can immobilize both arsenic and lanthanides. Lanthanides are shown as effective precipitants of arsenic (V) from aqueous solutions.

# 1. INTRODUCTION

Arsenic has received a great deal of public attention because of its link to certain types of cancers and its high levels in some drinking water supplies (Nordstrom, 2002; Hopenhayn, 2006). Studies on long-term human exposure have shown that arsenic in drinking water is associated with liver, lung, kidney, bladder cancers and skin cancers as well (Wu et al., 1989). In vitro studies have indicated that arsenic can interfere with DNA replication, DNA repair and cell division. Earlier studies have shown that arsenic can inhibit ligation of DNA strand breaks and if present during DNA synthesis, it can induce chromosomal aberrations, sister chromatid exchanges and mal-segregation of chromosomes (Natarajan et al., 1996). After a few years of continued low level of arsenic exposure, many skin ailments such as hypo-pigmentation (white spots), hyper-pigmentation (dark spots), collectively called Melanosis or dyspigmentaion (as called by some physicians) and keratosis (breakup of the skin on hands and feet) appear. After a latency of about 10 years, skin cancers appear. After a latency of 20 - 30 years, internal cancers - particularly bladder and lung appear. Death is mainly due to internal cancer. Some of the skin ailments caused by arsenic poisoning are shown in Figure 1. The compliance standard for arsenic is around 10 (μg/L) ppb and is enforced by EPA and other regulatory agencies. However, the maximum permissible limit of arsenic in drinking water of Bangladesh is 0.05 mg/l. The main source of arsenic in drinking water is arsenic-rich aquifer rocks in which the water is stored. It may also occur because of mining or industrial activities in some areas (WHO, 2001). Arsenic poisoning due to excessive exposure to natural and anthropogenic arsenic in drinking water has been reported in several countries including Bangladesh, Argentina, China, Taiwan, Thailand, India, Mexico, USA, Ghana, Hungary, Romania, United Kingdom, Chile, New Zealand and Vietnam (Smedley and Kinniburgh, 2002). The number of people approximately affected in many of these countries is listed in Table 1. (http://www.newscientist.com/news/). The latest statistics available on the arsenic contamination in groundwater indicate that 52 districts around 80% of the total area of Bangladesh an estimate, which around 40 million people are at risk.

Arsenic contamination is very dangerous because it is impossible to detect the presence of arsenic without special chemical tests. In water arsenic is tasteless, colorless and odorless even at high concentrations. Inorganic arsenic can occur in the environment in several forms but in natural waters it is mostly found as trivalent arsenite (As(III)) or pentavalent arsenate (As(V)) depending on the oxidation capacity of the water. As(III) species are more toxic than As(V) species. Arsenate species are predominant at moderate and high redox potentials, while arsenite species occur under more reducing conditions. Organic arsenic species, abundant in seafood, are very much less harmful to health, and are readily eliminated by the body (WHO, 2001).

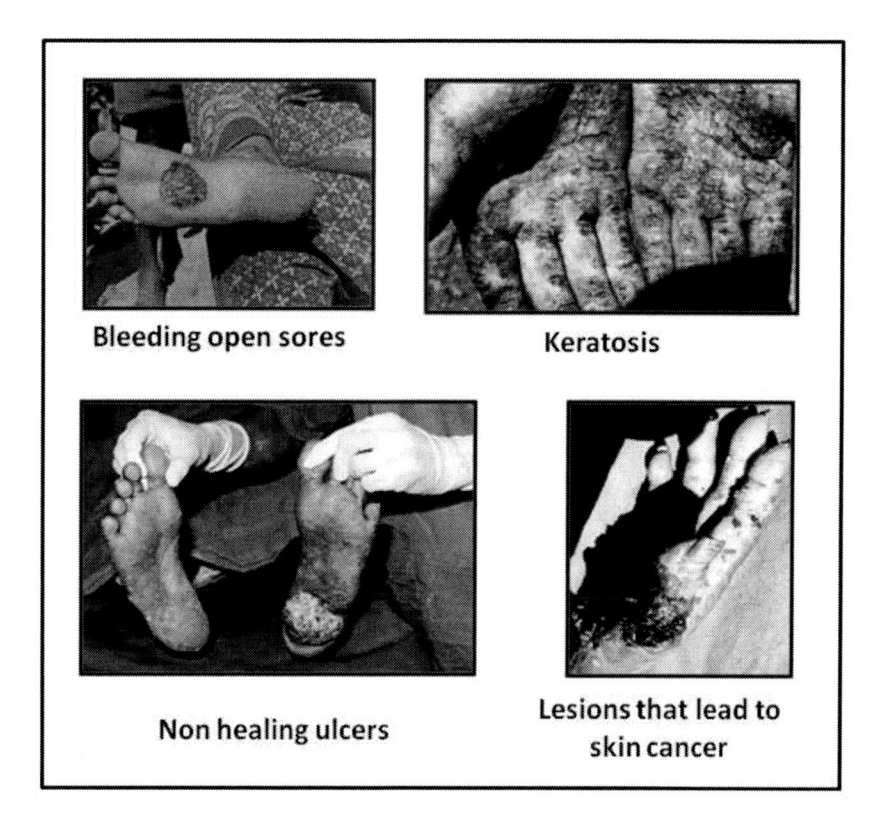

**Bleeding open sores**          **Keratosis**

**Non healing ulcers**          **Lesions that lead to skin cancer**

Figure 1. Images of bleeding open sores, keratosis, non-healing ulcers and lesions that lead to skin cancers caused by arsenic poisoning (WHO, 2001).

## 1.1. Sources of Arsenic in Ground Water

Arsenic (As) is a ubiquitous trace element found throughout the environment. Arsenic is released into ground water due to weathering of rock, and geochemical reactions as well as due to industrial waste discharges (from coal fired thermal fire plants, petroleum refining processes and ceramic industries), agricultural activities (fertilizers, arsenical pesticides, arsenical herbicides) and domestic discharges. Arsenic concentrations in groundwater vary greatly due to the heterogeneous distribution of source materials and subsequent geochemical controls on aqueous As mobility in aquifers (Cullen and Reimer, 1989). The causes of elevated As concentrations in groundwater, including the complex interactions between water, geologic substrate and biological processes, are not yet completely understood. Recent work around the globe, in such locations as West Bengal (Nickson et al., 2000), Bangladesh (Ahmed et al., 2004), Vietnam (Berg et al., 2001), Spain (Garcia-Sanchez et al., 2005), Argentina, Mexico, Chile, China, Ghana, Hungary, Taiwan and Thailand (Smedley and Kinniburgh, 2002) underscore how little was known about As concentrations in groundwater around the world until recently. While for many years it was believed that the only sources of arsenic contamination in the USA were arsenic in mine tailings, and the runoff from agricultural land which had used arsenical pesticides, it has now been realized that there is arsenic in many ground water aquifers some of which are now being tapped for drinking water supplies. Indeed hydrogeologists argue that there is as much arsenic in soil surrounding the aquifers in Massachussets as in Bangladesh; it is merely not easily available in the water, and moreover few people drink water from wells; but there is still a lot of arsenic around (gsw_nawqa_info@usgs.gov). Elevated arsenic (As) in groundwater has been identified within the United States in the West and south west (SW) (Savage et al., 2000; Welch and Lico, 1998), upper Midwest (Schreiber et al., 2000), and New England (Lipfert et al., 2006; Peters and Blum, 2003; Peters et al., 1999). Dissolution and desorption of As from naturally occurring As-containing minerals, geothermal water, and mining activity appear to be the key contributors to high-As ground water provinces within the United States. The fact that long term medium level arsenic exposure leads to chronic health problems was known in 1986, it was basically ignored by EPA until 2000. On a pessimistic calculation using a linear dose response curve the Office of Health and Hazard Assessment (OEHHA) of the California EPA calculated that a level for a one-in-a-million risk is 1.5 parts per trillion - 3000 times less than the implemented level. But it would be impossible to set a regulatory limit this low since no water supply could meet it. After much travail, the US Environmental Protection Agency, finally, in January 2001, issued a regulation with a new standard for drinking water of 10 ppb to take effect in 2006 to protect consumers served by public water systems from the effects of long-term, chronic exposure to arsenic. Unlike nuclear waste that decays or chlorinated hydrocarbons that break down, arsenic lasts forever, so that the disposal is an important issue. Since risks of waste disposal are risks of exposure of a lot of people to small amounts, it becomes very important to discover whether or not the effects on health are linear with exposure at low doses or are non-linear. So far these web masters see no reason why arsenic should not be treated as carefully as nuclear waste. The regulations should not be similar. Inversely if arsenic is let "off the hook" the regulations for nuclear waste should be correspondingly relaxed.

## 1.2. Removal of Arsenic through Adsorption on Solid Surfaces

Arsenate is the predominant species of arsenic in oxidized aqueous systems (Myneni et al., 1998). The concentration of arsenate in natural waters is strongly influenced by adsorption on oxide surfaces (Fuller and Davis, 1989; Fukushi et al., 2003). Consequently, the liquid/solid transition of arsenic is carried out through adsorption (Edwards, 1994). Adsorption is the process by which dissolved ions accumulate at the solid liquid interface. As solids are introduced into water containing dissolved species of arsenic, dissolved ions migrate from bulk solution towards the solid surface due to differences in: i) ion concentration gradient or charge gradient and ii) chemical potential gradient until the differences in concentration, charge and/or chemical potentials disappear. Consequently, ion complexes are formed on the solid surfaces and the adsorbed ions become unavailable for migration. The complexes formed between the adsorbate (toxic metal ions/ trace elements) and the adsorbent (solid surface site ions) that accumulate (Figure 2) at the interface are filtered out or allowed to settle and removed by coagulation and filtration processes.

### Table 1. Number of people currently affected worldwide by arsenic poisoning by country

| Country | # People affected |
| --- | --- |
| United States | Unknown |
| Mexico | 400,000 |
| Chile | 437,000 |
| Bolivia | 20,000 |
| Argentina | 2,000,000 |
| Hungary | 20,000 |
| Romania | 36,000 |
| India | 1,000,000 |
| Bangladesh | 50,000,000 |
| Thailand | 1,000 |
| Vietnam | Millions |
| Taiwan | 200,000 |
| China | 720,000 |
| Nepal | Unknown |

(source:www.newscientist.com).

There are several water quality parameters, which have been demonstrated to affect adsorption process such as: i) type and dosage of adsorbent, oxidation state of arsenic, ii) type and dosage of oxidant, iii) pH (Montiel and Welté,1996), iv) organic matter content (Kelemen, 1991), v) phosphorous (Holm, 2002; Manning and Goldberg, 1996), vi) inorganic carbon (carbonate/bicarbonate) (Holm, 2002), vii) silicate concentration (Liu et al., 2006; Swedlund and Webster, 1999) in the water and the viii) surface area of the solid (Goldberg et al., 2005). In spite of the large number of water quality parameters affecting arsenic removal efficiency, in water treatment systems often the needed arsenic:adsorbent ratio is used as the only parameter to achieve the goal of 10µg/L arsenic removal concentration. To predict the migration and long term fate of arsenate in natural environments the behavior and nature of adsorbed arsenate species must be known for a wide variety of minerals and over a full range of environmental conditions.

On the other hand the removal of trace amounts of arsenic from industrial process water and drinking water is a challenging problem. Overall, the X-ray studies on powders and single crystals provide strong indications of a variety of adsorbed arsenate species, presumably present under different conditions of pH, ionic strength and surface coverage, as well as being present on different solids and crystallographically different surfaces. Even on a single crystal surface, a great variety of possible surface species coordination geometries have been proposed, e.g. monodentate, bidentate and tridentate possibilities (Waychunas et al., 2005). A variety of possible protonation states adds even further potential complexity to the speciation of adsorbed arsenate. It is now widely recognized that the evidence of arsenic speciation from spectroscopic studies should be integrated with models describing macroscopic adsorption and electrokinetic data (Suarez et al., 1997; Hiemstra and van Riemsdijk, 1999; Blesa et al., 2000; Goldberg and Johnston, 2001). Subsequently the adsorption of arsenic should incorporate predictive models using a full range of environmental factors with spectroscopic measurements. Incorporating molecular calculations, as well as the reaction stoichiometries in surface complexation models can possibly help. Previous researchers who have developed adsorption models for metal oxides have used simplistic steady state conditions. Examples of these include the triple layer, Langmuir, Stern, Grahame, and Stern-Grahame models (Davis and Leckie, 1980; James and Leckie, 1978). The Grahame model is limited for use with surface excesses less than mono layer coverage and the Langmuir isotherm will not account for the formation of binuclear bidentate inner-sphere surface complexes (Davis and Misra, 1997) which are the predominant mechanism by which arsenate adsorbs. Currently there is a need for a better transport model for arsenic adsorption applicable for all solids under a variety of environmental conditions for most process water systems.

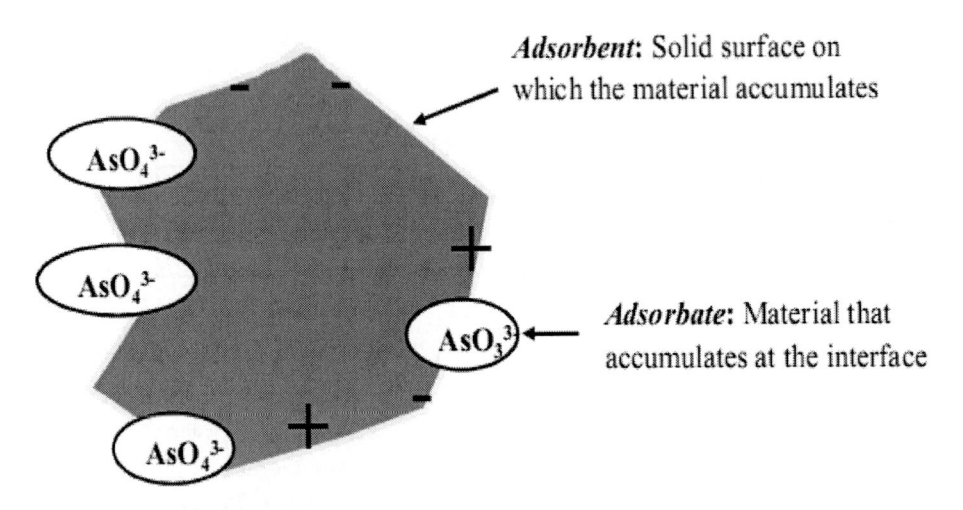

Figure 2. Process of accumulation of arsenic ion species on solid surface during adsorption.

In addition, the best adsorbent for arsenic removal is yet to be discovered. One of the major uses of activated metal oxides is for adsorbing metal or non metal ions from aqueous waste water streams. For instance activated alumina is suggested by EPA as the best available treatment (BAT) for removing arsenic ions from water (SHHWT, 1989). Recently it was observed that lanthanum oxide is a better adsorbent than activated alumina for removal of arsenic from water and waste water systems (Davis and Misra, 1997). In order to meet the imposed standard, it is essentially required to develop a new kind of adsorbent for arsenic removal from drinking and waste water systems (Davis and Misra, 1997).

## 1.3. Objectives of This Chapter

In order to point out the deficiency of general considerations (often recommendations) regarding arsenic removal from water and waste water systems and to analyze adsorption of arsenic as a function of multiple environmental factors and soil mineral properties, the present chapter as the major goal builds a very simple and robust predictive model, indicating the significant contributions of several parameters that affect arsenic adsorption in ground water such as: pH, organic matter content, cation exchange capacity, inorganic carbon content, iron and aluminum concentration in water and the surface area of solid minerals. The predictive model will be able to predict the amount of arsenic adsorbed under a wide range of natural environmental conditions with respect to technological optimization. The results are compared to arsenic adsorption measured spectroscopically on these soil minerals. Ten soil types were used in the development of the model. The second goal of this chapter is to develop a co-precipitation model for the adsorption of arsenic on rare earth hydroxides as a solid solution of hydrous rare earth oxides and rare earth arsenates.

# 2. BACKGROUND

## 2.1. Factors Influencing Arsenic Migration in Natural Waters

### Inorganic Carbon Concentration

Inorganic carbon ($H_2CO_3$, $HCO_3^-$ and $CO_3^{2-}$) is found in all natural waters. Depending on the geochemical conditions in nature, total inorganic carbon concentration (IC) can vary. The carbonate/bicarbonate system plays an essential role in the adsorption process (Stumm and Morgan, 1962). When iron or aluminum adsorbents are available in the water source, due to hydrolyses, metal-hydroxides are formed if the buffering capacity of the water is high enough. Relatively high buffering capacity of the ground/ subsurface waters can ensure the complete transformation of the metal ions to metal-hydroxides (Licskó, 1990). The IC concentration influences the pH after adsorption. Research work by Holm (Holm, 2002) with 1mM and 10mM of IC concentration has demonstrated that IC can have adverse effects on arsenic removal, due to competition between carbonate/ bicarbonate and arsenic for the adsorption sites on ferric-hydroxide.

### Phosphate and Silicate Concentration

If the subsurface water sources contain $PO_4^{3-}$ ions up to 0.8-1.0 mg/L concentration, which may reduce the availability of solid surface for adsorption of arsenic. Since the structure of arsenate and phosphate ions is similar, and they have similar sorption behavior (Dzombak and Morel, 1990), there is a competition between these ions for the adsorption sites at ferric-hydroxide (Holm, 2002; Manning and Goldberg, 1996). Moreover, the $PO_4^{3-}$ concentrations in groundwater can be 10–100 times greater than As. Holm (Holm, 2002) found significantly less As(V) sorption at all pH values for systems containing 0.9 mg/L $PO_4^{3-}$ compared to $PO_4^{3-}$ free systems. Previous research work has also demonstrated that silicate can also significantly inhibit arsenic removal by ferric hydroxides (Liu et al., 2006; Meng et al., 2000; Swedlund and Webster, 1999). According to Liu et al. (2006) the adsorption and surface complexation of silicate on ferric hydroxide results in decreased zeta potential. More significant zeta potential decrease was observed at elevated pH conditions. Decreasing zeta potential increases the repulsive forces between the ferric hydroxide precipitates which inhibits further particle agglomeration and result in the formation of small ferric-hydroxide colloids. Arsenate cannot be associated with these small flocs. They also pass through 0.45μm pore-size membrane filters in treatment systems for arsenic. The presence of $Ca^{2+}$ could buffer the adverse effects of silicate to some extent.

### Organic Matter Content

According to literature data (Kelemen, 1991; Laky and László, 2007) organic matter (OM) concentration of water is one of the most important influencing factors of arsenic removal (Kelemen, 1991). A difference of one order of magnitude in the requirement of coagulant dosages (to reach the 10μg/L standard arsenic concentration) was demonstrated for waters with low ($COD_{Mn}$ =1 mg/L) compared to high ($COD_{Mn}$ =13 mg/L) organic matter content (Kelemen, 1991; Laky and László, 2007).

## 2.2. Aqueous Speciation of Arsenic

Inorganic arsenic (AsIII) is the most toxic form and is commonly associated with groundwater where, according to US Geological Survey (USGS) and other data sources, it predominantly coexists with iron, (Welch et al. 2000). The changes between As(III) and As(V) depends on the redox potential of the subsurface environment and kinetically controlled. Arsenate species are predominant at moderate and high redox potentials, while arsenite species occur under more reducing conditions. Metal arsenites are much more soluble than the corresponding metal arsenates, and the concentrations of arsenic in waters increase with the reduction of As(V) as a consequence of more-soluble arsenite solid phases and lower extent of adsorption. Manganese oxides are very effective in the oxidation of As(III) to As(V) and play an important role controlling arsenic toxicity in the environment as arsenite is around 60 times more toxic for humans than arsenate. As(V) is the most stable species under normal atmospheric conditions, but the oxidation of As(III) by atmospheric oxygen is slow. Mine drainage waters can contain more than 30 % of their total arsenic as arsenite in moderate oxidation conditions. Arsenic can be either oxidized or reduced by bacterial activity in mine waters, aquatic sediments, raw and activated sewages, and soils. Arsenic sulfides are the less-soluble, arsenic-containing solid phases, but they become very soluble due to the oxidation of sulfide that occurs at very low redox potentials and the consequent generation of acid drainage (Clara and Magalhães, 2002).

## 2.3. Surface Speciation of Arsenic

The surface speciation of arsenate on oxide powders and on single crystal surfaces has been studied through X-ray spectroscopic and infrared investigations as well as theoretical molecular calculations. On powders, a variety of surface species have been inferred from extended X-ray absorption fine structure (EXAFS) studies. The majority of these studies have been concerned with iron oxides and a very limited range of ionic strengths. For arsenate on ferrihydrite and FeOOH polymorphs (goethite, akaganeite and lepidocrocite) at pH 8 and I = 0.1, it was inferred that arsenate adsorbed mainly as an inner-sphere bidentatebinuclear complex (Waychunas et al., 1993). Monodentate- mononuclear complexes were also inferred. The monodentate/bidentate ratio decreased with increasing arsenate surface coverage. In contrast, Manceau (1995) reinterpreted the EXAFS spectra as indicating bidentatemononuclear species. The possible existence of such a species, in addition to the ones described by Waychunas et al. (1993) was described in Waychunas et al. (2005). All three of the above species were subsequently inferred for arsenate adsorption on goethite from EXAFS results obtained at pH 6–9 and I = 0.1 (Fendorf et al., 1997). The monodentate-mononuclear species was inferred to dominate at higher pH values and lower surface coverages, whereas the bidentate-binuclear species dominated at lower pH and higher surface coverages. The bidentate-mononuclear species was detected in only minor amounts at high surface coverage. For arsenate on goethite and lepidocrocite at pH 6, a bidentate-binuclear species was inferred (Farquhar et al., 2002). The same species was inferred for arsenate on goethite, lepidocrocite, hematite and ferrihydrite at pH 4 and 7 and I = 0.1 through EXAFS combined with molecular calculations (Sherman and Randall, 2003). For arsenate on hematite, as a function of pH from 4.5 to 8 and surface coverage, Arai et al. (2004) interpreted EXAFS data to infer a predominant bidentate-binuclear and a minor amount of a bidentate-mononuclear species. In the only powder studies of aluminum oxides, EXAFS and XANES of arsenate adsorption on b-Al(OH)$_3$ at pH 4, 8, and 10 and I = 0.1 and 0.8 by Arai et al. (2001) showed that arsenate adsorbs to b-Al(OH)$_3$ as a bidentate-binuclear innersphere species (their XANES data did not indicate any outer- sphere species regardless of pH and ionic strength), and an EXAFS study of arsenate on a-Al(OH)$_3$ at pH 5.5 (in dilute Na-arsenate solutions) also showed that arsenate forms an inner-sphere bidentate-binuclear species (Ladeira et al., 2001). In situ ATR-FTIR studies of arsenate on ferrihydrite and amorphous hydrous ferric hydroxide have inferred that adsorption of arsenate occurred as inner-sphere species (Roddick-Lanzilotta et al., 2002; Voegelin and Hug, 2003). An ex situ FTIR study (Sun and Doner, 1996) has inferred binuclear-bridging complexes of arsenate on goethite. However, the latter noted that their results could be affected by not having bulk water present in their spectroscopic experiment. In addition, IR studies of arsenate adsorption may only show small or no shifts with changes of pH (Goldberg and Johnston, 2001). As a consequence, the assignment of arsenate surface species from infrared and Raman spectra is sufficiently difficult (Myneni et al., 1998) that a definitive in situ study has yet to be done.

## 2.4. Surface Complexation Models for Arsenic Adsorption

Numerous studies have focused on surface complexation modeling of arsenate on oxides (Goldberg, 1986; Dzombak and Morel, 1990; Manning and Goldberg, 1996; Grossl et al., 1997; Manning and Goldberg, 1997; Hiemstra and van Riemsdijk, 1999; Gao and Mucci, 2001; Goldberg and Johnston, 2001; Halter and Pfeifer, 2001; Dixit and Hering, 2003; Arai et al., 2004; Hering and Dixit, 2005). In surface complexation models it is assumed that the adsorption behavior of arsenic in soil can be described by surface complexation reactions written for generic surface functional groups that represent average properties of the soil as a whole rather than of specific mineral phases. Surface complexation modeling has the capability to predict the behavior and the nature of adsorbed arsenate species as functions of environmental parameters. However, the complexation reactions have not always been consistent with in situ spectroscopic results. Instead, with the exception of the study by Hiemstra and van Riemsdijk (1999), regression of macroscopic adsorption data has often been carried out merely to fit the data with the minimum number of surface species (Hering and Dixit, 2005). This is due to the reason that outer-sphere and inner-sphere complexes of the same arsenic are formed simultaneously. Although they are structurally similar, have different chemical characteristics, and reactive properties. Their equilibrium stability constants are significantly different. Currently their sum or just one of the equilibrium constants is used in chemical modeling. To accurately characterize the reactivity of ions at molecular level the two types of complexes need be separated at equilibrium. Their separation is currently done by spectroscopy (EXAFS, ATR-FTIR) under a wide range of reaction conditions with a range of minerals. Sample preparation and drying techniques of the currently available spectroscopic methods affect the number of hydrogen bonds and the structure of the complex formed hence has been found to give discrepant results. Spectroscopy is also expensive, tedious and time consuming.

In addition, surface complexation models (SCM) uses concentrations of ions and treat activities of solid surface ions ideally (activity of surface ions are set equal to unity [=1]) while using mass action to quantify adsorption of ions on solid surfaces. Mass action is strictly valid only when actual activities of surface site ions are used instead of concentrations. SCMs also represent the activities of the arsorbates by electrostatic potentials of the Boltzmann equation which is equivalent to assuming that the surface charge density is independent of pH. Activity coefficients depend on the difference between actual and ideal electrostatic potential not just the electrostatic potential of the Bolzmann equation. The above invalid assumptions made in the SCMs lead to erroneous calculation of adsorption.

## 2.5. Solubility of Rare Earth Arsenates Compared to Arsenates of Other Ions

Metal arsenites are much more soluble than the corresponding metal arsenates, and arsenites are adsorbed less by solid phases. Remediation techniques must consider the available information on solubility and adsorptive properties of As(III) and As(V). New remediation methods must consider solubility data for arsenic-containing materials and minerals since largely soluble arsenates are not suitable for arsenic decontamination. Lanthanum and actinium arsenates are examples of less-soluble metal arsenates that control lanthanides, actinides and arsenate concentrations in natural aquatic systems and can be used for remediation techniques under most conditions. On the other hand, the less soluble lead and barium arsenates are less suitable for arsenic decontamination (Magalhães, 2002).

### *Suitability of Calcium and Magnesium Arsenates*

Calcium oxide and hydroxides are commonly used in soils to treat acid mine drainage and industrial waste waters as they increase final pH and reduce the amount of dissolved matter discarded into aquatic systems. During this process some dissolved arsenic can be precipitated as calcium arsenate solid whose composition is related to the composition of the aqueous solutions (Nishimura et al., 1985). These solids are moderately soluble (Table 2) to be effective in the reduction of arsenic mobility in the environment. According to Bothe and Brown (1999) the lowest value for arsenic concentration in equilibrium with solid calcium arsenates is 0.01 mg/L arsenic, at pH 12.6 in closed systems. In addition, introduction of air with carbon dioxide can cause significant instability of calcium arsenates at pH > 8.3 in the $Ca–As–H_2O$ system (Nishimura et al., 1985) as calcium carbonate is the stable solid phase above this pH value. At pH values between 4.5 and 8.5, total arsenate concentrations around 200 times higher than the maximum contaminant level for total arsenic in potable water and treated sewage effluents and wastes, have been reported for aqueous solutions in equilibrium with calcium arsenates, as indicated by the presence of the rare calcium arsenate minerals such as weilite (CaHAsO4), pharmacolite ($CaHAsO_4·2H_2O$), haidingerite ($CaHAsO_4·H_2O$), and phaunouxite ($Ca_3(AsO_4)_2·11H_2O$) (their presence indicate high concentrations of total dissolved arsenate, calcium concentrations in the environment are usually controlled by equilibria with other less-soluble, calcium-containing solid phases). The change of calcium arsenate to calcium carbonate will also release arsenic in the environment (Magalhães, 2002).

Magnesium salts are used like calcium arsenates to promote arsenic fixation in soils, sediments, and wastes, although to a lesser extent. In the point of view of solubility they are similar to the calcium arsenates (Table 2). Hoernesite [$Mg_3(AsO_4)_2·8H_2O$] has been reported in arsenic-contaminated soils and in toxic waste sites, that contained high magnesium content in the ground water Voigt et al. (1996).

**Table 2. Total concentration of arsenate in aqueous solutions in equilibrium with calcium arsenates, magnesium arsenate, and iron(III) arsenates**

| Solid Phase | Temperature (K) | pH[a] | Arsenate$_{total}$/(mol/L)[a] |
|---|---|---|---|
| $CaHAsO_4·H_2O$ | 308 | acid | 0.12–1.2 |
| $Ca_3(AsO_4)_2$ | 293 | 6.90–8.35 | $1.5 \times 10^{-2} – 3.5 \times 10^{-3}$ |
| $Ca_3(AsO_4)_2·4.25H_2O$ | 296 | 7.32–7.55 | $1.1 \times 10^{-2} - 6.5 \times 10^{-3}$ |
| $Ca_{10}(AsO_4)_6(OH)_2$ | 310 | 5.56–7.16 | $7.5 \times 10^{-3} - 4.4 \times 10^{-4}$ |
| $Ca_{10}(AsO_4)_6Cl_2$ | 310 | 4.67–7.42 | $1.9 \times 10^{-3} - 3.7 \times 10^{-5}$ |
| $Mg_3(AsO_4)_2$ | 293 | 6.50–7.40 | $1.5 \times 10^{-2} - 4.6 \times 10^{-3}$ |
| $FeAsO_4$ | 293 | 1.90–2.95 | $3.7 \times 10^{-3} - 8.5 \times 10^{-5}$ |
| $FeAsO_4·2H_2O$ | 298 | 5.53–6.36 | $1.4 \times 10^{-4} - 2.5 \times 10^{-5}$ |

Source: Magalhães, 2002.

## *Suitability of Iron (III) Arsenates*

Iron(III) arsenates are too soluble, to provide an adequate decrease of arsenic in natural waters from precipitation processes (Table 2). Iron(III) arsenates are also more soluble than calcium arsenates when redox potentials are lower than zero (0) volts (Rochette et al., 1998; Meng et al., 2001) due to iron and arsenic reduction. Formation of iron (III) arsenates are very pH- and pE-dependent. The redox equilibria of iron and arsenic species (pH and pE dependence of the reduced and oxidized iron species and its relation with dissolved As(III) and As(V) species) are shown in Figure 3 (from Magalhães, 2002). Iron(III) oxides and oxyhydroxides are the stable solid phases at pH > 3. At pH less than 3, scorodite ($FeAsO_4·2H_2O$) has been reported to be widespread in arsenic-bearing ore deposits with high total concentrations of iron and arsenic (iron total activity around $10^{-3}$, with arsenate activity greater than $10^{-1.29}$) (Dove and Rimstidt, 1985). Their solubility could control the concentration of arsenates in natural waters. Scorodite was stable under oxidized conditions. Incongruent dissolution of scorodite with its transformation into iron(III) hydroxides has been observed (Dove and Rimstidt, 1985). Increased arsenic concentrations of tailings solutions with time from the breakdown of mainly scorodite has been reported too (Langmuir et al., 1999). This phase change reaction was very rapid at first and approached zero after 72 h at 298 K (Langmuir et al., 1999). According to the information presented above, concentrations of total arsenates in aqueous solutions, in equilibrium with iron(III) arsenates can be around 200 times higher than the maximum contaminant level for total arsenic in potable water.

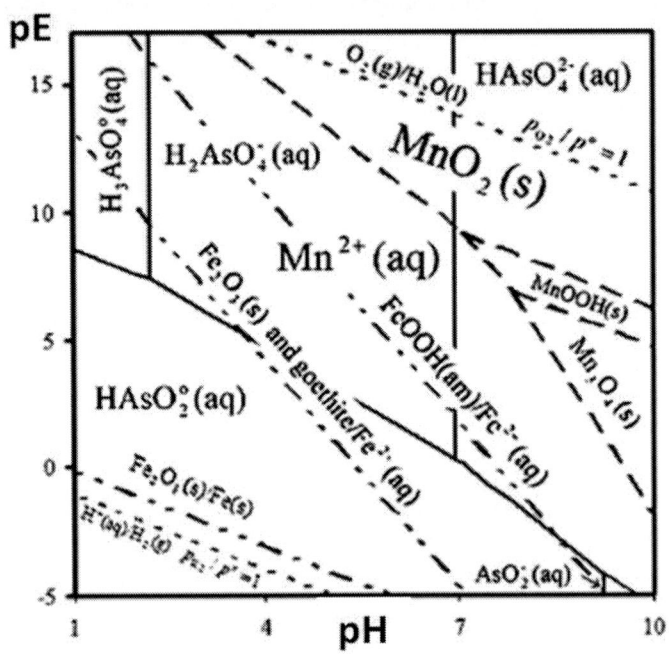

Source: Magalhães, 2002.

Figure 3. Arsenic (⊢——) iron (———) and manganese (--- ----) redox equilibria (pH-pE) diagram in environmental arsenic related processes (298.15K, TOTFe = TOTMn = 10-5 mol/L).

### Suitability of Lead Arsenates

Soils and sediments containing lead arsenates will pose high environmental problems of lead and arsenate bioavailability if they are in contact with phosphate-containing solutions. Lead arsenates are very low in solubility and will assure trace concentrations of these two elements in the aquatic systems under oxidizing conditions. However, the use of phosphates (as fertilizers or in soil amendments) induces the arsenate release and increases its mobility. It is observed that the arsenate released to aqueous solutions by phosphate amendment reverts to sparingly soluble solid phases (Peryea, 1991). Arsenate–phosphate co-precipitation has been usually observed. Aqueous systems with pH > 7 with lead and arsenate total concentrations lower than $10^{-7}$ mol/L have been observed. In very acidic solutions (pH < 2.5) and at chloride concentrations lower than $10^{-3}$ mol/L, lead and arsenate concentrations can be controlled by the equilibrium of aqueous solutions with schultenite ($PbHAsO_4$). An aqueous solution in equilibrium with schultenite at low pH (3.2) have been found to contain around $5 \times 10^{-5}$ mol/L of total arsenate (Magalhães and Jesus, 1988). These considerations show the importance of solubility in the design of programs for environmental remediation related to lead and arsenate. The crystallization of aluminum arsenophosphates solid solutions is also reported (Peryea, 1991).

Mimetite is the most stable lead arsenate phase in the pH range of natural waters. The crystallization of mimetite is a method already in use to remove dissolved arsenic from aqueous solutions to final arsenic concentrations lower than 0.2 µg/L (Bothe and Brown, 1999). Mimetite ($Pb_5(AsO_4)_3Cl$) is isostructural with pyromorphite ($Pb_5(PO_4)_3Cl$). All members of the pyromorphite-mimetite solid solution have the same value for the solubility constant, at 298.15 K, within experimental error (Inegbenebor et al., 1989). The composition of the solid solution is congruent with the composition of the aqueous solution in relation to the relative total amounts of dissolved phosphates and arsenates.

### Suitability of Arsenates of Lanthanides and Actinides

The rare-earth and trivalent actinide arsenates are very insoluble in water. Saturated solutions of rare-earth arsenates in pure water would contain a rare-earth concentration of about $10^{-11}$ mol dm$^{-3}$. In the saturated solution there is only one mathematically, significant form of the rare-earth ion and that is $Ln^{3+}$. Calculations indicate that other possible species such as $Ln(OH)^{2+}$ are not present in significant quantities at the pH conditions in the saturated solutions. $H_3AsO_4$ is the most significant form of arsenate in the saturated solutions. A smaller fraction of the total concentration of arsenate is present as $H_2AsO_4^-$. The concentration of the free arsenate anion $AsO_4^{3-}$ cannot be analytically determined, and therefore it cannot be used directly in the $K_{sp}$ expression. However, the concentration of the arsenate anion can be calculated by using the total concentration of arsenate and the weak acid dissociations of arsenic acid. An equilibrium exists between the rare-earth cation and the forms of arsenate present in solution: $AsO_4^{3-}$, $HAsO_4^{2-}$, $H_2AsO_4^-$, and $H_3AsO_4$. The equilibrium between solid rare-earth and actinide arsenate and the dissolved ion can be described as:

$$LnAsO_4(s) = Ln^{3+}(aq) + AsO_4^{3-}(aq) \tag{1}$$

where Ln is any rare-earth or trivalent actinide element. The solubility product ($K_{sp}$) expression for this equilibrium is given as:

$$K_{sp} = [Ln^{3+}][AsO_4^{3-}] \tag{2}$$

Using the following dissociation constants of arsenic acid (Fukushi and Sverjenski, 2007) and analytical concentrations of total arsenate, total Ln and total acidity obtained from Firsching, (1992) and Firsching, (1991) the solubility products of arsenates and of the structurally similar phosphate species of rare earth species were calculated in this paper. The model equations used in the calculations of solubility products are listed below.

$$H^+ + H_2AsO_4^- = H_3AsO_4^0, K_1 = 6.00 \text{ x } 10^{-3} \tag{3}$$

$$H^+ + HAsO_4^{2-} = H_2AsO_4^-, K_2 = 1.05 \text{ x } 10^{-7} \tag{4}$$

$$H^+ + AsO_4^{3-} = HAsO_4^{2-}, \quad K_3 = 3.00 \times 10^{-12} \tag{5}$$

The analytical concentration of total arsenate and similar phosphates are given by:

$$CT_{AsO_4} = [H_3AsO_4] + [H_2AsO_4^-] + [HAsO_4^{2-}] + [AsO_4^{3-}] \tag{6}$$

$$CT_{PO_4} = [H_3PO_4] + [H_2PO_4^-] + [HPO_4^{2-}] + [PO_4^{3-}] \tag{7}$$

$$\alpha_3 = \frac{MO_4^{3-}}{CT_{MO_4}} = \frac{K_1K_2K_3}{[H^+]^3 + [H^+]^2K_1 + [H^+]K_1K_2 + K_1K_2K_3} \tag{8}$$

where M is either As or P species. Suitable algebraic manipulations yield the solubility product ($K_{sp}$) of rare earth arsenates or equally phosphates to be equal to:

$$K_{sp}(MO_4) = \frac{[Ln^{3+}] \, CT_{MO_4}}{\frac{[H^+]^3}{K_1K_2K_3} + \frac{[H^+]^2}{K_2K_3} + \frac{[H^+]}{K_3} + 1} \tag{9}$$

The dissociation constants of phosphoric acid selected were as follows (Firsching, 1991):

$$H^+ + H_2PO_4^- = H_3PO_4^0, \quad K_1 = 7.11 \times 10^{-3} \tag{10}$$

$$H^+ + HPO_4^{2-} = H_2PO_4^-, \quad K_2 = 6.34 \times 10^{-8} \tag{11}$$

$$H^+ + PO_4^{3-} = HPO_4^{2-}, \quad K_3 = 4.17 \times 10^{-13} \tag{12}$$

Results of the above calculation indicate that the solubility of rare earth arsenates can be very low. The solubilities of the rare-earth arsenates calculated in this chapter are very similar to the solubilities of the rare-earth phosphates. The arsenates are slightly more soluble than the phosphates, as would be expected from charge density considerations of the phosphate and arsenate anions. The solubilities of the rare-earth arsenates are so low that the solubility product and the activity products are very close in value (Table 3). The activity products of the rare earth arsenates and rare earth phosphates reported in Table 3 were obtained from Firsching (1992) and Firsching (1991) respectively to be used for comparison.

Rare earth arsenates can be excellent scavengers of both toxic rare earth elements and arsenic in solution. Trivalent rare earth actinide and arsenates can occur as solid solutions with trivalent rare earth and actinide phosphates since the two solids are polymorphs. Rare earth hydroxides are considered slightly more soluble than the rare earth arsenates (Firsching, 1992) which, provides an advantage in using hydrated oxides of trivalent lanthanides and actinides for the immobilization of rare earth arsenates. The less soluble rare earth arsenates will form surface precipitates in the form of solid solution with rare earth arsenates and hydrous rare earth oxides. A co-precipitation model was developed in this chapter to analyze the co-precipitation of rare earth arsenates with hydrous oxides of trivalent rare earth and actinide elements using the experimental Gibb's free energies of formation of trivalent rare earth hydroxides. The co-precipitation model was used to develop the precipitation constants for the formation of surface precipitates of trivalent rare earth and actinide arsenates onto hydrated oxides of trivalent rare earth and actinide oxides. Davis and Misra (1997) has reported Lanthanum oxide as a better adsorbent than activated $\alpha$- and $\gamma$-alumina for oxyanions such as arsenates and biselenites. The University of Nevada, Reno, has been granted a patent protection on using LA(T), (a mixture of 10% lanthanum oxide and 90% activated alumina as an adsorbent for wastewater remediation (Misra and Nayak, 1996). LA(T) will be competitive with SORBPLUS, a mixed-metal oxide developed by ALCOA as an adsorbent for metals from aqueous streams (Davis and Misra, 1997).

**Table 3. Solubility products and activity products of rare earth arsenates and rare earth phosphates shown as pK$_{sp}$, and pK$_{ap}$**

| Rare earth element | [a]pK$_{sp}$ (arsenates) | [a]pK$_{sp}$ (phosphates) | [b]pK$_{ap}$ (arsenates) | [b]pK$_{ap}$ (phosphates) |
|---|---|---|---|---|
| Y | 21.64 ±0.06 | - | 22.60 ± 0.06 | 24.76 ± 014 |
| La | 21.18 ±0.08 | 21.63±0.07 | 21.45 ± 0.12 | 26.15 ± 0.52 |
| Pr | 21.50 ±0.02 | 21.50± 0.09 | 22.03± 0.02 | 26.06± 0.18 |
| Nd | 21.20 ±0.06 | 21.55± 0.21 | 21.86 ± 0.11 | 25.95 ± 0.06 |
| Sm | 21.79 ±0.04 | 21.56± 0.05 | 22.73 ±0.08 | 25.99 ±0.05 |
| Eu | 21.71 ±0.02 | 21.44 ±0.11 | 22.53 ± 0.03 | 25.75 ± 0.27 |
| Gd | 20.99 ±0.02 | 21.26 ±0.09 | 21.67 ±0.04 | 25.39 ±0.23 |
| Tb | 22.05 ±0.07 | 21.11 ±0.03 | 23.07 ± 0.09 | 25.07 ± 0.03 |
| DY | 21.86 ±0.05 | 21.12 ±0.04 | 23.83 ± 0.06 | 25.15 ± 0.07 |
| Ho | 21.90 ±0.05 | 21.48 ±0.35 | 22.87 ± 0.08 | 25.57 ± 0.46 |
| Er | 21.57 ±0.03 | 21.38 ±0.29 | 22.47 ± 0.04 | 25.78 ± 0.45 |
| Tm | 22.06 ±0.01 | 21.62 ±0.04 | 23.08 ± 0.01 | 26.05 ± 0.06 |
| Yb | 21.75 | 21.69 ±0.03 | 22.72 | 26.17 ±0.01 |
| Lu | 21.72 ± 0.01 | 21.27 ±0.04 | 22.66 ± 0.01 | 25.39 ± 0.03 |

Note: All calculations are at 25± 1OC.

[a] Rare-earth concentration, total arsenate concentration and total acidity data [for saturated solutions at 25± 1°C ] used in the calculations of pKsp were obtained from Firsching (1992) and Firsching (1991) respectively .[b] pKap values of arsenates and phosphates are from Firsching (1992) and Firsching (1991) respectively.

# 3. METHODS

## 3.1. Predictive Model for Arsenic Removal from Soil

Continuous removal of arsenic from water bodies is essential and requires careful measurement of arsenic adsorption. Arsenic adsorption is highly pH dependent and dependent on other soil properties such as cation exchange capacity, organic carbon content, inorganic carbon content, solid content and soil mineralogy such as surface area available for adsorption. It is useful to know how much arsenic is adsorbed on soil surfaces in natural environment. This will give insight into the availability and release of arsenic into ground waters during geochemical and weathering activities in subsurface environment such as soil. On the other hand since arsenic is present and adsorbed in trace amounts, measurement of arsenic in solution requires stringent methods of measurement such as spectroscopy which is expensive, tedious and time consuming.

This problem can be resolved if predictive equations for adsorption of arsenic can be developed considering a wide range of soil properties with easily measured soil parameters. As a first goal, a predictive equation is developed in this chapter to model the amount of arsenic adsorbed onto soil surfaces due to natural geochemical processes, using 10 different soil types as a function of several easily measurable soil parameters affecting adsorption of arsenic. The predictive model was developed using data from 5 soil types (first 100 data points). The fitted model was tested using the data from 5 soil types (113 data points).

### 3.1.1. Integrity of Data Used for Developing and Testing the Predictive Model

A total of 213 adsorption measurements of arsenic, as a function of six soil parameters obtained from the United States Salinity Laboratory (Goldberg et al., 2005) were used to develop and to test the predictive model. The soils were sampled from ten (10) soil types at two different depths at utility sites within from the state of California in the United States where ash landfills are presently in use or are planned. The soils chosen represent a wide range of chemical characteristics. For each site, soils were sampled down gradient from the landfill areas at depths ranging from 0 to 51 cm. Soil surface areas were determined using ethylene glycol monoethyl ether adsorption (Cihacek and Bremner, 1979). Soil pH values were measured in deionized water (1:25 soil/ water ratio) (Goldberg et al., 2005). Cation exchange capacities were measured by Na saturation and Mg extraction. Free Fe and Al were extracted with a Na citrate/citric acid buffer and Na hydrosulphite and measured using inductively coupled plasma (ICP) emission spectrometry (Goldberg et al., 2005). Carbon contents had been determined using a UIC Full Carbon System 150 with a C coulometer (UIC, Inc., Joliet, IL) (Goldberg et al., 2005). Mean, minimum, maximum and the standard deviation values of soil properties such as Cation exchange capacity (CEC), inorganic carbon content (IOC) organic carbon content (OC), pH and aluminum (Al) and iron (Fe) content for the ten different soil types (Burns et al., 2006; Hyun et al., 2006) along with surface areas (SA) are summarized in Table 4 through Table 13. Five soil types are collected at two depths and the rest of them were collected at one depth only (Goldberg et al., 2005).

Data collected from arsenate adsorption experiments performed in batch systems, were used to develop, adsorption envelopes (amount of As(V) adsorbed as a function of time), surface plots of adsorption as a function of six measured soil parameters, and adsorption contours. The surface plots of arsenic adsorption are shown with adsorption on the z axis of the surface, with grids created for the x (soil parameter 1) and y (soil parameter 2) variables ( Figures 4 through 9). Spline interpolation was used to create smooth surface profiles. Fine grid spacing was chosen to generate continuous adsorption surfaces. Contour plots of adsorption were developed for the same variables as in the adsorption surfaces to understand the behavior of adsorption on the ten soils more clearly (Figure 10 through Figure 13). Contour intervals of 0.1 mmol/kg for the adsorption of arsenic were chosen. Adsorption envelops were plotted for the 10 soil types by the soil depth (Figure 14). Data integrity testing (testing for normality of distribution, missing value correction) were performed before fitting the predictive model.

**Table 4. Selected physical and chemical properties of the soil type 1**
**(coarse-loamy, mixed thermic, Haplic, Durixeralf)**

| Soil Name | Soil Property | N | Mean | Standard deviation | Minimum | Maximum |
|---|---|---|---|---|---|---|
| Arlington | pH | 25 | 6.55 | 2.09 | 3.10 | 10.00 |
| | Surface Area ($km^2/kg$) | 25 | 0.08 | 0.02 | 0.06 | 0.10 |
| | Cation Exchange capacity (mmol/kg) | 25 | 146.84 | 42.32 | 107.00 | 190.00 |
| | Inorganic Carbon Content (g/kg) | 25 | 0.23 | 0.07 | 0.16 | 0.30 |
| | Organic carbon Content (g/kg) | 25 | 3.79 | 0.97 | 2.80 | 4.70 |
| | Iron Content (g/kg) | 25 | 9.11 | 0.97 | 8.20 | 10.10 |
| | Aluminum Content (g/kg) | 25 | 0.54 | 0.06 | 0.48 | 0.60 |

**Table 5. Selected physical and chemical properties of the soil type 2**
**(fine, smectitic, thermic, Natric, Palexeralf)**

| Soil Name | Soil Property | N | Mean | Standard deviation | Minimum | Maximum |
|---|---|---|---|---|---|---|
| Bonsall | Surface Area ($km^2/kg$) | 32 | 6.08 | 2.88 | 1.90 | 11.70 |
| | Cation Exchange capacity (mmol/kg) | 32 | 0.02 | 0.01 | 0.02 | 0.03 |
| | Inorganic Carbon Content (g/kg) | 32 | 88.00 | 34.54 | 54.00 | 122.00 |
| | Organic carbon Content (g/kg) | 32 | 0.10 | 0.03 | 0.07 | 0.13 |
| | Iron Content (g/kg) | 32 | 3.50 | 1.42 | 2.10 | 4.90 |
| | Aluminum Content (g/kg) | 32 | 13.05 | 3.81 | 9.30 | 16.80 |
| | Aluminum Content | 32 | 0.68 | 0.23 | 0.45 | 0.91 |

**Table 6. Selected physical and chemical properties of the soil type 3**
**(fine-loamy, mixed, thermic, Typic, Haploxeralf)**

| Soil Name | Soil Property | N | Mean | Standard deviation | Minimum | Maximum |
|---|---|---|---|---|---|---|
| Fallbrook | Surface Area ($km^2/kg$) | 32 | 6.43 | 2.98 | 2.00 | 11.70 |
| | Cation Exchange capacity (mmol/kg) | 32 | 0.05 | 0.02 | 0.03 | 0.07 |
| | Inorganic Carbon Content (g/kg) | 32 | 95.00 | 17.27 | 78.00 | 112.00 |
| | Organic carbon Content (g/kg) | 32 | 0.13 | 0.11 | 0.02 | 0.24 |
| | Iron Content (g/kg) | 32 | 3.30 | 0.20 | 3.10 | 3.50 |
| | Aluminum Content (g/kg) | 32 | 5.90 | 1.02 | 4.90 | 6.90 |
| | Aluminum Content | 32 | 0.29 | 0.08 | 0.21 | 0.36 |

**Table 7. Selected physical and chemical properties of the soil type 4
(coarse-loamy, mixed, superactive, nonacid, thermic, Typic, Xerorthent)**

| Soil Name | Soil Property | N | Mean | Standard deviation | Minimum | Maximum |
|---|---|---|---|---|---|---|
| Hanford | Surface Area ($km^2$/kg) | 20 | 6.84 | 2.25 | 3.00 | 11.70 |
| | Cation Exchange capacity (mmol/kg) | 20 | 0.03 | 0.00 | 0.03 | 0.03 |
| | Inorganic Carbon Content (g/kg) | 20 | 111.00 | 0.00 | 111.00 | 111.00 |
| | Organic carbon Content (g/kg) | 20 | 10.10 | 0.00 | 10.10 | 10.10 |
| | Iron Content (g/kg) | 20 | 28.70 | 0.00 | 28.70 | 28.70 |
| | Aluminum Content (g/kg) | 20 | 6.60 | 0.00 | 6.60 | 6.60 |
| | Aluminum Content | 20 | 0.35 | 0.00 | 0.35 | 0.35 |

**Table 8. Selected physical and chemical properties of the soil type 5
(very-fine, smectitic, calcareous, isohyperthermic, Cumulic, Endoaquoll)**

| Soil Name | Soil Property | N | Mean | Standard deviation | Minimum | Maximum |
|---|---|---|---|---|---|---|
| Nohili | Surface Area ($km^2$/kg) | 9 | 6.27 | 1.72 | 4.00 | 9.40 |
| | Cation Exchange capacity (mmol/kg) | 9 | 0.03 | 0.00 | 0.03 | 0.03 |
| | Inorganic Carbon Content (g/kg) | 9 | 467.00 | 0.00 | 467.00 | 467.00 |
| | Organic carbon Content (g/kg) | 9 | 2.70 | 0.00 | 2.70 | 2.70 |
| | Iron Content (g/kg) | 9 | 21.30 | 0.00 | 21.30 | 21.30 |
| | Aluminum Content (g/kg) | 9 | 49.00 | 0.00 | 49.00 | 49.00 |
| | Aluminum Content | 9 | 3.70 | 0.00 | 3.70 | 3.70 |

**Table 9. Selected physical and chemical properties of the soil type 6
(coarse-loamy, mixed, thermic, Mollic, Haploxeralf)**

| Soil Name | Soil Property | N | Mean | Standard deviation | Minimum | Maximum |
|---|---|---|---|---|---|---|
| Pachappa | Surface Area ($km^2$/kg) | 26 | 6.19 | 1.94 | 3.20 | 9.80 |
| | Cation Exchange capacity (mmol/kg) | 26 | 0.03 | 0.01 | 0.02 | 0.04 |
| | Inorganic Carbon Content (g/kg) | 26 | 45.50 | 6.63 | 39.00 | 52.00 |
| | Organic carbon Content (g/kg) | 26 | 0.02 | 0.01 | 0.01 | 0.03 |
| | Iron Content (g/kg) | 26 | 2.45 | 1.38 | 1.10 | 3.80 |
| | Aluminum Content (g/kg) | 26 | 7.40 | 0.20 | 7.20 | 7.60 |
| | Aluminum Content | 26 | 0.51 | 0.16 | 0.35 | 0.67 |

**Table 10. Selected physical and chemical properties of the soil type 7 (fine-loamy, mixed, superactive, thermic, Typic, Haploxeralf)**

| Soil Name | Soil Property | N | Mean | Standard deviation | Minimum | Maximum |
|---|---|---|---|---|---|---|
| Ramona | Surface Area ($km^2$/kg) | 20 | 6.28 | 2.30 | 3.20 | 9.90 |
| | Cation Exchange capacity (mmol/kg) | 20 | 0.03 | 0.01 | 0.03 | 0.04 |
| | Inorganic Carbon Content (g/kg) | 20 | 49.35 | 18.89 | 29.00 | 66.00 |
| | Organic carbon Content (g/kg) | 20 | 0.02 | 0.00 | 0.02 | 0.02 |
| | Iron Content (g/kg) | 20 | 3.41 | 1.12 | 2.20 | 4.40 |
| | Aluminum Content (g/kg) | 20 | 5.13 | 0.71 | 4.50 | 5.90 |
| | Aluminum Content | 20 | 0.41 | 0.01 | 0.40 | 0.42 |

**Table 11. Selected physical and chemical properties of the soil type 8 (fine-silty, mixed, superactive, mesic, Xeric Natridurid)**

| Soil Name | Soil Property | N | Mean | Standard deviation | Minimum | Maximum |
|---|---|---|---|---|---|---|
| Sabree | Surface Area ($km^2$/kg) | 16 | 6.37 | 3.07 | 1.90 | 11.90 |
| | Cation Exchange capacity (mmol/kg) | 16 | 0.02 | 0.00 | 0.02 | 0.02 |
| | Inorganic Carbon Content (g/kg) | 16 | 27.00 | 0.00 | 27.00 | 27.00 |
| | Organic carbon Content (g/kg) | 16 | 0.01 | 0.00 | 0.01 | 0.01 |
| | Iron Content (g/kg) | 16 | 2.20 | 0.00 | 2.20 | 2.20 |
| | Aluminum Content (g/kg) | 16 | 6.00 | 0.00 | 6.00 | 6.00 |
| | Aluminum Content | 16 | 0.46 | 0.00 | 0.46 | 0.46 |

**Table 12. Selected physical and chemical properties of the soil type 9 (coarse-loamy, mixed, superactive, nonacid, thermic, Typic, Torriorthent)**

| Soil Name | Soil Property | N | Mean | Standard deviation | Minimum | Maximum |
|---|---|---|---|---|---|---|
| Wasco | Surface Area ($km^2$/kg) | 16 | 6.82 | 3.38 | 2.00 | 11.70 |
| | Cation Exchange capacity (mmol/kg) | 16 | 0.03 | 0.00 | 0.03 | 0.03 |
| | Inorganic Carbon Content (g/kg) | 16 | 71.00 | 0.00 | 71.00 | 71.00 |
| | Organic carbon Content (g/kg) | 16 | 0.01 | 0.00 | 0.01 | 0.01 |
| | Iron Content (g/kg) | 16 | 4.70 | 0.00 | 4.70 | 4.70 |
| | Aluminum Content (g/kg) | 16 | 2.40 | 0.00 | 2.40 | 2.40 |
| | Aluminum Content | 16 | 0.42 | 0.00 | 0.42 | 0.42 |

**Table 13. Selected physical and chemical properties of the soil type 10
(fine-loamy, mixed, superactive, thermic, Mollic, Haploxeralf)**

| Soil Name | Soil Property | N | Mean | Standard deviation | Minimum | Maximum |
|---|---|---|---|---|---|---|
| Wyo | Surface Area (km²/kg) | 17 | 6.54 | 2.89 | 2.20 | 11.70 |
| | Cation Exchange capacity (mmol/kg) | 17 | 0.05 | 0.00 | 0.05 | 0.05 |
| | Inorganic Carbon Content (g/kg) | 17 | 155.00 | 0.00 | 155.00 | 155.00 |
| | Organic carbon Content (g/kg) | 17 | 0.01 | 0.00 | 0.01 | 0.01 |
| | Iron Content (g/kg) | 17 | 19.90 | 0.00 | 19.90 | 19.90 |
| | Aluminum Content (g/kg) | 17 | 9.50 | 0.00 | 9.50 | 9.50 |
| | Aluminum Content | 17 | 0.89 | 0.00 | 0.89 | 0.89 |

### 3.1.2. Integrity of Data Used in Model Fitting

Data from five soil types (Arlington, Ramona, Pachayappa, Sabree, Wasco, Wyo) were used to fit the predictive model for adsorption. The data used represented a broad range of chemical characteristics. Ranges of the properties were: pH, 1.92 to 11.9; surface area, 0.02 to 0.1 km²/kg; cation exchange capacity, 27 to 190 mmol/kg; organic carbon content, 1.9 to 19.9 g/kg; inorganic carbon content, 0.01 to 0.3 g/kg; iron content 2.4 to 10.1 g/kg and aluminum content 0.35 to 0.89 g/kg.

### 3.1.3. Predictive Model Fitting

A non-linear regression model for arsenic adsorption was developed to identify the best relationship of the six measured soil parameters affecting arsenate adsorption (surface area [SA], ph, cation exchange capacity [CEC], organic carbon content [OC], inorganic carbon content [IOC], iron content [Fe] and aluminum content [Al] to the measured adsorption. The NLIN subroutine with NLMIIXED procedure with SAS® software (SAS Institute Inc., Cary, NC) was used to fit the non linear mixed model. A significance level of 0.05 was used for all hypothesis tests.

### 3.1.4. Measured Adsorption Surfaces as a Function of Soils Parameters

The adsorption of arsenic increases initially as the pH and solid content were increase, reachees a maximum value and decreases thereafter (Figure 4 and Figure 5). Adsorption of arsenic increases initially at a faster rate with the pH, solid content reaching a maximum and decreases thereafter. It appears that there are optimum values of pH, solid content and surface area of solid at which maximum adsorption occurs.

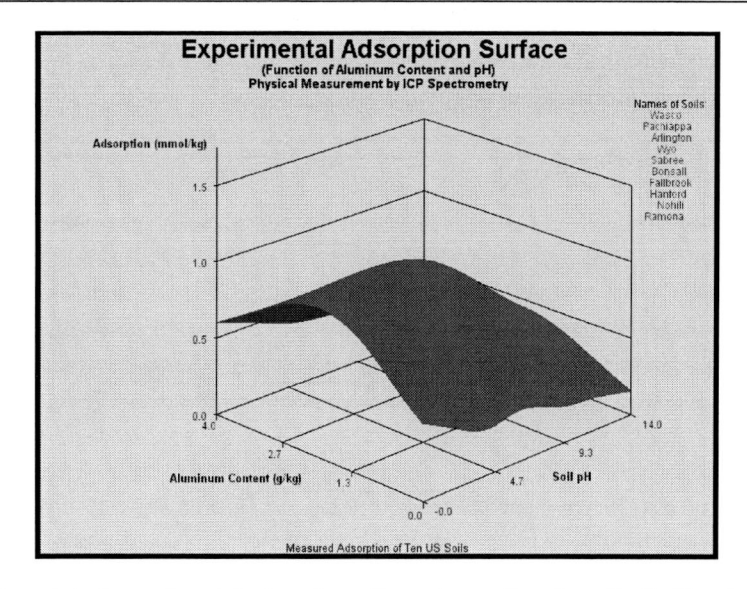

Figure 4. Surface plot of experimental adsorption of arsenic as a function of soil pH and aluminum content.

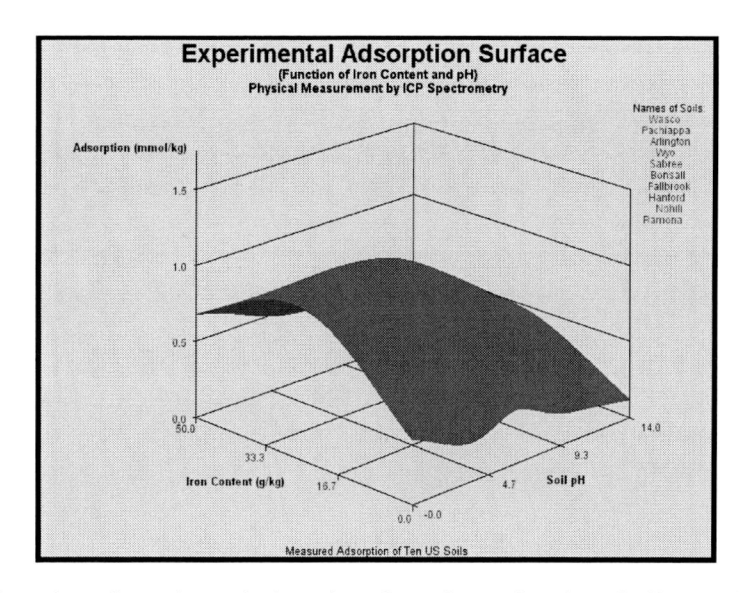

Figure 5. Surface plots of experimental adsorption of arsenic as a function of soil pH and iron content.

Adsorption increases as the organic carbon content in soil increases in the range of organic carbon content studied (Tables 3 through 13) but changes only slightly with the inorganic carbon content. A possible explanation for this could be that the equilibria between solid and dissolved ion concentration (e.x: $Al_2O_3(s)/Al(aq)$, $Fe_2O_3(s)/Fe(s)$) in the zone of representative redox potentials for many ground and soil waters where oxygen is consumed by degradation of organic matter. The inter-conversion between iron(II) and iron(III) oxidation states is very dependent on the redox potential of the media, which for soils is largely related to the organic matter content. Adsorption increases initially as the cation exchange capacity (CEC) of the soil increases (Figure 8) and reaches a steady level at low values of CEC, which indicates that CEC of the soil can improve the adsorption at low values and does not influence adsorption at moderate to high values.

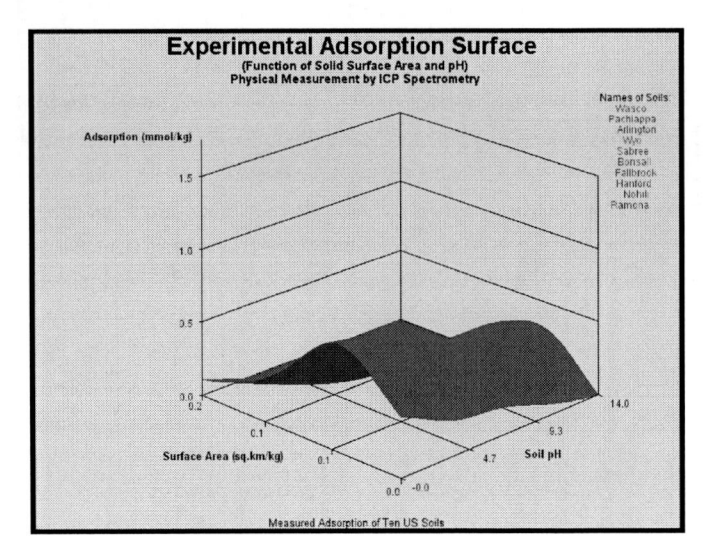

Figure 6. Surface plot of experimental adsorption of arsenic as a function of soil pH and surface area.

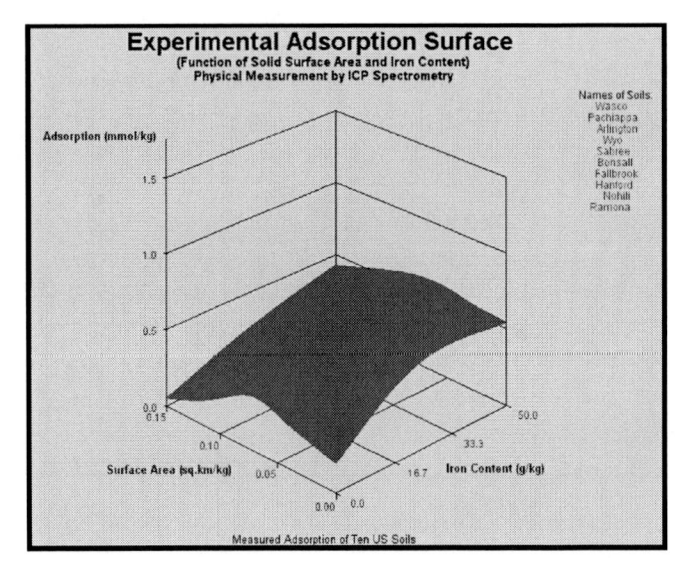

Figure 7. Surface plot of experimental adsorption of arsenic as a function of surface area and iron content.

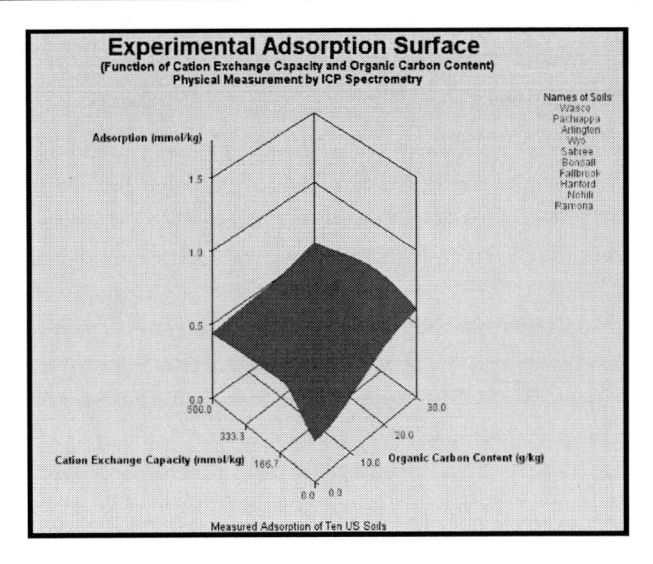

Figure 8. Surface plot of experimental adsorption of arsenic as a function of soil organic carbon content and cation exchange capacity.

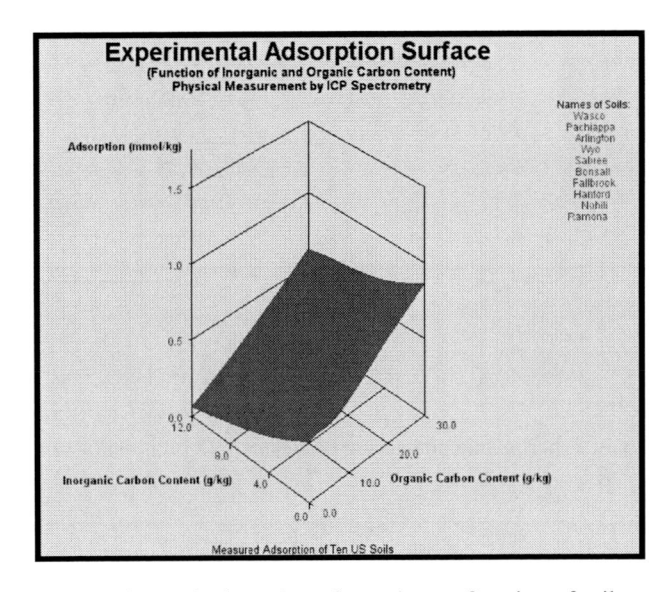

Figure 9. Surface plot of experimental adsorption of arsenic as a function of soil organic and inorganic carbon content.

### 3.1.5. Adsorption Contours of Arsenic as a Function of Soil Parameters

Adsorption of arsenic can be observed to increase initially to a maximum value (Figure 10 through Figure 13, circles shown in the middle) and to decrease thereafter as pH and solid content is increased. Largest adsorption is associated with medium values of pH and solid (Fe, Al) content (Figure 10 and Figure 11) and with relatively small values of surface areas (Figure 12). Adsorption appears to be negatively correlated with inorganic carbon content (Figure 13); adsorption decreases as inorganic carbon content in the soil increases, but increases significantly as the organic carbon content increases.

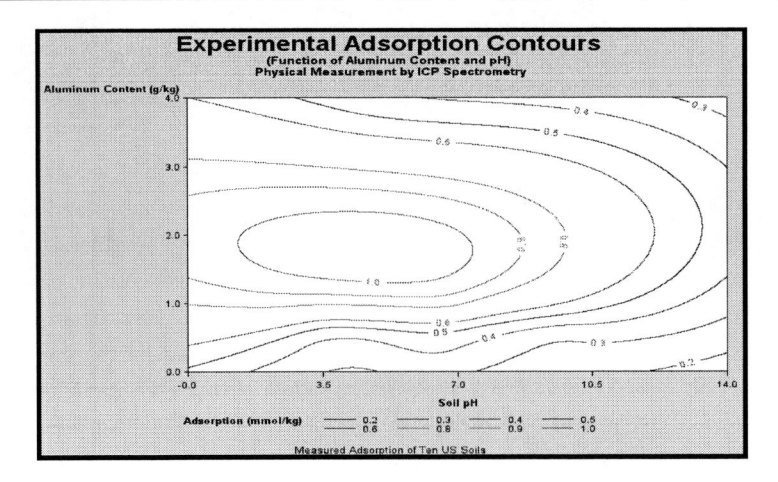

Figure 10. Contour plot of experimental adsorption of arsenic as a function of soil pH and aluminum content.

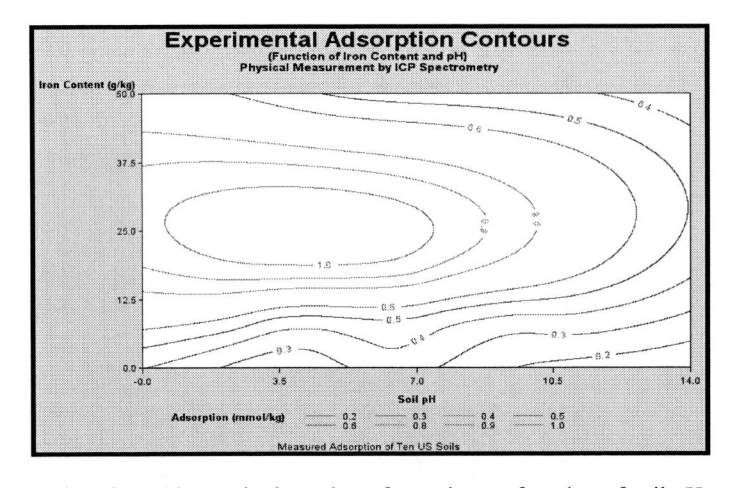

Figure 11. Contour plot of experimental adsorption of arsenic as a function of soil pH and iron content.

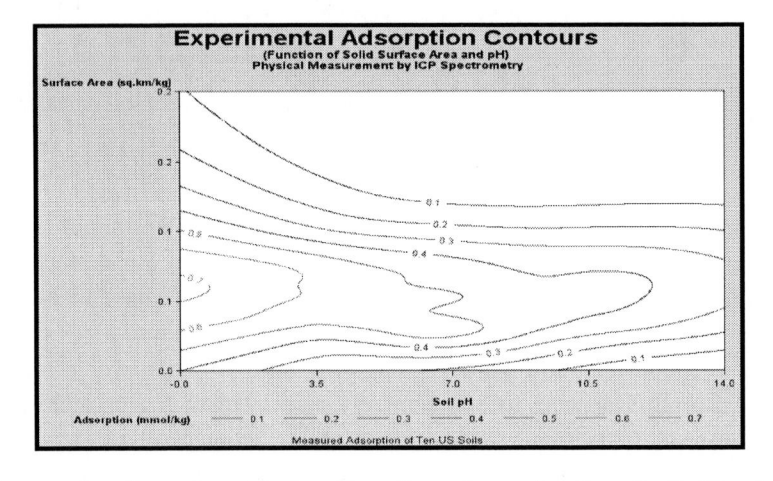

Figure 12. Contour plot of experimental adsorption of arsenic as a function of soil pH and surface area.

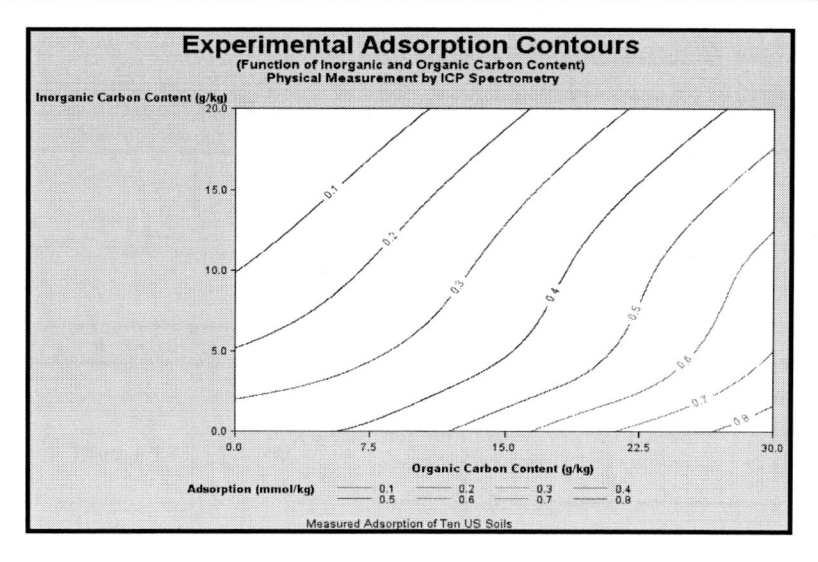

Figure 13. Contour plot of experimental adsorption of arsenic as a function of soil organic and inorganic carbon content.

### 3.1.6. Adsorption Envelops of Arsenic

Adsorption of arsenic increases with pH reaching a maximum between pH =5 and pH=7 for most soils and decrease thereafter (Figure 14). The adsorption envelops (adsorption as a function of soil pH) is shown for the soils separated by depth (in cm) of the soils. No significant variation is visible between depths from soil type to soil type, except for the soil, Pachyappa, which shows a sudden decrease in adsorption at pH=9.0 at depth between 25cm and 51cm, which is not present at depth between 0cm and 25cm.

## 3.2. Surface Precipitation Model for Arsenic Adsorption

According to Twidell et al., (1999), the injection of ozone in treatment systems leads iron to form a hydroxide, which is a filterable precipitate. The same ozonation process oxidizes As(III) to As(V), which binds by adsorption to iron hydroxides and therefore can be removed with it by filtration. This process, called "co-precipitation", is a Best Demonstrated Available Technology (BDAT) for mine water treatment, but has had limited application to drinking water in the United States. It is possible that an intermediate solid solution of $Fe(OH)_3$-$FeAsO_4$ is formed that results in the surface precipitation of arsenic in the form of $FeAsO_4$-$Fe(OH)_3$ that is responsible for the co-precipitation of arsenic onto iron hydroxides. Incongruent dissolution of scorodite ($FeAsO_4.2H_2O$) with its transformation in iron(III) hydroxides has been observed (Dove and Rimstidt, 1985). The mechanism by which arsenic is adsorbed onto hydroxides of iron is yet to be explained. On the other hand, studies have observed lanthanum oxide to be a better adsorbent than activated alumina and iron hydroxides for removal of arsenic from water and waste water systems (Davis and Misra, 1997). In this chapter we propose a surface precipitation model for the adsorption of arsenic onto hydrated oxides of trivalent lanthanides and actinides with lanthanum arsenate (LA) structure.

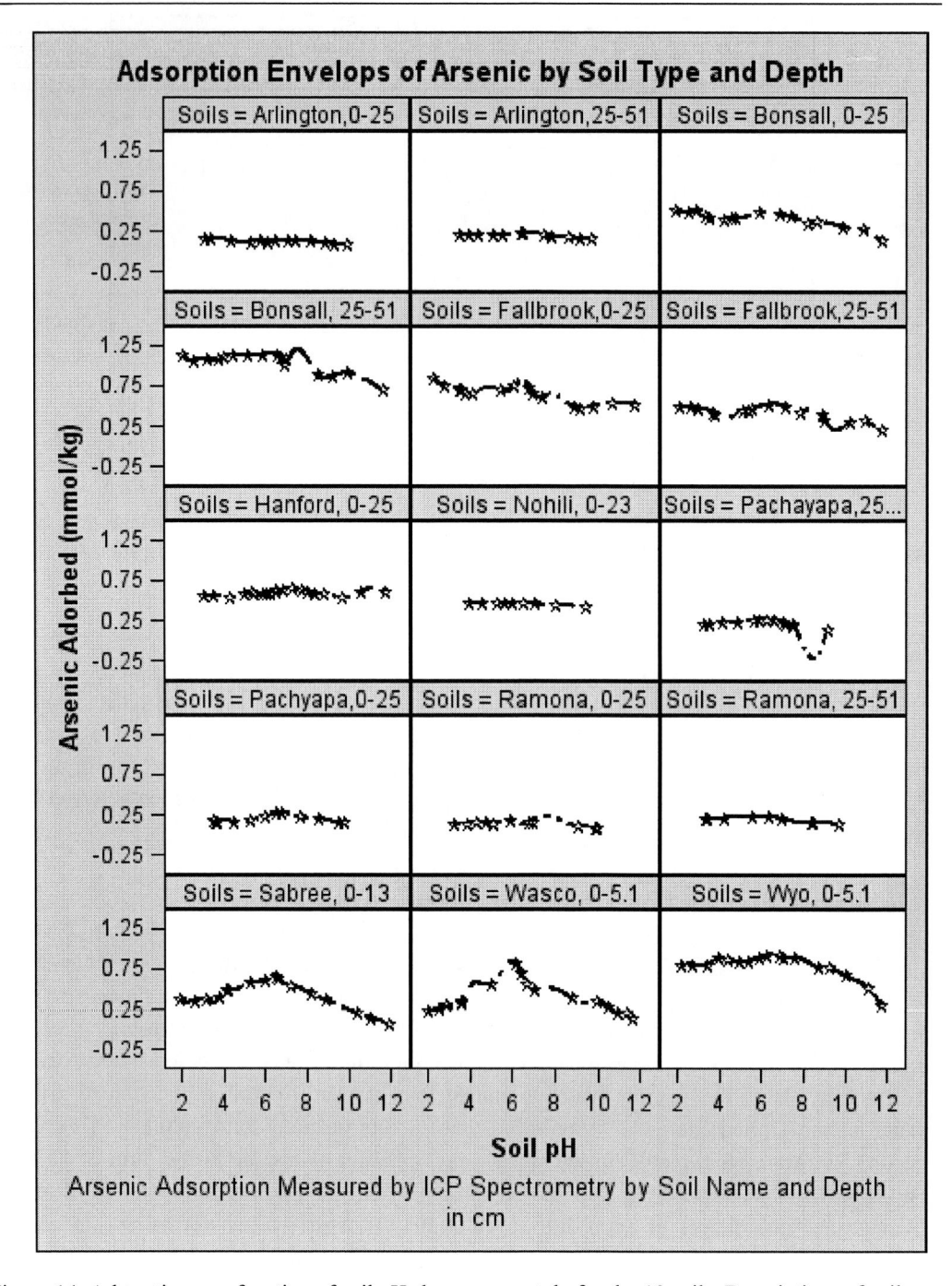

Figure 14. Adsorption as a function of soil pH shown separately for the 10 soils. Descriptions of soils are given in Table 4 through Table 13.

### 3.2.1. Linear Free Energy Model

Sverjensky and Molling (1992) proposed an empirical linear free energy correlation for crystalline solids of divalent cations. The linear free energy formula of Sverjensky and Molling (1992) is analogous to the well-known Hammett relationship developed for functional group substitution in organic compounds (Wells, 1968; Exner, 1988). Sverjensky Molling (1992) correlation has been successfully applied to various crystalline structure families of divalent and tetravalent cations to predict surface reaction controlled dissolution rates and formation energies of end members (Xu and Wang, 1999a, b, c; Xu et. al., 1999; Wang and Xu, 2000; Wang and Xu 2001).

This linear free energy equation quantifies cation substitution in crystalline solids hence can be used to predict the rates at which solids react (Sverjensky and Molling, 1992). The standard state free energy parameters in the equation were found to be remarkably close to the dissolution rate parameters in several studies (Sverjensky and Molling, 1992; Zhu, 2002). The original Sverjensky and Molling (1992) linear free energy relationship was modified in this study for trivalent cations with lanthanum arsenate (LA) structure and was applied to derive standard state Gibbs free energies of formation ($\Delta G^0_{f,MvX(ss)}$) of surface precipitates with LA structure. The modified Sverjensky and Molling (1992) equation for trivalent cations is given as:

$$\Delta G^0_{f,MvX} - \beta_{MvX} r_{M^{3+}} = a_{MvX} \Delta G^0_{n,M^{3+}} + b_{MvX} \tag{13}$$

where $\Delta G^0_{f,MvX}$ denotes the standard state Gibbs free energies of formation of the end member solids ($MvX$) in the solid solution (surface precipitate), $r_{M^{3+}}$ represents the Shannon-Prewitt radius (Shannon, 1976) of the $M^{3+}$ cation in a given coordination state, and $\Delta G^0_{n,M^{3+}}$ denotes the non-solvation contribution to the Gibbs free energy of formation of the aqueous $M^{3+}$ ion. The coefficients $a_{MvX}, b_{MvX}$, and $\beta_{MvX}$ are regression parameters. Values of coefficient $a_{MvX}$ are related only to the stoichiometry of the crystal solid formed and are very close for all polymorphs of the composition $MvX$ (Sverjensky and Molling, 1992). The coefficient $\beta_{MvX}$ is only related to the effect of nearest neighbors (or coordination numbers: CN) of the cation in the solid formed. In polymorphs, the structure family with small CN (e.g., CN=6 in calcite structure family) has higher value of $\beta_{MvX}$ than does the family with large CN (e.g., CN=9 in aragonite structure family) (Sverjensky and Molling, 1992). The coefficient $b_{MvX}$ reflects characteristic of the reaction type and conditions under which solid formation took place regardless of the valence of the cation or the stoichiometry of the solid.

Although Eq. (13) applies to the free energies of crystalline solids, it was initially developed as a family of linear free energy equations for aqueous and mixed solvent organic reactions (Wells, 1968; Exner, 1988). One of these is the Hammett equation, which has the form:

$$Log(K/K_0) = \sigma\rho \tag{14}$$

where $K$ and $K_0$ represent the equilibrium (or rate) constants for substituted and unsubstituted reactions involving a series of substituent functional group. Each substituent functional group has a characteristic group $\sigma$. $\rho$ is a regression coefficient characteristic of the reaction conditions. The substituent parameter $\sigma$ is derived from equilibrium constants of substituted ($K^b$) and unsubstituted ($K_0^b$) species of the functional group according to the equation:

$$\sigma = Log(K_b / K_0^b) \tag{15}$$

The relationship of Eq. (13) to the Hammett equation can be established by converting Eq. (13) to equilibrium constants and by including a reference reaction for a specific cation ( $N$ ) as follows:

$$Log(K_{MvX} / K_{NvX}) = \sigma_{M^{3+}} \rho_{MvX} + \beta_{MvX} r * \tag{16}$$

where

$$\sigma_{M^{3+}} = Log(K_{M^{3+}} / K_{N^{3+}}) \tag{17}$$

$$\rho_{MvX} = a_{MvX} \tag{18}$$

and

$$r* = -(r_{M^{3+}} - r_{N^{3+}}) / 2.303RT \tag{19}$$

Addition of $\beta_{MvX} r *$ in Eq. (16) enables the application of Eq. (16) accurately to crystalline solids. The success of Eq. (13) in describing the standard state Gibbs free energies of formation of numerous iso-structural families of crystalline solids (Sverjensky and Molling, 1992; Sverjensky, 1992), and the analogy with the Hammett equation as described above, suggests that linear free energy relation for crystalline solids might be applicable to the rate at which solids react. Hence this chapter explores the possibility of applying the linear free energy relation for calculating surface precipitation constants of iso-structural families of trivalent metal arsenates of lanthanides for which no reliable thermodynamic data are currently available.

According to the hypothesis of surface reaction control and transition state theory, rate of reaction ( $r'$ ) for arsenates far from equilibrium can be related to the rate constant ( $k$ ) by the following equation:

$$\log r' = \log k + n \log( a_{H^+}^c ) \tag{20}$$

where $a_{H^+}^c$ represents the activity of aqueous $H^+$ ions and $n$ the order of the reaction (Murphy and Helgeson, 1987). Experimental surface reaction controlled dissolution rates have been discussed in Casey and Westrich (1992) in detail. For a constant $n$ ($n$ is roughly constant for each structure), it is possible to develop linear free energy relations involving the function $2.303RT \log(r')$. One such relation employs involving the trivalent cation parameters $\Delta G_{n,M^{3+}}^0$ and $r_{M^{3+}}$ as:

$$2.303RT \log(r') + \beta r_{M^{3+}} = a' \Delta G_{n,M^{3+}}^0 + b' \tag{21}$$

Eq. (21) yields strong linear free energy correlations, using the $\beta$ values defined by the regression of the standard Gibbs free energies. The standard state free energy of reaction ($\Delta G_{R,MvX(ss)}^0$) forming a surface precipitate on a surface of a solid as a solid solution can be defined by:

$$\Delta G_{R,MvX(ss)}^0 = -2.303RT \log(r') \tag{22}$$

Thus the linear free energy correlation of Sverjensky and Molling (1992) permits the prediction of the surface precipitation constants at equilibrium.

### 3.2.2. Model Adaptation for Surface Precipitation of Arsenic

The standard state Gibbs free energy of formation of the end member of the solid solution on the surface of the solid can be defined as follows (Eq. (23), derivation of which for lanthanum arsenate(LA) is given in sections 3.2.3):

$$\Delta G_{f,MvX(ss)}^0 = \Delta G_{R,MvX(ss)}^0 + \Delta G_{f,M^{3+}}^0 + \Delta G_{f,AsO_4^{3-}}^0 \tag{23}$$

where $\Delta G_{f,MvX(ss)}^0$ is the standard state Gibbs free energy of formation of the pure end member component of the solid solution with $MvX$ structure. The $\Delta G_{f,M^{3+}}^0$, and $\Delta G_{f,AsO_4^{3-}}^0$ are the conventional apparent free energy of formation of the trivalent cation ($M^{3+}$), and arsenate ($AsO_4^{3-}$) ions. Definition of the empirical standard state Gibbs free energy of formation of the end member component of solid solution permits the most direct analogy to Eq. (13) described through the following linear relation:

$$\Delta G_{f,MvX(ss)}^0 - \beta'_{MvX} r_{M^{3+}} = a'_{MvX} \Delta G_{n,M^{3+}}^0 + b'_{MvX} \tag{24}$$

The regression parameters $\beta'_{MvX}, a'_{MvX}, b'_{MvX}$ are analogous to regression parameters ($\beta_{MvX}, a_{MvX}, b_{MvX}$) of the bulk solid precipitates in Eq. (13). Since the end member components of the surface precipitates and the bulk precipitates are polymorphs with stoichiometry equal to $MvX$, the regression parameters $a_{MvX}$ and $a'_{MvX}$ are similar or very close to each other (Sverjensky and Molling, 1992). In addition since the end member components of the surface precipitates and the bulk solid precipitates possess the same crystal structure (LA), they have same cationic CN in the solids. Therefore, the regression coefficients $\beta_{MvX}$ and $\beta'_{MvX}$ are similar. The regression coefficients $a_{MvX}$, $\beta_{MvX}$ obtained through regressing the standard state Gibbs free energies of formation of bulk solid precipitates through Eq. (13) can be used to develop linear free energy correlations for the end member components of the solid solutions with same crystal structure (LA).

In this study the standard state, Gibbs free energies of formation of the end member solids of solid solutions ($\Delta G^0_{f,MvX(ss)}$) with LA structures were calculated using the regression coefficients ($a_{MvX}$ and $\beta_{MvX}$) obtained from the regression relation developed for the bulk solid precipitates. The regression coefficient $b'_{MvX}$ for this free energy relation was obtained from similar free energy equations developed for the solid solution formation in surface precipitation reactions of trivalent lanthanide and actinide cations onto hydrated oxides of lanthanides and actinides under similar reaction conditions (Ragavan and Adams, 2009). The regression coefficient $b'_{MvX}$ depends only on reaction type (Sverjensky and Molling, 1992). According to Sverjensky (1992) within experimental uncertainties the same regression parameters apply to pure solids formed by surface precipitation onto solids with polymorphic crystalline structure. In both cases the surface precipitation is onto hydrated oxides of lanthanides and actinides.

The $\Delta G^0_{R,MvX(ss)}$ values for surface precipitates were calculated from the empirical correlations (Eq. (23)) and were used to estimate the equilibrium constants for surface precipitation reactions of lanthanides and actinides with LA structure, for the following trivalent cations: $Dy^{3+}$, $Er^{3+}$, $Tm^{3+}$, $Lu^{3+}$, $Ho^{3+}$, $La^{3+}$, $Nd^{3+}$, $Ce^{3+}$, $Pr^{3+}$, $Gd^{3+}$, $Sm^{3+}$, $Tb^{3+}$, $Yb^{3+}$, $Y^{3+}$, $Eu^{3+}$, $Am^{3+}$, $Np^{3+}$, $U^{3+}$, and $Pu^{3+}$, for which reliable experimental thermodynamic data are currently not available. The surface precipitation reactions were formulated based on the Farley et. al. (1985) solid solution model linked to the surface complexation model of Dzobmbak and Morel (1990). The calculated surface precipitation constants need to be verified against experimentally derived surface precipitation constants. Applicability of the model to individual metal ions due to differences of ionic radius and other ionic properties also need to be tested experimentally for respective individual cations.

### 3.2.3. Surface Precipitation Reactions

Surface precipitation reactions of trivalent cations onto hydrous oxides of lanthanides and actinides can be defined following Farley et. al. (1985) and, Dzombak and Morel (1990) as follows:

$$\equiv LnOH^0 + Ln^{3+} + 3H_2O \Leftrightarrow Ln(OH)_3(ss) + = LnOH^0 + 3H^+ \qquad (25)$$

where $Ln$ is any trivalent ion in the lanthanide (lanthanum (La) to lutetium (Lu)) and actinides (actinium (Ac) through lawrencium (Lw)) series, subscript 'ss' stands for the end member component of the solid solution $LnAsO_4 - Ln(OH)_3$, symbol '$\equiv$' (triple) and '=' (double) denote the bonds of metal atoms at the solid surface. The $\equiv LnAsO_4^0$ represents concentrations of $LnAsO_4$ surface sites and $= LnOH^0$ represents the concentrations of $Ln(OH)_3$ surface sites (Farley eta al., 1985). The $\Leftrightarrow$ sign was used to indicate equality. At high surface coverage the sorption of $AsO_4^{3-}$ and $Ln^{3+}$ onto the sorbent surface forms a solid solution of $LnAsO_4 - Ln(OH)_3$ (Farley eta al., 1985; Dzombak and Morel, 1990) as shown below.

$$= LnOH^0 + Ln^{3+} + 3H_2O \Leftrightarrow Ln(OH)_3(ss) + = LnOH^0 + 3H^+ \qquad (26)$$

$$\equiv LnOH^0 + AsO_4^{3-} + Ln^{3+} \Leftrightarrow LnAsO_4(ss) + \equiv LnOH^0 \qquad (27)$$

The components of the solid solution obey the mass balance constraint as:

$$X_{Ln(OH)_3(ss)} + X_{Ln(AsO_4)(ss)} = 1 \qquad (28)$$

where $X$ stands for the mole fraction of an end member component in the solid solution and are calculated from:

$$X_{Ln(AsO_4)_3(ss)} = \frac{[LnAsO_4(ss)]}{[Ln(OH)_3(ss)] + [LnAsO_4(ss)]} \qquad (29)$$

$$X_{Ln(OH)_3(ss)} = \frac{[Ln(OH)_3(ss)]}{[Ln(OH)_3(ss)] + [LnAsO_4(ss)]} \qquad (30)$$

where the brackets, [], stand for the mole concentrations of components per liter of aqueous solution.

For an ideal solid solution, the activity of the component 'i' ($a_i^{ac}$) in the solid solution is equal to $X_i$. Most $Ln(OH)_3$ solid (pure end members and bulk solids) have a structure similar to HLO. In addition the substitution is isovalent, for most applications. Hence mixing in this type of solid solution can be considered ideal. As a result, the activity coefficient for the end-member $Ln(OH)_3(ss)$ is unity in the dilute solution regions where Henry's law is obeyed for the $Ln(OH)_3(ss)$ component (Zhu, 1992). Henrian standard state for $Ln(OH)_3(ss)$ assumes a hypothetical pure $Ln(OH)_3(ss)$ end-member of the solid solution with an HLO structure. In such systems the mole balance equation for modeling surface precipitation can be described by the following construct (Farley eta al., 1985; Dzombak and Morel, 1990).

$$TOTLn \Leftrightarrow \sum [Ln^{3+}]_{aq} + ([LnOH^0] - [LnAsO_4^0]) + [Ln(OH)_3(ss)] + [LnAsO_4(ss)] \quad (31)$$

$$TOTAs \Leftrightarrow \sum [AsO_4^{3-}]_{aq} + [LnAsO_4^0] + [LnAsO_4(ss)] \quad (32)$$

$$TOT(\equiv LnOH) \Leftrightarrow [\equiv LnOH^0] + [\equiv LnOH_2^+] + [\equiv LnO^-] + [LnAsO_4^0] \quad (33)$$

$$TOTH \Leftrightarrow [H^+] - [OH^-] + [LnOH^{2+}] + 2[Ln(OH)_2^+] + 2[\equiv LnOH^0] - 3[Ln(OH)_3(ss)] \quad (34)$$

where the symbols $TOTM, TOTAs, TOT(\equiv LnOH)$, and $TOTH$ denote the total concentrations computed with $Ln^{3+}, As, \equiv LnOH,$ and $H^+$ as components in the system in a tableau representation of the system. The summation sign stands for the sum of all aqueous species and complexes of $Ln^{3+}$ or $AsO_4^{3-}$ ion and $\equiv LnOH$ stands for surface site. Following the assumption of Farley et al. (1985) and Dzombak and Morel (1990), all precipitated $Ln$ is accounted for in the calculation of mole fractions in Eqs. (29) and (30).

### 3.2.4. Calculating Equilibrium Surface Precipitation Constants

The standard Gibbs free energies of formation for the pure bulk $LnAsO_4(s)$ solids with lanthanum arsenate structure was obtained through the Sverjensky and Molling (1992) equation modified for trivalent cations (Eq. 13). The regression coefficients obtained from this correlation were used to calculate the standard state Gibbs free energies of formation of the "fictitious" pure end members of the solid solutions onto HLO. From the standard state Gibbs free energies of formation of the 'fictitious' pure end-members of the solid solution the standard state Gibbs free energies of reactions of the solid solution onto HLO were calculated as:

$$\Delta G^0_{R, LnAO_4^{3-}(ss)} = \Delta G^0_{f, LnAO_4^{3-}(ss)} - \Delta G^0_{f, Ln^{3+}} - \Delta G^0_{f, AsO_4^{3-}} \quad (35)$$

where $\Delta G^0_{f, LnAO_4^{3-}(ss)}$ is the standard state Gibbs free energy of formation of the $LnAsO_{4,}(ss)$ end member component of the solid solution onto HLO, $\Delta G^0_{R, MvX(ss)}$ is the standard state Gibbs free energy of the solid solution reaction for the solid ($MvX$) with lanthanum arsenate (LA) structure. The thermodynamic construct of derivation of Eq. (35) are given above in sections 3.2.2 and 3.2.3 respectively.

For aqueous species, unit activity of the species in a hypothetical one molal ideal solution referenced to infinite dilution at the temperature and pressure of interest defines the standard state. The standard state, Gibbs free energies of formation of $AsO_4^{3-}$ ion is from Krause and Ettel (1988). Values of $\Delta G^0_f$ of the cations of lanthanides and actinides are from Brookins (1988). Finally the equilibrium solubility products ($K$) for the solid solution reactions were calculated using the standard state Gibbs free energies of the solid solution reactions as:

$$\Delta G^0_{R,M_vX(ss)} = -2.303RT \ln(K) \tag{36}$$

where $K$ are the values of the equilibrium constants for reactions given in Eq. (13), $R$ is the gas constant and $T$ is the temperature in Kelvin. The $K$ values thus calculated are the solubility products of the pure end-member components of the solid solution with LA structure.

## 4. RESULTS AND DISCUSSION

### 4.1. Fitted Predictive Model for Arsenic Adsorption

Results of non-linear model are shown in Table 14. All soil parameters tested significantly influences adsorption of arsenic onto soil. Adsorption is larger on aluminum surface in the presence of soil. Adsorption is also negatively influenced by pH in the range of pH studied indicating that adsorption is larger at low values of pH and decreases as pH increases. Adsorption is also larger at low values of inorganic carbon content and cation exchange capacity of soil. Separate predictive models may be required for low and high values of pH to better explain the relationship of pH with adsorption of arsenic. The fitted model can be described as follows:

$$Y_{ij} = \beta_0 + \beta_j g_j + u_i + \epsilon_{ij} \tag{37}$$

where $i$ indexes the measurement observation and $j$ indexes the treatment effects. The $u_i$ indicates the measurement level normal random effect. $\epsilon_{ij}$ s are iid normal errors. $\beta_j$ are unknown coefficients to be estimated. The model is continuous. The estimated values of the coefficients for the six treatment variables (parameters) are shown in Table 14 along with the probabilities of estimates. The parameter estimates table indicates significance of all the parameters at 5% level of significance. Surface area and solid content are the largely, positively influencing variables of arsenic adsorption in soil. Separate analyses for aluminum and iron content are required to study the influence of each solid individually along with other soil parameters on adsorption of arsenic. Adsorption of arsenic was positively correlated (p=0.0087, Table 14) with aluminum content in solution.

According to the above results from the model, arsenic adsorption increases linearly as surface area of the solid (p=0.0052) and organic carbon content (p=0.0011) increases. Adsorption is larger, smaller the values of pH of soil solution, cation exchange capacity of soil, and inorganic carbon content of soil. Adsorption is also negatively correlated with iron content in the soil and positively correlated with aluminum content. Krause and Ettel (1988), observed congruent dissolution of scorodite ($FeAsO_4.2H_2O$) to occur in the pH range between 0.97 and 2.43 according to the following equilibrium reaction:

$$FeAsO_4. 2H_2O + 3H^+ = Fe^{3+} + H_3AsO_4 + 2H_2O \tag{38}$$

**Table 14. Solution for non linear mixed model fixed effects of arsenic adsorption as a function of six soil parameters at convergence. A probability (Pr, column 5 from left) value less than 0.05 indicates that the parameter studied significantly influences adsorption of arsenic. Negative or positive estimates indicate that the parameter tested negatively or positively influences adsorption of arsenic respectively.**

| Parameter | Estimate | Standard Error | t Value | Pr > |t| |
|---|---|---|---|---|
| Intercept | -0.451 | 0.181 | -2.49 | 0.0674 |
| Surface Area | 61.207 | 11.077 | 5.53 | 0.0052 |
| Cation Exchange Capacity | -0.030 | 0.005 | -5.46 | 0.0055 |
| Organic Carbon Content | 0.089 | 0.011 | 8.35 | 0.0011 |
| Inorganic Carbon Content | -0.915 | 0.150 | -6.11 | 0.0036 |
| pH | -0.018 | 0.004 | -4.10 | 0.0148 |
| Iron Content | -0.162 | 0.030 | -5.44 | 0.0055 |
| Aluminum Content | 2.765 | 0.578 | 4.79 | 0.0087 |

The experimentally measured solubility of scorodite decreased with pH in the range between 1 and 5 (Table 15), where adsorption of arsenic remains significantly active (Figure 14). The authors further observed arsenic solubility to pass through a minimum at the pH of about 4.0 (Table 15). Above pH of 5.0 arsenic and iron solubility increased with pH linearly although the values were lower than the solubility at pH less than 5. In other studies, uptake of arsenic has been found to be at a maximum at initial pH values within a range of 3.5 to 7.0 (Shevade and Ford, 2004). Adsorption of arsenic reported in this chapter reached its maximum values at pH of about 5 and decreased after pH of about 5 for most soils (Figure 14). The congruent dissolution of iron and arsenic in the pH range between 1 and 5 observed by Krause and Ettel (1988) indicates high affinity of iron for arsenic in this pH range, which decreases at pH values greater than 5 by the incongruent dissolution of iron and As. According to Pal (nodate: from Dufendach, 2002) adsorption of As(V) by $Fe(OH)_3$ decreases as pH rises, and increases as arsenic concentration rises. Their results indicate that adsorption of arsenic is linearly, negatively correlated to concentration of iron in solution, which is similar to the results observed in this chapter. Further, an approximate value of 1 mg/L Fe will adsorb about 60 ppb (16.7:1) of arsenic from a water source with a pH between 5 and 8 and an arsenic concentration between 0 and 750 ppb (Dufendach, 2002). The use of iron hydroxides for the coprecipitation of arsenic in industrial wastewater (in which arsenic is in the mg/L range) requires an iron dosage 4 to 8 times higher than that of the soluble arsenic; a greater iron dosage yields no further benefit, (Vance, 2001).

Dufandach (2002) have tested the portable ground water from wells in Alaska, with 237 ppb of arsenic and 9.43 mg/L iron contents and found that for redox potentials between 471 and 849 mV the arsenic content in the water reduces to 10ppb. The system reduced iron content to below secondary maximum contaminant goal and total organic carbon by 52%. Trivalent arsenite ($AsO_3^{3-}$) and pentavalent arsenate ($AsO_4^{3-}$) anion forms are both common in ground water. The above redox potential appears to be the most feasible for arsenic immobilization by iron. The oxidation of arsenite to arsenate is necessary to achieve effective arsenic removal, (Gupta, 1978). The reaction of ozone with arsenite is shown as:

$$AsO_3^{3-} + O_3 = AsO_4^{3-} + O_2 \qquad (39)$$

**Table 15. Solubility of crystalline scorodite (FeAsO₄.2H₂O) at 25°C**
**(Krause and Ettel, 1988)**

| Original pH | Equilibrium pH | Analytical Concentration (mg/L) | |
|---|---|---|---|
| | | Fe | As |
| | | Congruent dissolution | |
| 0.90 | 0.97 | 58.00 | 54.00 |
| 1.04 | 1.08 | 41.00 | 34.00 |
| 1.20 | 1.24 | 21.00 | 19.00 |
| 1.38 | 1.41 | 13.00 | 14.00 |
| 1.64 | 1.67 | 5.30 | 5.10 |
| 2.48 | 2.05 | 0.95 | 1.50 |
| 2.85 | 2.43 | 0.26 | 0.33 |
| | | Incongruent dissolution | |
| 3.45 | 2.55 | 0.043 | 0.19 |
| - | 2.64 | 0.053 | 0.11 |
| 5.57 | 2.65 | 0.058 | 0.20 |
| 10.00 | 2.69 | 0.035 | 0.093 |
| 11.05 | 2.86 | 0.04 | 0.08 |
| - | 5.42 | 0.013 | 0.59 |
| - | 6.53 | 1.50 | 3.80 |
| 11.76 | 7.08 | 7.80 | 31.00 |
| 12.06 | 7.79 | 12.00 | 129.00 |
| 12.36 | 7.92 | 52.00 | 463.00 |

The organic carbon content and presence of other oxidizable ion species in solution such as manganese will reduce the availability of ozone for the above reaction. Ozone reacts with ferrous iron in ground water to form the ferric hydroxide a gelatinous precipitate in two steps:

$$2Fe^{2+} + O_3 + H_2O = 2Fe^{3+} + O_2 + 2OH^- \tag{40}$$
$$2Fe^{3+} + 6H_2O = 2Fe(OH)_3 + 6H^+ \tag{41}$$

According to Twidell (1999) "arsenic is effectively removed from solution by ferric precipitation but the removal is via adsorption rather than ferric arsenite precipitation", (Twidwell, 1999). However, statements by other researchers such as: "ferric arsenate is not a suitable compound for As control in mine effluents while nanocrystalline scorodite that can be easily precipitated at ambient pressure and temperature conditions would be satisfactory in meeting the regulatory guidelines at pH 3 to 4 (Paktrunc and Buggerman, 2010; Krause and Ettel, 1985) corroborate the statement by Twidell (1999). $FeAsO_4$ precipitate indicates an arsenic solubility of 160mg/L at pH 5.0 (Krause and Ettel, 1987). It is possible that a solid solution in the form of $FeAsO_4$-$Fe(OH)_3$ is formed during the adsorption process. Further studies on the effect of iron content on arsenic adsorption including the interference by total organic and inorganic carbon is required. Dufendach, (2002) also recommends further study to determine the reliability of iron as a surrogate for arsenic in his report. These results would be particularly timely as it is anticipated that 12% of public water systems in the Western

United States (US) will be out of compliance with the new arsenic standard in 2006, (US Environmental Protection Agency 2000).

### 4.1.1. Validation of Prediction Model

The non linear mixed model fitted for the first half of the data was validated using data from the remaining five soil types that were not used to fit the prediction model. Surface plots showing measured and model predicted adsorption for these soils (validation data set) as a function of aluminum content, surface area and soil pH are shown in Figures 15 through 18.

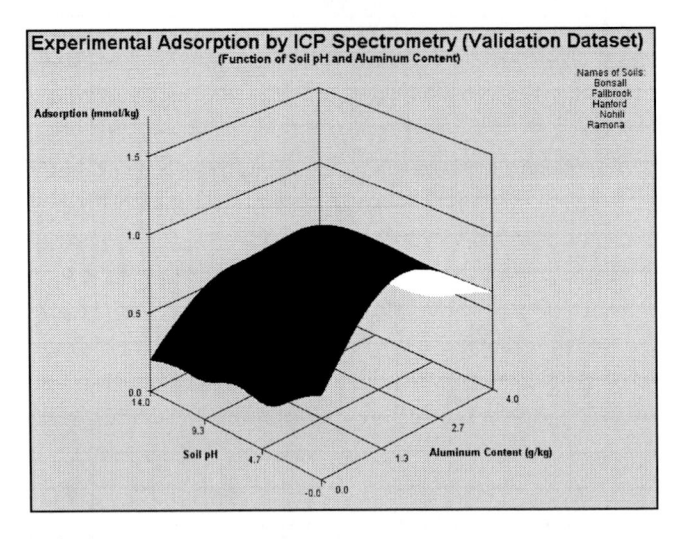

Figure 15. Surface plot of experimental adsorption of arsenic as a function of soil pH and aluminum content for validation data set.

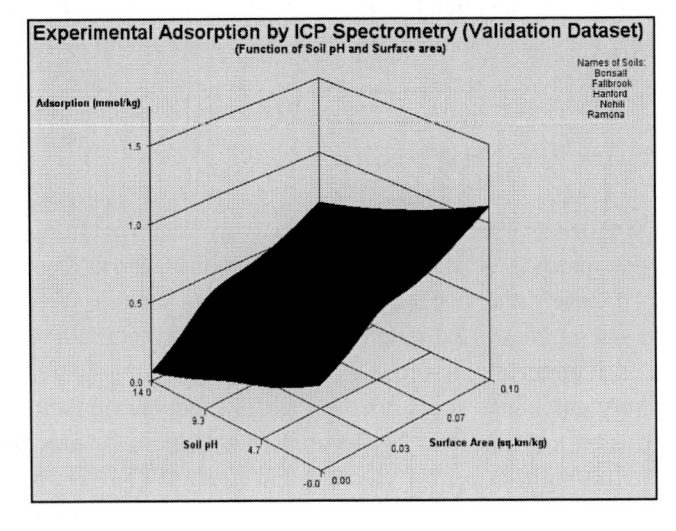

Figure 16. Surface plot of experimental adsorption of arsenic as a function of soil pH and surface area for validation data set.

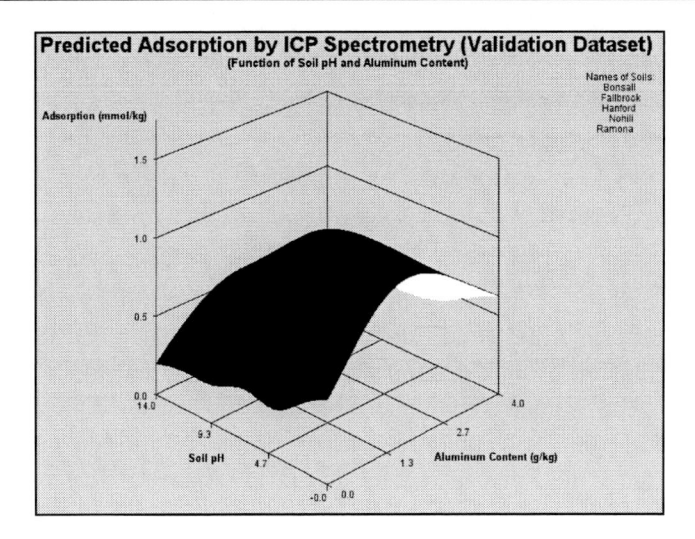

Figure 17. Surface plot of model predicted adsorption of arsenic as a function of soil pH and aluminum content for validation data set.

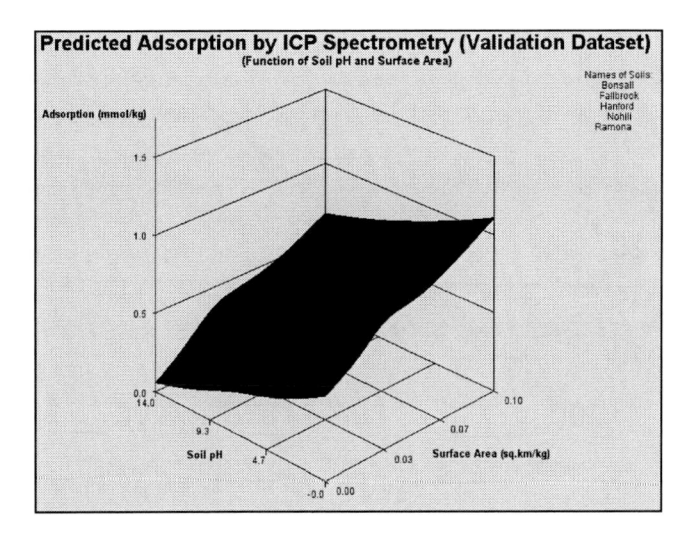

Figure 18. Surface plot of model predicted adsorption of arsenic as a function of soil pH and surface area for validation data set.

The adsorption predicted by the model fits the measured data well indicating that the fitted model is adequate to predict arsenic adsorption in soils. Model diagnostic statistics indicated adequacy of the models as well (AIC (smaller the better) =-253.9; BIC(smaller the better) =-258.2).

## 4.2. Surface Precipitation of Arsenic onto Hydrated Oxides of Lanthanides and Actinides

The following linear free energy correlations were obtained for the pure bulk solid precipitation of arsenates of lanthanides and actinides by performing a multiple linear regression analysis using experimental standard state Gibbs free energies of formation and cationic radii of the trivalent cations (Table 16) based on Eq. (13).

$$\Delta G^0_{f,MAsO_4(s)} - 35.1\,r_{M^{3+}} = 0.85\,\Delta G^0_{n,M^{3+}} - 2595.90 \qquad (42)$$

The above regression equation was validated using the Gibbs free energy of formation of the bulk solid precipitate of arsenates of lanthanides and actinides calculated from first principles as described in Krause and Ettel (1988). The calculated and the estimated Gibbs free energies of formation of bulk precipitates of arsenates of lanthanides and actinides are shown in Table 16. The estimated correlations between the estimated precipitates against the independent variables (non solvation free energy of the cation and cationic radii) were statistically significant at the 5 percent significance level (Figure 19, $R^2$=1.0).

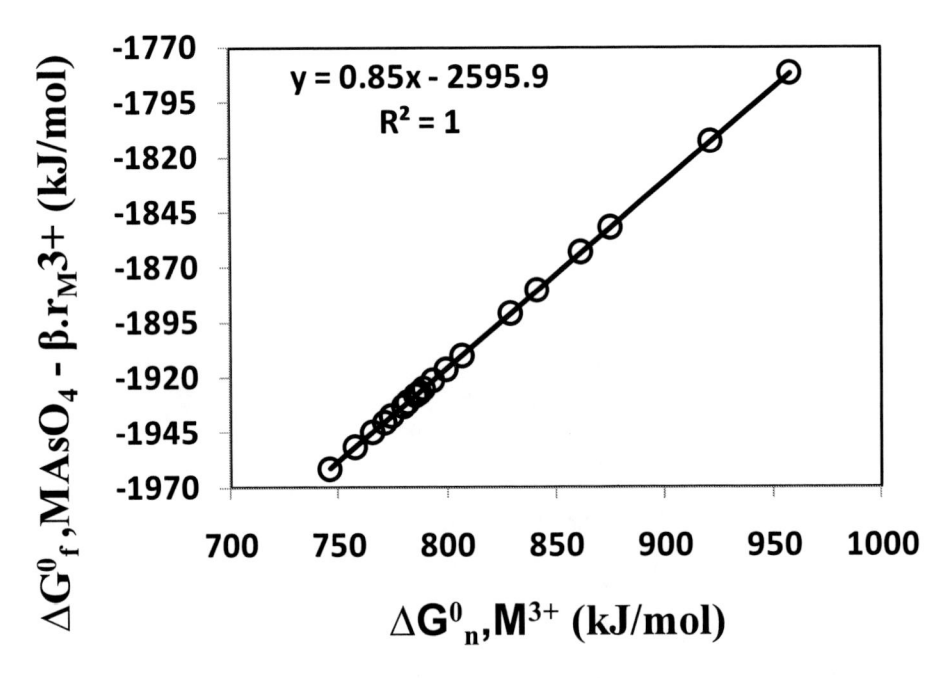

Figure 19. Multiple linear regression. Graphical representation of Eq. (13) for the standard state Gibbs free energies of formation of the bulk solid trivalent metal arsenates with lanthanum arsenate structure. Regression calculations were computed using Eq. (13) and using the data from Table 16. The vertical axis shows the left side of Eq. (13).

The calculated and the estimated bulk precipitates are shown in Figure 20 ($R^2$=0.99). The estimated Gibbs free energies are within 0.41 % of the calculated values. The standard state Gibbs free energies of formation of the bulk solid precipitate estimated from the above correlations are shown in Table 16. The discrepancies between the calculated and the estimated Gibbs free energies are acceptable considering the metastable nature of the solids. These correlations from Eq. (42) allow prediction of standard state Gibbs free energies of formation of the pure end-members of solid solution, and the equilibrium constants for solid solution reactions for which there are no reliable experimental data currently available.

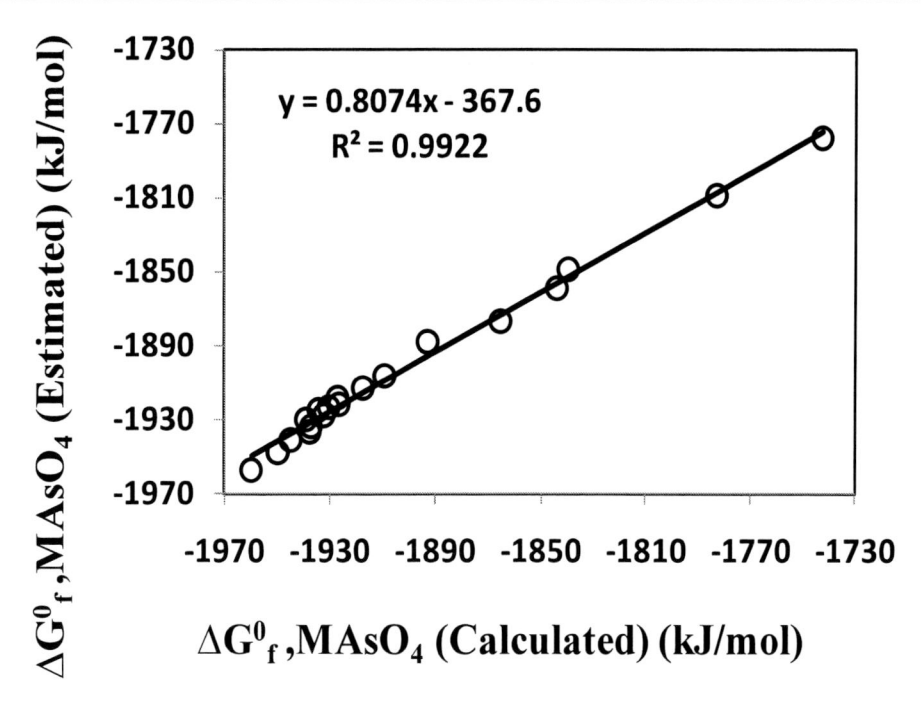

Figure 20. Graphical representation of the calculated and the estimated standard state Gibbs free energies of formation of the bulk solid trivalent metal arsenates with lanthanum arsenate structure. The estimated values are by Eq. (13) and using the data from Table 16. The calculated values are as described in Krause and Ettel (1988) and using data from Table 16.

The following regression correlations (Eq. (43)) were developed for the Gibbs free energies of formation of the pure end member components of the solid solutions, using the coefficients $a_{MvX}, \beta_{MvX}$ obtained from the correlations developed for the bulk solids with LA structure (Eq. (42)) and the coefficient $b_{MvX}$ obtained from a previous calculation for the surface precipitation of trivalent lanthanides and actinides onto hydrated lanthanum oxides.

$$\Delta G^0_{f,MAsO_{43}(ss)} - 35.1 r_{M^{3+}} = 0.85 \Delta G^0_{n,M^{3+}} - 1273.07 \tag{43}$$

The standard state Gibbs free energies of formation of the pure end members of the solid solution with LA structure obtained using Eq. (43) are given in Table 16 (column 7). The $\Delta G^0_{R,MvX(ss)}$ values calculated from Eq. (35) for the solid solution reactions for LA (column 8) and the equilibrium surface precipitation constants (column 9) calculated from Eq. (36) for the respective solid end members are listed in Table 16. The equilibrium constants (solubility products) of the bulk $M(OH)_3(s)$ solid precipitates were also calculated in similar way using the standard state Gibbs free energies of formation of pure bulk solids in Eq. (35) for the respective solids and the equilibrium constants calculated from Eq. (36). The equilibrium constants of the bulk solid precipitation reactions thus calculated are shown in column 10 of Table 16 (for LA structure) for comparison. The equilibrium constants for the solid solution reactions were much smaller than that of the bulk solid precipitation reactions for LA structure (see columns 9 and 10, Table 16). The results are consistent with the results from Farley et. al. (1985) surface precipitation model, which states that the solubilities of the solid solution components are far lower than those of the bulk solid precipitates which allow surface precipitation to occur at much lower concentrations of the cations in the solutions than that required for bulk solid precipitation to occur.

**Table 16. Experimental, calculated and estimated thermodynamic data for pure bulk and solid solution of arsenates of trivalent lanthanides and actinides with hydrous oxides of lanthanides and actinides**

| $M^{3+}$ | $r_{M^{3+}}$ | $\Delta G^o_{f,M^{3+}}$ | $\Delta G^o_{n,M^{3+}}$ | $\Delta G^o_{f,MAsO_4(s)}$ | $\Delta G^o_{f,MAsO_4(s)}$ | Difference Calc(2)-Cal (1) | $\Delta G^o_{f,MAsO_4(ss)}$ | $\Delta G^o_{R,MAsO_4(ss)}$ | $Log(K)$ | $Log(K)$ |
|---|---|---|---|---|---|---|---|---|---|---|
| | (nm) | (kJ/mol) | (kJ/mol) | (kJ/mol) | (kJ/mol) | (kJ/mol) | (kJ/mol) | (kJ/mol) | Bulk Precip. | Surface Precip. |
| $Dy^{3+}$ | 0.09 | -665.70 | 787.21 | -1927.47 | -1923.59 | -3.88 | -600.76 | 713.33 | 106.76 | -124.96 |
| $Er^{3+}$ | 0.09 | -669.47 | 785.41 | -1931.23 | -1925.21 | -6.02 | -602.38 | 715.48 | 106.38 | -125.33 |
| $Tm^{3+}$ | 0.09 | -662.35 | 793.33 | -1924.12 | -1918.52 | -5.60 | -595.69 | 715.05 | 106.46 | -125.26 |
| $Lu^{3+}$ | 0.09 | -628.02 | 829.12 | -1889.81 | -1888.16 | -1.65 | -565.33 | 711.08 | 107.15 | -124.56 |
| $Ho^{3+}$ | 0.09 | -674.07 | 779.85 | -1935.83 | -1929.89 | -5.95 | -607.06 | 715.41 | 106.40 | -125.32 |
| $La^{3+}$ | 0.11 | -684.12 | 757.70 | -1945.88 | -1948.13 | 2.26 | -625.30 | 707.21 | 107.83 | -123.88 |
| $Nd^{3+}$ | 0.10 | -671.98 | 774.61 | -1933.74 | -1933.99 | 0.25 | -611.16 | 709.21 | 107.48 | -124.23 |
| $Ce^{3+}$ | 0.10 | -672.40 | 771.37 | -1934.16 | -1936.61 | 2.45 | -613.78 | 707.01 | 107.87 | -123.85 |
| $Pr^{3+}$ | 0.10 | -679.52 | 765.77 | -1941.27 | -1941.44 | 0.17 | -618.61 | 709.30 | 107.47 | -124.25 |
| $Gd^{3+}$ | 0.09 | -661.51 | 789.21 | -1923.28 | -1921.78 | -1.50 | -598.95 | 710.96 | 107.18 | -124.54 |
| $Sm^{3+}$ | 0.10 | -666.96 | 781.88 | -1928.72 | -1927.92 | -0.80 | -605.09 | 710.26 | 107.30 | -124.42 |
| $Tb^{3+}$ | 0.09 | -652.30 | 799.51 | -1914.08 | -1913.07 | -1.00 | -590.24 | 710.45 | 107.26 | -124.45 |
| $Yb^{3+}$ | 0.09 | -644.35 | 806.96 | -1906.13 | -1906.72 | 0.59 | -583.89 | 708.85 | 107.55 | -124.17 |
| $Y^{3+}$ | 0.11 | -694.17 | 746.28 | -1955.92 | -1957.77 | 1.85 | -634.94 | 707.62 | 107.76 | -123.95 |
| $Eu^{3+}$ | 0.10 | -574.43 | 875.42 | -1836.25 | -1848.45 | 12.20 | -525.62 | 697.19 | 109.59 | -122.13 |
| $Am^{3+}$ | 0.11 | -599.51 | 841.66 | -1861.32 | -1876.73 | 15.41 | -553.90 | 694.00 | 110.15 | -121.57 |
| $Np^{3+}$ | 0.11 | -517.45 | 921.57 | -1779.31 | -1808.71 | 29.40 | -485.88 | 679.96 | 112.61 | -119.11 |
| $U^{3+}$ | 0.12 | -476.50 | 957.85 | -1738.39 | -1777.64 | 39.25 | -454.81 | 670.08 | 114.34 | -117.38 |
| $Pu^{3+}$ | 0.11 | -578.41 | 862.05 | -1840.23 | -1859.37 | 19.14 | -536.54 | 690.26 | 110.80 | -120.91 |

Cationic radii are from Shock and Helgeson, (1988) and Shannon, (1976). Values of $\Delta G^o_f$ of the cations are from Brookins, (1988). The calculated $\Delta G^o_{f(s)}$ values of the bulk precipitate solid crystals are from Eq. (42). Calculated $\Delta G^o_{R(ss)}$ and $\Delta G^o_{f(ss)}$ values of the solid solution component are from Eq. (35) and Eq. (43). Calculated Log $K$ values for the solid reactions (column 11 from left) and the bulk precipitation (column 10 from left) are from Eq. (36) using the $\Delta G^o_{f(ss)}$ of the component of the solid solution (column 8 from left) and that of bulk precipitate (column 6 from left). All calculations are at 25°C and 1 bar.

# 5. CONCLUSIONS

Hydroxides of lanthanides may be used as precipitants for removing arsenic from water and waste water. Such compounds are abundant in the environment and have been found to bind with arsenic in the environment. The reactions of lanthanide ions with arsenic (V) ions indicate that lanthanides can be used to immobilize arsenic successfully. Hydroxides of lanthanides appear to remove arsenic more efficiently under the same conditions, than iron (III), aluminum, calcium and magnesium hydroxides. Arsenates of lanthanides are extremely insoluble and form stable precipitates under a wide range of pH. The solubility products of 15 lanthanide arsenates in aqueous solution at $25 \pm 1°C$ indicates that lanthanum arsenate (pK, = 21.45) in the most soluble while scandium arsenate (pK, = 26.72) is the least soluble. Calculated equilibrium surface precipitation constants (Log(K) between -125.33 and -117.38) of lanthanum arsenates were much smaller than that of their bulk counterparts (Log(K) between 106.38 and 114.34 ) indicating that surface precipitation could be the predominant mechanism of arsenic removal in aqueous solutions of oxides of lanthanides. Surface precipitation of arsenic (V) using lanthanides may immobilize both arsenic and lanthanides effectively. A predictive model developed for the effect of soil parameters on adsorption of arsenic from ground water indicates significant negative correlation of iron concentration (p=0.0055) and pH (p=0.0149) of solution with the amount of arsenic adsorbed, while significant positive correlations exist between the concentration of aluminum in solution and the amount of arsenic adsorbed (0.0087) and between surface area of solid and the amount of arsenic adsorbed (0.0052).

# REFERENCES

Ahmed SA, et al. (2004) Arsenicosis in two villages in terai, lowland Nepal. *Environmental Science.* 11(3):179-88.

Arai Y, Sparks DL, and Davis JA. (2004) Effects of dissolved carbonate on arsenate adsorption and surface speciation at the hematite-water interface. *Environ. Sci. Technol.* 38(3):817–824.

Berg M, Tran HC, Nguyen TC, Pham HV, Schertenleib R, Giger W. (2001) Arsenic contamination of ground and drinking water in Vietnam: A human health threat. *Environmental Science and Technology.* 35: 2621-2626

Blesa MA, Weisz AD, Morando PJ, Salfity JA, Magaz GE, and Regazzoni AE. (2000) The interaction of metal oxide surfaces with complexing agents dissolved in water. *Coord.Chem. Rev.* 196:31–63.

Bothe JV, and Brown PW. (1999) *Environ. Sci. Technol.* 33:3806-3811.

Brookins DG. *Eh-pH Diagrams for geochemistry*, Springer, Berlin, 1988.

Burns PE, Hyun S, Lee LS, and Murarka I. (2006) Characterizing As(III, V) adsorption by soils surrounding ash disposal facilities, *Chemosphere.* 63:1879-1891.

Casey WH, Westrich H. (1992) *Nature.* 355:157.

Chukhlantsev VG. *J. Inorg. Chem.* USSR. (1965) 1:41-48.

Cihacek LJ, and Bremner JM. (1979) A simplified ethylene glycol monoethyl ether procedure for assessing soil surface area. *Soil Sci. Soc. Am. J.* 43:821–822.

Cullen ER, and Reimer KJ. (1989) Arsenic speciation in the environment. *Chemical Reviews.* 89:713-764.

Davis SA, and Misra M. (1997) Transport Model for the Adsorption of Oxyanions of Selenium ( IV) and Arsenic (V) from Water onto Lanthanum and Aluminum-Based Oxides. *Journal of Colloid and Interface Science.* 188:340–350.

Davis JA, and Leckie JO. (1978) Surface ionization and complexation at the oxide/water interface. II. Surface properties of amorphous iron oxyhydroxide and adsorption of metal ions. *J. Colloid Interf. Sci.* 67:90–107.

Davis JA, James RO, and Leckie JO. (1978) *J. Colloid Interface Sci.* 63(3).

Dixit S, and Hering JG. (2003) Comparison of arsenic(V) and arsenic(III) sorption onto iron oxide minerals: Implications for arsenic mobility. *Environ. Sci. Technol.* 37(18):4182-4189.

Dove PM, and Rimstidt JD. (1985) *Am. Miner.* 70:838–844.

Dufendach, JW, 2002: http://campwater.com/alaska/docs/CoprecipitationPaper2002.pdf.

Dzombak DA, Morel FMM. Surface Complexation Modeling: Hydrous Ferric Oxide, John Wiley and Sons, New York, 1990.

Edwards M. (1994) Chemistry of Arsenic Removal During Coagulation and Fe-Mn Oxidation. *Journal AWWA.* 86(9):64–78.

Exner O. *Correlation Analysis of Chemical Data*, Plenum, New York, 1988.

Farley KJ, Dzombak DA, and Morel FMM. (1990) Surface precipitation model for the sorption of cations on metal oxides. *J. Colloid Interface Sci.* 106:226-242.

Farquhar ML, Charnock JM. Livens FR, and Vaughan DJ. (2002) Mechanisms of arsenic uptake from aqueous solution by interaction with goethite, lepidocrocite, mackinawite, and pyrite: an X-ray absorption spectroscopy study. *Environ. Sci. Technol.* 36(8):1757-1762.

Fendorf S, Eick MJ, Grossl P, and Sparks DL. (1997) Arsenate and chromate retention mechanisms on goethite. 1. Surface structure. *Environ. Sci. Technol.* 31:315–320.

Firsching FH. (1992) Solubility Products of the Trivalent Rare-Earth Arsenates. *J. Chem. Eng. Data.* 37(4):497-499.

Firsching FH. (1991) Solubility Products of the Trivalent Rare-Earth Phosphates. *J. Chem. Eng. Data.* 36(1):93-95.

Fukushi K, Sasaki M, Sato T, Yanase N, Amano H, and Ikeda H. (2003) A natural attenuation of arsenic in drainage from an abandoned arsenic mine dump. *Appl. Geochem.* 18(8):1267-1278.

Fukushi K, and Sverjensky DA. (2007) A surface complexation model for sulfate and selenate on iron oxides consistent with spectroscopic and theoretical molecular evidence. Geochim. *Cosmochim. Acta.* 71:1-24.

Fuller CC, and Davis JA. (1989) Influence of coupling of sorption and photosynthetic processes on trace-element cycles in natural-waters. *Nature.* 340(6228):52-54.

Gao Y, and Mucci A. (2001) Acid base reactions, phosphate and arsenate complexation, and their competitive adsorption at the surface of goethite in 0.7 M NaCl solution. *Geochim. Cosmochim. Acta.* 65:2361-2378.

Garcia-Sanchez A, Moyano A, and Mayorga P. (2005) High arsenic contents in groundwater of central Spain. *Environ. Geol.* 47:847-54.

Goldberg S, Lesch SM, Suarez DL, and Basta NT. (2005) Predicting arsenate adsorption by soils using soil chemical parameters in the constant capacitance model. *Soil Sci. Soc. Am. J.*, 69:1389-1398.

Goldberg S. (1986) Chemical modeling of arsenate adsorption on aluminum and iron oxide minerals. *Soil Sci. Soc. Am. J.* 50:1154-1157.

Goldberg S, and Johnston CT. (2001) Mechanisms of arsenic adsorption on amorphous oxides evaluated using macroscopic measurements, vibrational spectroscopy, and surface complexation modeling. *J. Colloid Interface Sci.* 234:204-216.

Grossl PR, Eick M, Sparks DL, Goldberg S, and Ainsworth CC. (1997) Arsenate and chromate retention mechanisms on goethite. 2. Kinetic evaluation using a pressure-jump relaxation technique. *Environ. Sci. Technol.* 31(2):321-326.

Halter WE, and Pfeifer HR. (2001) Arsenic (V) adsorption onto alpha-Al2O3 between 25 and 70 degrees C. *Appl. Geochem.* 16(7–8):793-802.

Hering JG, and Dixit S. (2005) Contrasting sorption behavior of arsenic(III) and arsenic(V) in suspensions of iron and aluminum oxyhydroxides. *Adv. Arsenic Res.* 915: 8-24.

Hiemstra T, and van Riemsdijk WH. (1999). Surface structural ion adsorption modeling of competitive binding of oxyanions by metal hydroxides. *J. Colloid Interface Sci.* 210:182–193.

Holm TR. (2002) Effects of CO32−/bicarbonate, Si and PO43- on arsenic sorption to HFO, *Journal AWWA.* 94(4):174–181.

Hopenhayn C. (2006) Arsenic in drinking water: impact on human health. *Elements* 2(2): 103-107.

Hyun S, Burns PE, Murarka I, and Lee LS. (2006) Selenium(IV) and (VI) sorption by soils surrounding fly ash management facilities. *Vadose Zone J.* 5:1110-1118.

Inegbenebor AI, Thomas JH, and Williams PA. (1989) *Miner. Mag.* 53:363–371.

Kelemen B. (1991) Arsenic removal from water containing humic substances by ferric-salt. Water Research Institute (VITUKI), Budapest, 1991. in Hungarian.

Krause E, and Ettel VA. (1985) Ferric arsenate compounds: Are they environmentally safe? Solubilities of basic ferric arsenates. In A.J. Oliver, Ed., Impurity control and disposal. Proceedings, *15th Annual Hydrometallurgy Meeting CIM*, Vancouver, Canada, August 18-22, 1985

Krause E, and Ettel VA. (1987) Solubilities and stabilities of ferric arsenates. In G.L. Strathdee, M.O. Klein, and LA. Melis, Eds., Crystallization and precipitation. Proceedings I,S CAP'87, Saskatoon, Canada, October 5-7, 1987, 195-210.

Krause E, and Ettel VA. (1988). Solubility and stability of scorodite FeAsO4.2H2O: New data and further discussion. *American Mineralogist.* 73: 850-854.

Laky D, László B. (2007) Arsenic removal from drinking water by coagulation – laboratory and pilot-scale experiments. *Proceedings of the IWA – UK's Young Water Professionals Conference,* Guildford, United Kingdom, April 18-20, 2007, 44–52.

Ladeira ACQ, Ciminelli VST, Duarte HA, Alves MCM, and Ramos AY. (2001) Mechanism of anion retention from EXAFS and density functional calculations: Arsenic (V) adsorbed on gibbsite. *Geochim. Cosmochim. Acta.* 65(8):1211-1217.

Langmuir D, Mahoney J, MacDonald A, Rowson J. (1999) *Geochim. Cosmochim. Acta.* 63: 3379-3394.

Licskó I. Colloid Destabilization in the Treatment of Surface Waters Containing Colloid-Stabilizing Pollutants, *Proceedings of the 4th Gothenburg Symposium*, October 1-3, 1990, Madrid, Spain, 55–73.

Lipfert G, Reeve AS, Sidle WC, and Marvinney R. (2006) Geochemical patterns of arsenic-enriched ground water in fractured, crystalline bedrock, Northport, Maine, USA, *Applied Geochemistry*. 21(3): 528-545.

Liu R, Qu J, Xia S, Zhang G, and Li G. (2007) Silicate Hindering In Situ Formed Ferric Hydroxide Precipitation: Inhibiting Arsenic Removal from Water, *Environmental Engineering Science*. 24 (5): 707-715.

Magalhães MCF, de Jesus JDP, Williams PA. (1988) *Miner. Mag.* 52:679-690.

Magalhães MCF. (2002) Arsenic. An environmental problem limited by solubility. *Pure Appl. Chem.* 74(10):1843-1850.

Manceau A. (1995) The mechanism of anion adsorption on iron oxides: evidence for the bonding of arsenate tetrahedra on free FeO(OH)6 edges. *Geochim. Cosmochim. Acta.* 59: 3647-3653.

Manning BA, and Goldberg S. (1996) Modeling competitive adsorption of arsenate with phosphate and molybdate on oxide minerals. *Soil Sci. Soc. Am. J.* 60(1):121-131.

Meng X, Bang S, and Korfiatis GP. (2000) Effects of silicate, sulfate, and carbonate on arsenic removal by ferric chloride. *Water Research.* 34(4): 1255-1261.

Misra M, and Nayak DC. *Process for Removal of Selenium and Arsenic From Aqueous Streams.* U.S. patent approved in August 1996.

Montiel A, and Welté B. (1996) Removal of Arsenic from Drinking Water, Proceedings of IWSA-AIDE International Workshop: *Natural Origin Inorganic Micropollutants: Arsenic and Other Constituents.* Wien, May 6-7, 1996.

Murphy WM, and Helgeson HC. (1987) *Geochim. Cosmochim. Acta.* 51:3137.

Myneni SCB, Traina SJ, and Waychunas GA, and Logan TJ. (1998) Experimental and theoretical vibrational spectroscopic evaluation of arsenate coordination in aqueous solutions, solids, and at mineral-water interfaces. *Geochim. Cosmochim. Acta.* 62(19–20):3285-3300.

Narasaraju TSB, Lahiri P, Yadav PR, and Rai US. (1985) *Polyhedron.* 4:53-58.

Nickson RT, McArthur JM, Ravenscroft P, Burgess WG, and Ahmed KM. (2000) Mechanism of arsenic release to groundwater, Bangladesh and West Bengal. *Applied Geochemistry.* 15:403-413

Nishimura T, Ito CT, Tozawa K, and Robins RG. (1985) Impurity, Control and Disposal, *Proceedings 15th Annual Hydrometallurgical Meeting*, Vancouver, Canada, 1985, 2.1-2.19.

Nordstrom DK. (2002) Public health-Worldwide occurrences of arsenic in ground water. *Science.* 296(5576):2143-2145.

Pal, BN. (no date) Granular Ferric Hydroxide for Elimination of Arsenic from Drinking Water, M/S Pal Trockner, Ltd., Calcutta, India

Paktunc D, Bruggeman K. (2010) Solubility of nanocrystalline scorodite and amorphous ferric arsenate: Implications for stabilization of arsenic in mine wastes, *Applied Geochemistry.* 25:674-683.

Peryea FJ. (1991) *Soil Sci. Soc. Am. J.* 55: 1301-1306.

Peters SC, and Blum JD. (2003) The source and transport of arsenic in a bedrock aquifer, New Hampshire, USA, *Applied Geochemistry.* 18 (11):1773-1787.

Peters SC, Blum JD, Klaue B, and Karagas MR. (1999) Arsenic occurrence in New Hampshire ground water. *Environmental Science and Technology* 33(9):1328-1333.

Ragavan AJ, and Adams VD. (2009) Estimation of equilibrium surface precipitation constants for trivalent metal sorption onto hydrous ferric oxide and calcite, *J. Nucl. Mater.* 389(3):394-401.

Rochette EA, Li GC, and Fendorf SE. (1998) *Soil Sci. Soc. Am. J.* 62:1530-1537.

Roddick-Lanzilotta AJ, McQuillan AJ, and Craw D. (2002) Infrared spectroscopic characterisation of arsenate (V) ion adsorption from mine waters, Macraes mine, New Zealand. *Appl. Geochem.* 17(4):445-454.

Savage KS, Bird DK, and Ashley RP. (2000) Legacy of the California gold rush: environmental geochemistry of arsenic in southern mother lode gold district. *Int. Geol. Rev.* 42:5385-5415.

Schreiber ME, Simo JA, and Freiberg PG. (2000) Stratigraphic and geochemical controls on naturally occurring arsenic in groundwater, eastern Wisconsin, USA. *Hydrogeology Journal.* 8(2):161-176.

Shevade S, Ford RG. (2004) Use of synthetic zeolites for arsenate removal from pollutant water. *Water Research.* 38:3197-3204.

Sherman DM, and Randall SR. (2003) Surface complexation of arsenic(V) to iron(III) (hydr)oxides: structural mechanism from ab initio molecular geometries and EXAFS spectroscopy. *Geochim. Cosmochim. Acta.* 67(22):4223.

Shannon RD. (1976) *Acta Crystallogr.* A 32: 751.

SHHWT, Standard Handbook of Hazardous Waste Treatment and Disposal (Harry M. Freeman, Editor-In-Chief). McGraw-Hill, New York, 1989.

Shock EL, and Helgeson HC. (1988) Calculation of the thermodynamic and transport properties of aqueous species at high temperatures: correlation algorithms for ionic species and equation of state predictions to 5 Kb and 100oC. *Geochim. Cosmochim. Acta* 52: 2009-2036.

Smedley PL, and Kinniburgh DG. (2002) A review of the source, behavior and distribution of arsenic in natural waters. *Applied Geochemistry.* 17(5):517-628.

Stumm W, and Morgan JJ. (1962) Chemical aspects of coagulation. *Journal AWWA.* 54(8):971-991.

Suarez DL, Goldberg S, Su C, and Manning BA. (1997) Evaluation of oxyanion adsorption mechanisms on oxides using FTIR spectroscopy and electrophoretic mobility. *Abstr. Papers Amer. Chem. Soc.* 213, 59-GEOC.

Sun X, and Doner HE. (1996) An investigation of arsenate and arsenite bonding structures on goethite by FTIR. *Soil Sci.* 161:865-872.

Sverjensky DA, and Molling PA. (1992) A linear free energy relationship for crystalline solids and aqueous ions. *Nature.* 356:231-234.

Sverjensky DA, Molling PA. (1992) A linear free energy relationship for crystalline solids and aqueous ions. *Nature.* 356:231-234.

Swedlund PJ, and Webster JG. (1999) Adsorption and polymerization of silicic acid on ferrihydrite, and its effect on arsenic adsorption. *Water Research.* 33(16):3413-3422.

Twidwell LG, McCloskey J, and Gale M. (1999) Technologies and Potential Technologies for Removing Arsenic from Process and Mine Waste Water

Vance DB. (2001) Arsenic-Chemical Behavior and Treatment: http://www.2the4. net/arsenic.

Voegelin A, and Hug SJ. (2003) Catalyzed oxidation of arsenic(III) by hydrogen peroxide on the surface of ferrihydrite:an in situ ATR-FTIR study. *Environ. Sci. Technol.* 37(5):972-978.

Voigt DE, Brantley SL, Hennet RJC. (1996) *Appl. Geochem.* 11:633–643.

Wang Y and Xu H. (2000) in: Materials Research Society Symposium Proceedings, *Scientific Basis for Nuclear Waste Management.* 23:367.

Wang Y, Xu H. (2001) *Geochim. Chosmochim. Acta.* 65:1529.

Waychunas GA, Rea BA, Fuller CC, and Davis JA. (1993). Surface chemistry of ferrihydrite: Part 1. EXAFS studies of the geometry of coprecipitated and adsorbed arsenate. *Geochim. Cosmochim. Acta.* 57:2251-2269.

Waychunas GA, Trainor TP, Eng PJ, Catalano JG, Brown Jr. GR, Davis J, Rogers J. and Bargar J. (2005) Surface complexation studied via combined grazing-incidence EXAFS and surface diffraction: arsenate on hematite (0001) and (10–12). *Anal. Bioanal. Chem.* 383:12-27.

Welch AH, and Lico MS. (1998). Factors controlling As and U in shallow ground water, southern Carson Desert, Nevada. *Applied Geochemistry.* 13(4):521-539.

Wells PR. *Linear Free Energy Relationships.* Academic Press, London, 1968.

World Health Organization (2001) *Arsenic in drinking water, fact sheet.* No. 210.

Wu MM, Kuo TI, Hwang YH, and Chen CJ. (1989) Dose-response relation between arsenic well water and mortality from cancer. *American Journal of Epidemiology.* 130(6):1123-1132.

Xu H, and Wang Y. (1999) *J. Nucl. Mater.* 275:211.

Xu H, and Wang Y. (1999) *J. Nucl. Mater.* 275:216.

Xu H and Wang Y. (1999) *Radiochim. Acta.* 87:37.

Xu H and Ewang Y, Barton L. (1999). *J. Nucl. Mater.* 273:343.

Zavarin M, and Bruton CJ. A Non-Electrostatic Surface Complexation Approach to Modeling Radionuclide Migration: The Role of Iron Oxides and Carbonates. *Proceedings of Migration,* Incline Village, Nevada, September 26 - October 1, 1999.

Zhu C, and Sverjensky DA. (1992) *Geochim. Cosmochim. Acta* 56:3435-3467.

In: Nuclear Materials
Editor: Michael P. Hemsworth, pp. 47-80

ISBN 978-1-61324-010-6
© 2011 Nova Science Publishers, Inc.

*Chapter 2*

# EXPERIMENTAL STUDIES AND FIRST PRINCIPLES CALCULATIONS IN NUCLEAR FUEL ALLOYS FOR RESEARCH REACTORS

*P. R. Alonso[1,2][1], P. H. Gargano[1,2], L. Kniznik[1], L. M. Pizarro[1,2], and G. H. Rubiolo[1,2,3]*

[1]Departamento de Materiales – Centro Atómico Constituyentes – Comisión Nacional de Energía Atómica, San Martín, Buenos Aires, Argentina
[2] Instituto Sabato, Universidad Nacional de General San Martín-Comisión Nacional de Energía Atómica, Argentina
[3]Consejo Nacional de Investigaciones Científicas y Tecnológicas, CONICET, Argentina

## ABSTRACT

We present here a review of our recent years work and some new results in phase stability determination in uranium alloys, as a contribution to progress in nuclear fuel alloys for research reactors development. In the frame of RERTR Program (Reduced Enrichment for Research and Test Reactors) it is being developed a high density uranium based fuel that could remain stable in the body cubic centered (bcc) phase during fabrication and irradiation in the reactor. Research is focused in an U-Mo alloy dispersed fuel in aluminum matrix. The main problem focuses in an undesirable growth of the interface between fuel and Al matrix. This problem could be reduced with the addition of Si or/and Zr. Our efforts have been devoted to enlighten phase stability knowledge. U-Mo, U-Al and pseudo-binary $UAl_3$-$USi_3$ systems have been studied with DFT first principles methods to establish their ground states. U-Mo finite temperature bcc phases equilibria have been determined from first principles calculations applying the cluster variation method, while MonteCarlo simulations were applied for the finite temperature study of $UAl_3$-$USi_3$ system. U-Al system was also interest from a kinetic point of view since compounds formation and growth are crucial for nuclear fuel performance. Thus, phases equilibria have been described with a semiempirical CALPHAD method capable

---

[1] Corresponding author e-mail: pralonso@cnea.gov.ar, rubiolo@cnea.gov.ar.

of being used as input for kinetic calculations and diffusion parameters have also been evaluated so as to use DICTRA package to simulate interphase growth in a $UAl_3$-Al diffusion couple. Finally, the effect of the addition of Zr and Si has been analyzed from an experimental approach. Overall results have succeeded in reproducing known data in cases where it was available and to provide with novel insights of bonding interactions in the alloys formation as well as new valuable experimental data.

# 1. INTRODUCTION

The Materials Testing Reactors cores currently work with high-enriched, low-density $UAl_x$ and $U_3Si_2$ based dispersion fuel [1,2]. In the last four decades international effort has been devoted to develop a more dense fuel capable to meet the requirements of the nuclear non-proliferation treaty, with low or no modification in initial design. This treaty promotes peaceful nuclear issues and gives a value of 20% as the maximum permissible content of $^{235}$U enrichment (low enriched uranium (LEU)) in the fuel that could replace the existing one while having a high uranium density (>15.0 gU/cm3) and good irradiation behavior. Body-centered-cubic uranium alloys are considered as strong candidates.

The objective is that the fuel could remain stable in the body centered cubic phase ($\gamma$U bcc solid solution) during fabrication ($\approx$500°C) and irradiation ($\approx$250°C), i.e., at temperatures at which $\alpha$U is the equilibrium phase [3]. This can be achieved by alloying U with proper metals. Several transition metals, particularly 4d and 5d elements in Group IV through VIII, form high temperature solid solutions with $\gamma$U and this cubic phase can be retained in its metastable state upon cooling. Among them, Mo was recognized as a good candidate [4], showing a high solubility in $\gamma$U ($\approx$ 35 at %) and an acceptable amount ($\approx$ 20 at. %) needed to stabilize $\gamma$U during fuel element fabrication at a working temperature of $\approx$500°C. Small amounts of a third element such as transition metals 4d and 5d, from groups VII and VIII, have a powerful stabilizing effect when added to U-Mo alloys [5,6].

The development of fuel elements based in LEU(Mo) alloys is in progress and encouraged worldwide activities such as in pile experiments, out of pile experiments and computational simulations. The following paragraphs introduce the research areas where our effort has contributed to understand the performance of such nuclear fuel elements.

## 1.1. Role of Interaction Energies in $\gamma$U(Mo) Phase Stability

Hofmann and Meyer [7] have found an empirical relationship between the measured nucleation time for the decomposition of metastable $\gamma$U(Mo) phase in $\alpha$ U(Mo) + $U_2$Mo (the position of the "nose" in the TTT diagram) and the enthalpy of mixing of the $\gamma$ phase as estimate from semi-empirical method by Miedema et al [8]. From this, it appears that the activation energy of nucleation is proportional to the negative of the enthalpy of mixing. Moreover, they show that this correlation can be extended to the addition of a third element X (transition metals, 4d and 5d, from groups VII and VIII) that replaces atoms of uranium in the alloy U-Mo. Ternary enthalpy of mixing is estimated as the sum of that in the binary alloys U-Mo and U-X.

Besides, the ultimate effectiveness of third element substitution on $\gamma$ stabilization depends on the solubility limit of this element in $\gamma$U(Mo) alloy and on the value of enthalpy of mixing in the ternary system. Unfortunately, solubilities of these third elements in the $\gamma$U(Mo) phase are not well known.

A report was published in 2001 [9] which compiles thermodynamic data on the U-Mo alloy, enthalpy and Gibbs energy of formation, and presents them as a function of the uranium composition and temperature in the form of a suitable polynomial. The authors also stated that there were no previous data on the thermodynamic functions of the quenched metastable alloy below the stable $\gamma$ phase region, and measured the enthalpy increment of quenched $U_{0.823}Mo_{0.117}$ alloy from 299 to 820 K. From this data, together with the estimation of the enthalpy of formation at 298.15 K from Miedema model [8] and the entropy at the same temperature from the method of irreversible thermodynamics, they gave the expression for the Gibbs energy of formation. They finally compared these results with estimates obtained by applying the regular solution model and an empirical approach [10] to nearest neighbors' bond energy.

With the aim of emphasizing the role of bonding energy between atoms in the decomposition of the metastable $\gamma$ phase and the solubility of third elements in U-Mo alloy, we decided to calculate the thermodynamic functions of the binary U-Mo by using first principles thermodynamics [11]. This formalism establishes that the thermodynamic properties of an alloy can, in principle, be computed as accurately as desired through a technique known as the *cluster expansion*. A cluster is defined as a set of lattice points chosen in such a way that it contains the maximum correlation length to be considered. In earlier works on the U-Mo system [12] we have constructed a "minimal" start-up cluster expansion on the simplest six bcc-based ordered structures whose total energies of formation were calculated and its effective cluster interactions (ECIs) determined. The minimal cluster expansion consisted of the empty cluster, all the point clusters, all the nearest-neighbor pairs surrounding all atoms and the irregular tetrahedron. An U-Mo phase diagram was obtained in that approach, showing discrepancies with experimental one in the composition range where $\gamma$U(Mo) should be stable. That first approach thus remarked that the irregular tetrahedron interaction plays no significant role in stabilizing $\gamma$U(Mo) phase. In a later study [13], we performed first-principles electronic calculations of sixteen totally relaxed bcc-based ordered U-Mo structures. We decided which clusters to retain in the cluster expansion of the thermodynamic functions by comparing their ability to reproduce the experimental ground state while exhibiting a good convergence of energy cluster interactions.

We briefly discuss in this paper the dominant role of multisite interactions in the stabilization of disordered $\gamma$U(Mo) phase and how the platinum addition reinforces the stabilization modifying the values of these multisite interactions.

## 1.2. Interaction Layer between the $\gamma$U(Mo) Fuel and the Aluminum Matrix

Irradiation of U-Mo fuel dispersed in aluminum matrix results in the formation of an interaction layer through a diffusion reaction process that coats the fuel particle surface. Consequently, an increase in thickness is observed, weakening the fuel meat cohesion. Porosities observed as small voids occurring in-between the fuel meat particles have also been reported during irradiation experiments of these dispersion fuel elements [14]. This overall behavior can lead to failure of the fuel plates. In experiments with a U–7wt%Mo alloy, the composition of the interaction layer was classified in two major layers and measured as 77 or 82 at.% Al [14]. As the ratio of U/Mo content in both layers was unaltered by irradiation, the authors wrote the compositions as $(U,Mo)Al_3$ and $(U,Mo)Al_{4.4}$. Mirandou et al [15] found that the constituents of the interaction layer (IL) in unirradiated diffusion couples U–7wt%Mo/Al heat treated at 580°C were, accordingly, $(U, Mo)Al_4$ (oI20, space group 74), $(U, Mo)Al_3$ (cP4, space group 221) and also $UMo_2Al_{20}$ (cF184, space group 227).

It has been suggested that the IL is amorphized during irradiation and the structural instability of the IL is due to its amorphous character [16]. The same authors remark that in–pile test show that an IL with a high Al-content tends to be amorphized more easily. Maintaining the interaction layer of U-Mo/Al dispersion fuel as a stable low-Al content compound such as $(U-Mo)Al_3$ appears to be the key to avoiding massive pore formation. In brief, it has been proposed that the $(U,Mo)Al_3$ compound should be stabilized against formation of compounds with a higher aluminum content, such as $UAl_4$. $UAl_4$ has also been identified as an undesired product in the IL with different arguments. Nazarg and coauthors [17] have reported the results from irradiation experiments establishing the irradiation stability of $UAl_4$. More recently, Gan and coauthors [18] comment that $UAl_4$ compound remains crystalline under irradiation, but its fragility can be responsible for the breakaway swelling observed in the reactor-irradiated dispersion fuel.

Several potential remedies are available to correct these fuel performance problems, among them smaller changes in the fuel and matrix chemistry was proposed. Then, our research work was shifted towards understanding some issues about silicon and zirconium addition to aluminum matrix and U-Mo fuel respectively.

### *1.2.1. The Pseudo-Binary Phase Diagram $USi_3$-$UAl_3$ as a Basis for the Characterization of The Reaction Layer in U-Mo/Al-Si Alloys Diffusion Couples*

Previous studies have tried the Si-modification of the $UAl_3$ fuel in the $UAl_3$–Al dispersion fuel to suppress $UAl_4$ formation during the high-temperature fabrication process [19, 20, 21, 22]. In the case of the $U_3Si_2$ dispersions in Al the composition of the interaction layer is $U(Al_{0.75},Si_{0.25})_3$ [23] as it came out in post-irradiation examinations. More recently, the unstable behavior of the reaction product has not been observed in tested dispersion fuel systems such as $U_3Si/Al$ [24] and $U_3Si_2/Al$ [25]. On an analogous base, the Si-modification of the Al matrix in the U–Mo/Al dispersion fuel has been recently proposed to solve the stability of the IL [26]. Out-of-pile diffusion studies have shown that Si indeed accumulates in the IL between U–Mo and Al–Si [27,28]. In-pile tests have also shown the positive effect of Si: the IL thickness was much thinner than the one observed in the fuel plates using a pure Al matrix, and no porosity was formed [29,30].

The Si/Al ratio within $U(Al, Si)_3$ phase was found to be 1/3.5 in a $U_3Si/Al$ sample irradiated under a heat flux of 550 W/cm2 [31]. The IL composition thus lies at the intersection of the lines $U_3Si_2$—Al and $USi_3$—$UAl_3$ in the ternary U-Al-Si phase diagram. On the other hand, the formation of $(U,Mo)(Al,Si)_3$ has been shown by means of diffusion couples of U-7wt%Mo / Al-5.2wt%Si and U-7wt%Mo /Al-7.1wt%Si in the temperature range 340 to 580°C in out-of-reactor experiments [32]. The Si/Al ratio within this phase was found to be 1/3 and no $(U,Mo)Al_4$ was found. Although the 1/3.5 and 1/3 ratios are close enough the last one does not lay at the intersection of the lines U—Al-5.2wt%Si (or U—Al-7.1wt%Si) and USi$_3$—UAl$_3$. This could be due to the fact that the Si/Al ratio is related to the phase stabilities on the pseudo-binary USi$_3$-UAl$_3$.

This pseudo-binary was evaluated in the past by Dwight [33] through measurements in binary or ternary fields of the U-Al-Si phase diagram. He suggested a complete miscibility between UAl$_3$ and USi$_3$ at 900°C but indicated that a miscibility gap might exist at some lower temperature.

We present in this chapter our new results concerning U(Al,Si)$_3$ disordered phase. We have evaluated the formation energy at T=0K of the solid solution in the pseudo-binary UAl$_3$-USi$_3$ by using first principles calculated total energies in a cluster expansion method.

### 1.2.2. Low Silicon U(Al,Si)$_3$ Stabilization by Zr Addition

It was also noticed that minor silicon quantities would be required to solve the stability of the IL if a fourth element is present [26]. The early U–Al fuel developers have also tried the Zr-modification of the UAl$_3$ fuel [22]. The experimental evidence showed that the addition of 14wt.% zirconium as a third element is enough to inhibit UAl$_4$ formation.

In a recent publication [34], the results of diffusion-couple tests between U–Mo–XZr and Al–YSi alloys with various contents of Zr and Si were reported. The interaction product formed in a U–Mo–2Zr vs. Al–5Si (wt.%) diffusion couple tested at 580 °C showed that the composition of the IL goes from U(Al, Si)$_2$ on the U–Mo–Zr side to U(Al, Si)$_3$ on the Al-Si side, which indicates an increased chemical potential gradient for Si with the presence of Zr in the U-Mo alloy.

Modification of the UAl$_3$ phase by means of simultaneous addition of silicon and zirconium in order to suppress UAl$_4$ formation had not been reported. We have thus decided to determine UAl$_3$ phase stability in U-Al-Zr-Si alloys containing 50 – 47 wt.% uranium, 49.9 – 46.9 wt.% aluminum, 0 - 6 wt.% zirconium and 0.1 wt.% silicon [35]. Four alloys within the quaternary system U-Al-Si-Zr were made and heat treated at 600°C. The samples were analyzed by means of X-ray diffraction (XRD), electron-probe microanalysis (EPMA) and Energy Dispersive X-ray Spectrometry (EDS) techniques.

The main findings of this investigation and their implication about the stability of the IL are briefly described here.

### 1.2.3. UAl$_3$-Al Interaction Layer Growth

The more recent out of pile experiences [15, 36,37] in U(Mo)/Al diffusion couples show that the IL comprises four stable ordered intermediate phases: UAl$_3$, UAl$_4$, UMo$_2$Al$_{20}$ and U$_6$Mo$_4$Al$_{43}$ (hP106, space group P63/mcm). This reaction product is stratified in three main zones, two of which present a periodically layered morphology. In Ref. [37] the authors conclude that both growth kinetics and global energy of activation of reaction product are very close to that found for the U/Al binary system despite the fact that the UMoAl$_{20}$ and U$_6$Mo$_4$Al$_{43}$ ternary phases appear in the interaction zone. This means that the presence of these ternary phases does not have significant influence neither on the growth kinetics of the reaction product nor on its activation energy.

Out of pile experiences in diffusion couples U/Al [38] and U/UAl$_3$/Al at high pressures [39] show that the IL comprises three stable ordered intermediate phases in the system U-Al: UAl$_2$ (cF24) and the already mentioned UAl$_3$ and UAl$_4$. It was found that the growth of UAl$_2$

and UAl$_4$ phases is much slower than the growth of UAl$_3$ phase, being the growth of UAl$_2$ the slowest.

Computer simulation has become an important and effective tool for gaining insight into complex materials processes. Two recent works present simulations of the IL growth in U(Mo)/Al [40] and U/Al [41] interdiffusion systems by adjusting parameters from out of pile diffusion couple experiments [42] and in pile measurements. The main hypothesis in reference [40] were: (i) both U and Al are mobile species in the IL; (ii) the layer growth is a diffusion-controlled process; the chemical reactions kinetics at the interfaces of the IL with the fuel and the IL with the matrix are not considered; (iii) U and Al exhibit a continuous composition variation through the IL from (U,Mo)Al$_{4.4}$ at the layer–matrix boundary to (U,Mo)Al$_3$ at the layer–alloy boundary. The hypothesis (ii) is modified to include chemical reaction kinetics at the IL-matrix interface in reference [41].

Recently we have presented to pairs consideration a work [43] where (1) a thermodynamic database of the U-Al system suitable for the Thermo-Calc code was built following the thermodynamics parameters given in reference [44] and making corrections over some of them in order to fit the experimental balances reported in [45]; and (2) self-consistent atomic mobility parameters for both U and Al in the UAl$_4$ phase of the U–Al system were obtained via DICTRA simulation of interdiffusion in a UAl$_3$/Al diffusion couple and comparison against reported experimental data at different temperatures [46]. The successful description for the interdiffusion process and the obtained value of Al tracer diffusion coefficient in the UAl$_4$ are shown in this paper.

## 1.3. U-Al System and UAl$_4$ Stability

The preceding results concerning UAl$_4$ growth required a review of the Al/U system, since as it was already mentioned in the previous paragraph, some corrections had to be made to known thermodynamic parameters.

The system has been the subject of numerous studies since the middle of last century due to its out coming role in nuclear fuel development. Works of Gordon and Kaufmann [47] and R. F. Hills [48] contributed to establish stable compounds and formation reactions.

Enthalpies of formation for the intermetallic compounds were experimentally determined by Chiotti and Kateley [49] from electromotive force and calorimetric measurements. They also mention the uncertainty that hold on composition range of UAl$_4$ compound. Borie [50] had determined the crystallography as body-centered orthorhombic, spatial group I2ma or Imma, and had obtained by chemical analysis the composition values in this phase between 81.8 and 83.1 at.% Al. On this basis he postulated the existence of constitutional defects, U vacancies or Al sustitutional at U sites. Later, metallographic studies [51] showed a solubility range between 80 and 81.96 at% Al. Runnals and Boucher presented a study on UAl$_4$ phase composition based on measurements of density and X-ray diffraction [52]. They suggested that the composition could be specified as U$_{0.9}$Al$_4$ (81.63 at% Al), and the departure from stoichiometry could be due to the existence of U constitutional vacancies. Jesse et al [53] found that the range of homogeneity extends from UAl$_4$ (80 at% Al) to UAl$_{4.8}$ (82.76 at% Al). In 1989 Kassner et al. [54] presented an exhaustive review of thermodynamic data of the system and a reevaluation of phase equilibria. Their assessment was performed using Gibbs free energy minimization by means of a computer code. Intermetallic compounds were treated as line compounds except for $UAl_4$. This compound also exhibits a polymorphic transformation at 646°C attributed to a rearrangement of vacancies. Later, two contradictory investigations by X-ray diffraction methods have been presented. Zenou et al established that the concentration of U in $UAl_4$ is $18.5 \pm 0.5$ at% [55] by using scanning electronic microscopy (SEM) coupled with EDS, XRD analysis and density measurements, the Rietveld refinement assuming a random occupation of U sites by vacancies confirmed this result. However, Tougait and Noël [56] have used the same techniques and suggested that the compound is stoichiometric.

Researchers related to fuel development have also measured a nonstoichiometric $U_{1-x}Al_4$ phase as a product within the interaction layer in irradiated $UAl_3/Al$ fuel plates [57].

In a previous work [58] some of the present authors presented the calculated ground state of U-Al system by a first principles method and a many body potential for describing the whole range of compositions. The $UAl_2$ and $UAl_3$ resulted as stable compounds but the $UAl_4$ compound appeared as metastable at T=0K. Based on the discussion present in the literature about the role of vacancies in the stabilization of $UAl_4$, we reported later [59] an investigation of stoichiometry of $UAl_4$ through a calculation method. Supercells consistent of two adjacent unit cells were made up containing vacancies resulting less stable than the stoichiometric compound. We present here new results showing instability of larger supercells though a tendency is exhibit to reduce formation energy as composition of supercell approaches stoichiometry. Antisite structures $U_{1-x}Al_{1+x}$ are presented as a more reliable possibility of stabilizing the compound.

## 1.4. Monolithic Fuel

An alternative for fixing the problem of undesired interdiffusion between Al and UMo alloy has been proposed by the use of monolithic fuel, this is, by placing a thin sheet of monolithic fuel between Al cladding and thus reducing U-Mo alloy–Al interaction surface. The whole set is friction welded [60]. Due to differences in elastic constants and thermal expansion coefficients problems can arise from internal stresses produced by temperature changes during fabrication and later use inside reactor. If elastic constants are known, these internal stresses can be evaluated a priori.

Although elastic constants can be experimentally measured, we have not found that information in literature for U(Mo) alloys. We have made an effort to obtain estimations of these quantities from first principles thermodynamics [61]. We show here the results for lattice parameter and three independent elastic constants of the cubic solid solution U(Mo) as a function of uranium concentration. These result implies that a deformation of a retained $\gamma$ U(Mo) phase at low temperature can produce a phase transformation to a more stable structure, in agreement with experiment.

## 2. Modeling Tools

This section provides an overview of the ab-initio and computational-thermodynamic methods that we have applied to model the above mentioned basic features of the γ U(Mo)/Al dispersed nuclear fuel.

### 2.1. First-Principles Calculations of Thermodynamic Properties of Ordered Solids

In an ordered solid thermal fluctuations take the form of electronic excitations and lattice vibrations and, accordingly, the free energy can be written as $F = E_0 + F_{vib}$, where $E_0$ is the absolute zero total energy while $F_{vib}$ denote vibrational free energy contributions. Both quantities can be evaluated by methods involving applications of quantum mechanical total-energy calculations performed within the framework of electronic density-functional theory (DFT)[62]. DFT-based methods allow one to predict bulk and defect properties for elemental and multicomponent solids, starting from knowledge of only the atomic numbers of the constituent atomic species.

The total energies of the superstructures and the pure elements were calculated using DFT together with the full-potential linearized augmented-plane wave (FP-LAPW) method in the generalized gradient approximation(GGA) [63,64], including scalar relativistic corrections [65,66], and implemented in the WIEN 97.8 code [67] (U-Mo and U-Mo-Pt system) and in the WIEN2K code [68]. We have neglected here the correction introduced by the spin-orbit term taking into account that the aim of the work is to evaluate relative stability of compounds. As it has already been noticed [69], changes in total energies are canceled at a great extent when differences in total energies are made to obtain formation energies. All parameters were optimized so as to enhance accuracy of calculations. We searched for an optimum value for the kinetic energy cut-off for the plane-wave basis by testing the parameters RKmax and muffin-tin radii (rmt). They were set as rmt[U]=2.6, rmt[Mo]=2.3, rmt[Pt]=2.6, rmt[Al]=2.2, rmt[Si]=2.2 and RKmax=10. The core configurations [Xe] $4f^{14}$ $5d^{10}$, [Kr] $3d^{10}$, [Xe] $4d^{10}$, [He] $2s^2 2p^6$ and [He] $2s^2 2p^6$ were considered for U, Mo, Pt, Al and Si respectively. The calculations used a number of k points such that the total energy of the structures changed less than 0.1mRy/atom. Depending on the number of atoms in the cell, the number of k points in the calculation was 256 and 288 points in an irreducible wedge of the Brillouin zone. With regards to the cubic structures, the criterion to obtain the total energy was to minimize it as a function of lattice parameter. For non cubic structures the process was to minimize iteratively cell shape and volume till convergence was achieved. In the case of $U_2Mo$ and $UAl_4$ (with and without defects) structures, atomic positions were also relaxed.

The results of the vibrational free energy contributions are not included in the discussion of this chapter.

## 2.2. First-Principles Calculations of Thermodynamics of Disordered Solids. The Cluster Expansion Formalism

The cluster expansion is a generalization of the well-known Ising Hamiltonian [70]. Over the last twenty years, several publications have taken care of the formalism for obtaining thermodynamics of disordered solids alloy from quantum mechanical energy calculations involving a small number of ordered structures [71, 72, 73, 74]. Consider a binary system where the configurational variable at site $i$ ($\sigma_i$) takes values $\sigma_A$ or $\sigma_B$ depending on the type of atom, A or B, occupying the site $i$. Any configuration of the system will be completely specified by the $N$-dimensional discrete vector $\sigma = (\sigma_1, \sigma_2, \ldots, \sigma_N)$ for a lattice with $N$ sites and any function of the configuration $P(\sigma)$ can be expanded in terms of a complete set of basis functions as follows

$$P(\sigma) = \sum_{\alpha} p_{\alpha} \, \Phi_{\alpha}(\sigma) \qquad (1)$$

where the $\Phi_{\alpha}(\sigma)$ are the cluster functions, $p_{\alpha}$ the appropriate expansion coefficients and the index $\alpha$ signifying a given cluster of lattice sites. The summation in (1) runs over all possible clusters in the system. Since the basis functions $\Phi_{\alpha}(\sigma)$ are orthogonal, the coefficients are calculated in the usual way

$$p_{\alpha} = \rho_0 \sum_{\sigma} P(\sigma) \, \Phi_{\alpha}(\sigma) \qquad (2)$$

where the normalization factor $\rho_0$ is $2^{-N}$, i.e., the inverse of the total number of configurations in the system. The cluster functions $\Phi_{\alpha}(\sigma)$ can be expressed as a product of the configurational variables associated with a given cluster of lattice sites. In other words,

$$\Phi_{\alpha}(\sigma) = \prod_{k \in \alpha} \sigma_k \qquad (3)$$

The expectation value of a function of the configuration, such as the thermodynamic average of the configurational energy, can be expressed easily in terms of the multisite correlation function (MCF) $\zeta_{\alpha}$:

$$\langle P \rangle = \sum_{\alpha} e_{\alpha} \, \zeta_{\alpha} \qquad (4)$$

with

$$\zeta_{\alpha} = \langle \Phi_{\alpha}(\sigma) \rangle = \left\langle \sigma_{k_1} \, \sigma_{k_2} \cdots \sigma_{k_{|\alpha|}} \right\rangle \qquad (5)$$

where $k_i$ ($i = 1, 2, \ldots, |\alpha|$) denote the lattice points in a cluster $\alpha$ of $|\alpha|$ sites. The expansion coefficients $e_{\alpha}$ in (4) are known in the literature as the effective cluster interactions (ECI) for the physical property $P$. The complete cluster expansion of equation (4) is formally exact, however, the utility of this rests in the possibility of identifying a hierarchy of a small number of clusters whose contributions $e_\alpha$ to the physical property $P$ dominates those of the remaining clusters.

The formation energy, $\Delta E_F = E_T - \sum_i$, where the sum runs over all species $i$, of any ordered or disordered alloy lattice may be now described with a truncated expression of the bilinear form of equation (4). The unknown parameters ECIs of the cluster expansion can then be determined by fitting this expression to a set of formation energies of ordered compounds, for which the corresponding correlation functions are known. These energies can be obtained, for instance, through first-principles electronic calculations as mentioned in the last section. Because those calculations demand time, we have access to a finite number of structural energies. The number of $e_\alpha$ parameters to compute can be equal or smaller than the number of known structural energies. In case they are equal, ECIs can be obtained by the direct inversion method of Connolly – Williams [75]. However, it has been determined that a better cluster expansion is obtained if the system of equations (4) is over determined [73]. The choice of the number and hierarchy of the clusters is determined using a Cross-Validation (CV) score defined as

$$CV = \left( \frac{\sum_{n=1}^{N} \left( \Delta E_F^n - \Delta \hat{E}_F^n \right)^2}{N} \right)^{1/2} \tag{6}$$

where $\Delta E_F^n$ is the first-principles electronic calculated formation energy of the ordered compound $n$, whereas $\Delta \hat{E}_F^n$ is the value predicted by the cluster expansion with ECI's obtained with a least-squares fit to the (N-1) other structural energies. For a finite number of structural energies the CV score goes through a minimum when the number of clusters included in the expansion increases.

The three criteria used in this work for choosing the right clusters and the number of known structural energies that will enter in the cluster expansion has been implemented in the software Alloy Theoretic Automated Toolkit (ATAT)[73,76], and are the following:

i.    the ground state phase diagram should exhibit known stable phases and the predicted energies for structures other than the ones included in the cluster expansion should lie above the ground state tie lines;
ii.   the predicted cluster-expanded energies of ordered compounds in the expansion should agree with the corresponding formation energy calculation via first principles (the CV score is small);
iii.  the magnitude of the ECI decays as a function of the diameter of the corresponding cluster and as a function of the number of sites it contains.

In the random alloy $A_x$–$B_{1-x}$ the correlation functions take the values

$$\zeta_\alpha^{rand} = \langle \prod_{k \in \alpha} \sigma_k \rangle_{rand} = \prod_{k \in \alpha} \langle \sigma_k \rangle = \left( \sigma_A + \sigma_B (1 - x) \right)^{k_\alpha} = (2x - 1)^{k_\alpha}$$

where $k_\alpha$ is the number of points comprised in the cluster $\alpha$, and $x$ is the concentration of element $A$, and we have made use of the values $\sigma_A =$, $\sigma_B =$. Finally, the cluster expanded formation energy for the completely disordered alloy takes the simple form

$$E_f^{rand} = E^{rand} - xE^A - (1-x)E^B == \sum_\alpha e_\alpha (2x-1)^{k_\alpha} - x\sum_\alpha e_\alpha (+1)^{k_\alpha} - (1-x)\sum_\alpha e_\alpha (-1)^{k_\alpha} \quad (7)$$

## 2.3. CALPHAD Modeling of Thermodynamics

The CALPHAD method [77] is based on mathematically formulated models describing the thermodynamic properties of individual phases. The model parameters are evaluated from thermochemical data of the individual phases and phase equilibrium data between phases, as phase equilibria are a manifestation of the thermodynamic properties of the phases involved. More specifically, under typical experimental conditions of constant temperature and pressure, phase equilibrium is obtained by minimization of the Gibbs energy of a closed system.

The molar Gibbs free energy of a solid solution or liquid phase $\varphi$ is described by the following expression:

$$G^\varphi - H^{SER} = {}^0G^\varphi + {}^{id}G^\varphi + {}^EG^\varphi \qquad (8)$$

where the first term includes the description of the mechanical mixture of pure elements in phase $\varphi$ at temperature T. The second term is the ideal entropy mixing contribution and the third term is the excess energy, which is expressed in the Redlich-Kister formalism for an $A$-$B$ binary phase [78] as:

$$^EG^\varphi = x_A \, x_B \left[ \sum_{m=0}^{n} {}^mL^\varphi_{A,B} (x_A - x_B)^m \right] \qquad (9)$$

with ${}^mL^\varphi_{A,B} = a_m + b_m T$.

The reference enthalpy, $H^{SER} = \sum_i H_i^{SER}$ represents the mechanical mixing enthalpy of the pure elements in their stable states at standard temperature of 298.15K.

The intermediate phases with solubility $A_nB_m$ are assessed with the two-sublattice model. This description is equivalent to a crystalline structure which distinguishes the occupation of two types of sites (the sublattice $I$ with $n$ sites and the sublattice $II$ with $m$ sites) where species ($A$ and $B$) have no preferential location. The occupation of the sublattices by atoms of both $A$ and $B$ represents the widening of homogeneity range of phase $A_nB_m$.

The atom fraction of a species referred to the total number of sites in a sublattice is called "site fraction" or $y_j^{(n)}$, where (n) is the index of the sublattice (I or II) and $j$ labels the specie.

The molar Gibbs free energy is written with the three contributions:

$$G^{A_nB_m} - H^{SER} = {}^0G^{A_nB_m} + {}^{id}G^{A_nB_m} + {}^EG^{A_nB_m} \qquad (10)$$

where $^{0}G^{A_n B_m}$ contains the energies of formation of possible compounds in $A_n B_m$ structure with only one specie in each sublattice, times the site fractions of the species. The term $^{id}G^{A_n B_{mk}}$ includes the independent contributions from each sublattice to the entropy of mixing assuming a random mix in each sublattice. The excess term is expressed by Redlich-Kister polynomials which include the site fractions of species in each sublattice:

$$
^{E}G^{A,B_m} = y_A^I\, y_B^I \left( y_A^{II} \sum_{k=0} {}^{k}L_{A,B:A}\left(y_A^I - y_B^I\right)^k + y_B^{II} \sum_{k=0} {}^{k}L_{A,B:B}\left(y_A^I - y_B^I\right)^k \right) +
$$

$$
+ y_A^{II}\, y_B^{II} \left( y_A^I \sum_{k=0} {}^{k}L_{A:A,B}\left(y_A^{II} - y_B^{II}\right)^k + y_B^I \sum_{k=0} {}^{k}L_{B:A,B}\left(y_A^{II} - y_B^{II}\right)^k \right) +
$$

$$
+ y_A^I\, y_B^I\, y_A^{II}\, y_B^{II}\, L_{A,B:A,B} \tag{11}
$$

The $^{k}L_{A,B:*}$ $\left( {}^{k}L_{*:A,B} \right)$ is the interaction parameter between component $A$ and $B$ in the first (second) sublattice and $^{k}L_{A,B:A,B}$ the cross-interaction parameter, all of them depends only of the temperature.

Intermetallic compounds $A_n B_m$ are treated as stoichiometric phases and the Gibbs free energy depends only on temperature. Therefore, the energy of formation in the two-sublattice model is written as a linear function of T:

$$
\Delta G_f^{A_n B_m} = a' + b'\, T \tag{12}
$$

In this work, the application of the CALPHAD method was carried out employing the Thermo-Calc package [79].

## 2.4. CALPHAD Modeling of Atomic Mobility

The multicomponent extension of the Fick's first law, which enables the concentration gradient of one species to cause another species to diffuse, is expressed by the following relationships:

$$
J_k = -\sum_{j=1}^{n} D_{kj}\, \frac{\partial C_j}{\partial z} \qquad D_{kj} = \sum_{i=1}^{n} L'_{ki}\, \frac{\partial \mu_i}{\partial C_j} \tag{13}
$$

where $C_k$ is the concentration of component $k$, $\mu_i$ is the chemical potential of component $i$, $n$ is the number of components in the system and $D_{kj}$ are the interdiffusion coefficients. The interdiffusion coefficients are expressed in terms of pure thermodynamic quantities $\frac{\partial \mu_i}{\partial C_j}$, and of phenomenological coefficients $L'_{ki}$, which are merely kinetic quantities related to atomic mobility.

In the CALPHAD method the $L'_{ki}$ coefficient is described by the following expression [80]:

$$L'_{ki} = (\delta_{ik} - C_k V_i) C_i M_i \tag{14}$$

where $\delta_{ik}$ is the Kronecker delta, $V_i$ and $M_i$ are the partial molar volume and the mobility of component $i$ respectively. From absolute-reaction rate theory arguments, the mobility is proposed as

$$M_i = \frac{M_i^0}{RT} \exp\left(-\frac{Q_i}{RT}\right) = \frac{1}{RT} \exp\left(-\frac{\left(Q_i - RT \ln M_i^0\right)}{RT}\right) = \frac{1}{RT} \exp\left(-\frac{\Delta G_i}{RT}\right) \tag{15}$$

where $M_i^0$ is a frequency factor, $Q_i$ an activation energy and both will generally depend on the composition, the temperature, and the pressure. Finally the composition dependence of $\Delta G_i$ is represented with a Redlich-Kister expansion [78]. For a phase described by the model of sublattices it becomes:

$$\Delta G_i = \sum_p \sum_q y_p y_q \Delta G_i^{p,q} + \sum_p \sum_{q>p} \sum_s y_p y_q y_q \left[\sum_r (y_p - y_q)^r \Delta^r G_i^{p,q,s}\right] + \cdots \tag{16}$$

where $\Delta G_i^{p,q}$ is the value of $\Delta G_i$ in the phase with the first sublattice ocuppied with $p$ and the second with $q$ and $\Delta^r G_i^{p,q,s}$ represents the binary interaction parameter of order $r$ between $p$ and $q$. As in the case when evaluating thermodynamic data for a Thermo-Calc database, the model parameters ($\Delta G_i^{p,q}, \Delta^r G_i^{p,q,s}$, etc) are assessed by an optimization procedure where experimental information is taken into account.

Our research reported in this paper was accomplished by using the DICTRA (diffusion-controlled transformations) software package [81]. DICTRA has been developed with the purpose of simulating diffusion-controlled transformations in multicomponent systems with the CALPHAD method. The *DICTRA* methodology handles diffusion in multicomponent and multiphase systems based on the numerical solution of the diffusion equations with local thermodynamic equilibrium at the phase interfaces. The software is also suitable for treating problems involving a moving boundary. DICTRA is coupled with the Thermo-Calc code. Both software packages require databases assessed from experimental data, Thermo-Calc uses a database with thermodynamic parameters and DICTRA a database with tracer diffusion coefficients.

## 3. RESULTS AND DISCUSSION

In this section, we show the modeling results obtained with first-principles calculations of thermodynamic properties on U-Mo, U-Al and UAl$_3$–USi$_3$ systems, the experimental results describing the role of Zr in stabilizing the U(Al, Si)$_3$ phase, the DICTRA modeling of UAl$_4$

growth in an UAl3/Al diffusion couple and the first-principles calculations of elastic constant in the disordered $\gamma$U(Mo) alloy.

## 3.1. Cluster Expansion in the Bcc U-Mo System. $\gamma$U(Mo) Phase Stability

First principles total energy calculations were carried out for a total of 16 bcc-based ordered structures [13]. In selecting bcc-based structures a natural candidate was $U_2Mo$ structure, whose existence was experimentally shown in the U-Mo system [82].

Formation energy values for U-Mo system are shown in the ground state phase diagram of Figure 1 (circles). There are two features of this phase diagram that are worth emphasizing. The calculated energies of formation predict only one stable compound ($U_2Mo$) and an asymmetry that would stabilize the disordered alloy on the Uranium rich side as it is observed in the experimental phase diagram of U-Mo.

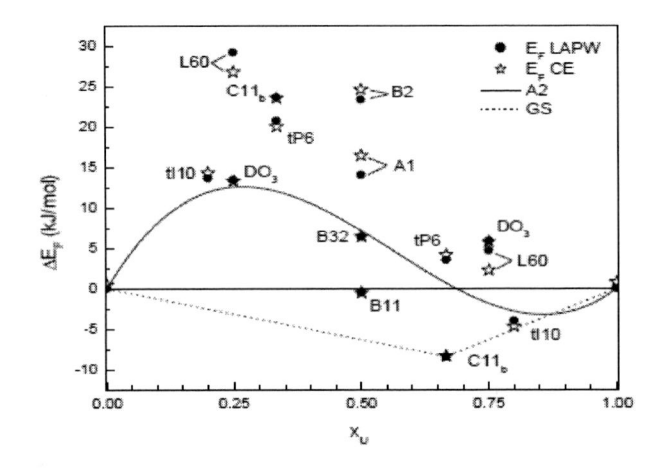

Figure 1. Ground state bcc phase diagram of U-Mo alloys. The ab-initio calculations are represented by circles while the cluster-expanded energies of formation are represented by stars. The filled and dashed curves stand respectively for the energy of formation of disordered $\gamma$U(Mo) alloy and for the ground state.

In order to find the minimal cluster expansion that provides the best possible fit based on the set $\{\Delta E_F\}$ calculated and also fulfills the three criteria mentioned in section 2.2, we proceed by iterative refinement considering all the clusters till a maximum cluster containing one pair of first order, one pair of second order, and one pair of fourth order. Figure 2 shows the selected set which consists of p1 + p2 + p3 + p5 + p6 + t112 + t113 + t115 (where: $p_i$ is the ith neighbor pair and $t_{ijk}$ is the triplet that involves $i^{th}$, $j^{th}$ and $k^{th}$ neighbors pairs).

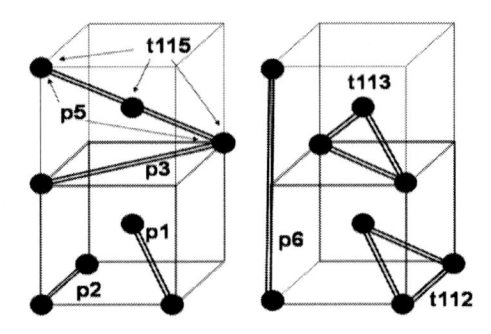

Figure 2. Real-space depiction of the clusters used in the expansion.

Some remarks can be done about the three criteria stated in section 2.2 and our approach for the best fit:

i)   The cluster-expanded energies of formation for 25 structures, including the sixteen ordered compounds considered in the calculation of ECI's, are shown in Figure 3. It can be seen that only one ground state compound is obtained, it is the body-centered tetragonal $C11_b$ $U_2Mo$ structure as has been experimentally observed [82].

ii)  A direct comparison of first principles calculated (circles) and cluster-expanded predicted (open stars) formation energies of the eleven ordered compounds considered in the expansion is shown in Figure 1. The average fitting error is 0.8 kJ/mol and the maximum fitting error is 2.4 kJ/mol. Calculation of CV parameter yields an acceptable low value (5.6 kJ/mol).

iii) The effective cluster interactions (ECI) as a function of their size are shown in Figure 4. We took into account the convergence criteria stating that within the set of clusters with the same number of atoms, the interactions should decay as size is incremented [83]. Pair and triplet interactions accordingly show this behavior. We could comment here that the inclusion of the pair p4 and the triplet t124 in the expansion causes an increment with size in the intensity of pair interactions, thus resulting in a not eligible choice.

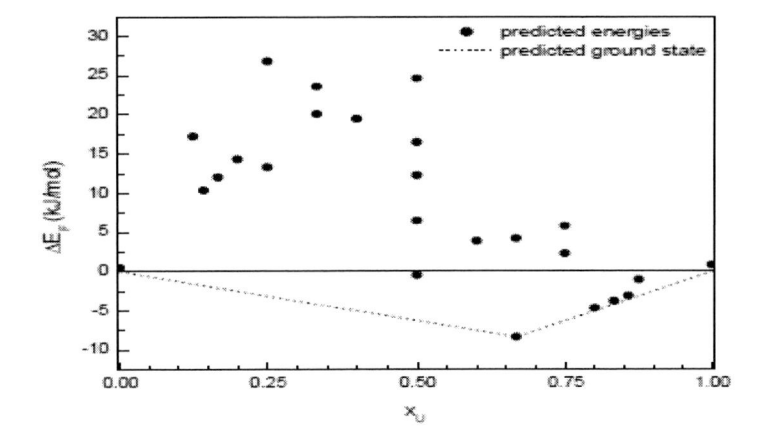

Figure 3. Cluster expansion predicted energies for structures not used for obtaining the ECIs.

The cluster-expanded predicted formation energy of random binary $U_x Mo_{1-x}$ alloys is then calculated analytically with equation (7) and is shown as solid line in Figure 1.

$$\langle \Delta E_F \rangle_{random} = \left( \Sigma e_{pair} \right) \left[ (2x-1)^2 - 1 \right] + \left( \Sigma e_{triplet} \right) \left[ (2x-1)^3 - (2x-1) \right]$$

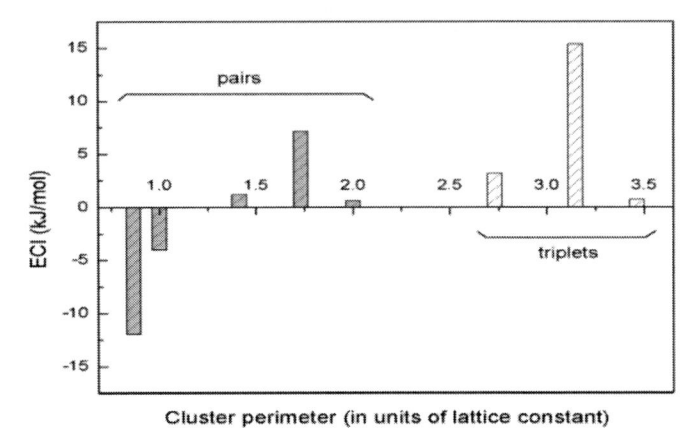

Figure 4. Effective cluster interactions for bcc-based Uranium-Molybdenum alloys. In the case of the three-body multisite ECIs, we have used the perimeter of the triplet as a measure of the range of the interactions.

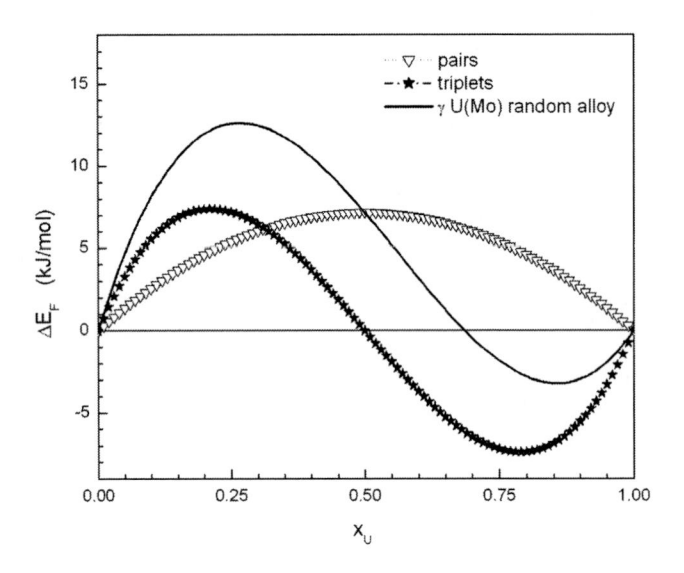

Figure 5. Contributions of pairs and triplets to the formation energy of γU(Mo) random phase.

A consideration of pairs and triplets separately may lead to a more comprehensive picture of triplet contributions to γU(Mo) stability. Result is drawn in Figure 5. We note that the short-range pair interactions up to second neighbours dominate the miscibility gap observed both in our calculation and in the experimental diagram between the terminal solid solution Mo(U) and the γU(Mo) phase but, the asymmetry of the gap and the stability of the gamma phase in an ample range of molybdenum concentration is due to the presence of triplet interactions, in particular that including first and third neighbours (t113).

Finally, since the triplet t113 has a fundamental role in stabilizing $\gamma$U(Mo) disordered phase, we can establish that the ultimate effectiveness of third element substitution on $\gamma$U(Mo) stabilization depends on how much this element modifies the t113 multisite interaction. Thus, in the frame of the Cluster Expansion, the stabilization of a structure that has a ternary occupation in t113, as $L2_1$ structures, suggest the stabilization of the disordered phase. Platinum was proposed as third element addition to stabilize $\gamma$U(Mo) at lower temperatures, our calculation of the formation energy of $L2_1$ $U_2PtMo$ (-14 kJ/mol) agrees with this suggestion because it is 5.8 kJ/mol more negative than the formation energy of $U_2Mo$.

## 3.2. UAl₃ Stabilization by Si Addition. UAl₃-USi₃ Pseudobinary System Ground State

In order to obtain a cluster expansion for the $UAl_3$-$USi_3$ pseudo binary system a model was assumed for $L1_2$ (spatial group 221) U(Al, Si)₃ structure in which U atoms are fixed at 1a Wickoff positions and Al and Si atoms are interchangeable species (mobile) at the 3c Wickoff positions in the $UAl_3$ structure. This is, sustitutional occupation of Al and Si was assumed at 3c sites.

First principles total energy calculations were carried out for a total of 24 fcc-based ordered structures. Formation energy values for $UAl_3$-$USi_3$ pseudo binary system are shown in the ground state phase diagram of Figure 6.

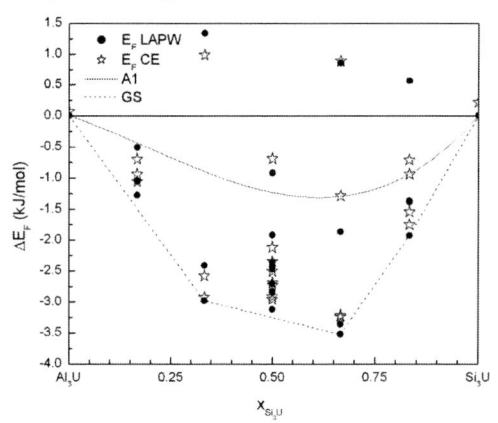

Figure 6. Ground state fcc phase diagram of Al₃U-Si₃U pseudobinary alloys. The ab-initio calculations are represented by circles while the cluster-expanded energies of formation are represented by stars. The filled and dashed curves stand respectively for the energy of formation of disordered Al₃U-Si₃U pseudobinary alloy and for the ground state.

The minimal cluster expansion that provides the best possible fit based on the set of calculated energies and also fulfills the three convergence criteria contains pairs till sixth order, and three triplets including pairs of first and second order:

i)  The cluster-expanded energies of formation for several structures not included in the calculation of ECI's, are shown in Figure 7. Predicted ground state agrees with calculated one and there are no predicted energies below predicted groundstate.

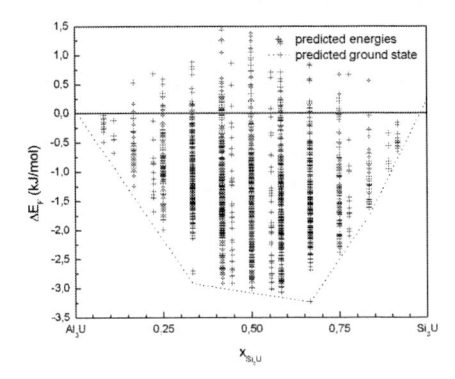

Figure 7. Cluster expansion predicted energies for structures not used for obtaining the ECIs.

i)  A good agreement was found between first principles calculated and cluster-expanded predicted formation energies of the twentyfour ordered compounds considered in the expansion. Calculation of CV parameter yields an acceptable low value (0.854 kJ/mol).

ii)  The effective cluster interactions (ECI) as a function of their size are shown in Figure 8. Intensity of the ECIs decay as a function of the number of atoms and also as size is incremented for a fixed number of atoms.

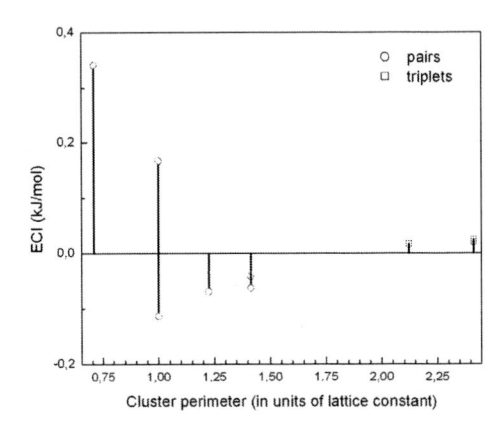

Figure 8. Effective cluster interactions for fcc $Al_3U$-$Si_3U$ pseudobinary alloys. In the case of the three-body multisite ECIs, we have used the perimeter of the triplet as a measure of the range of the interactions.

Finally, we show in Figure 9 the separate contribution to solid solution formation energy of pairs and triplets interactions. Triplets are responsible for the asymmetry that can lead to a miscibility gap in $U(Al,Si)_3$ phase. This result appears as a theoretical base for the previously suggested existence of the separation of the solid solution at low temperatures [33] and for the 1/3 Si/Al ratio found through experiments with diffusion couples for the formation of $(U,Mo)(Al,Si)_3$ phase [32]. If no miscibility gap was present, the composition of $(U,Mo)(Al,Si)_3$ phase would have lay at the intersection of the lines U—Al-5.2wt%Si (or

U—Al-7.1wt%Si) and USi₃—UAl₃. On the other hand, in the presence of the gap, (U,Mo)(Al,Si)₃ phase is only formed at its stable compositions.

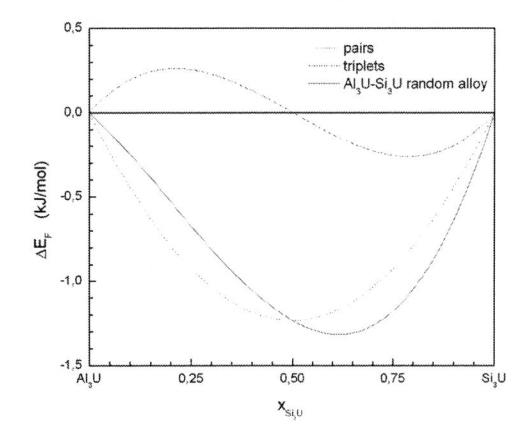

Figure 9. Contributions of pairs and triplets to the formation energy of $Al_3U$-$Si_3U$ pseudobinary random phase.

## 3.3. Low Silicon U(Al,Si)₃ Stabilization by Zr Addition. Experimental Results

Four alloys were made (Table 1) and submitted to heat treatments at 600 °C during 1000 hours in order to test increasing amounts of Zr addition to suppress the reaction between Al and primary (U,Zr)(Al,Si)₃ to form UAl₄ after solidification.

### Table 1. U-Zr-Al-Si alloys compositions

| Sample | U wt.% / at.% | Al wt.% / at.% | Si wt.% / at.% | Zr wt.% / at.% |
|--------|---------------|----------------|----------------|----------------|
| 0 | 50.00 / 10.18 | 49.90 / 89.65 | 0.1 / 0.17 | 0 / 0.00 |
| 1 | 49.47 / 10.12 | 49.43 / 89.18 | 0.1 / 0.17 | 1 / 0.53 |
| 3 | 48.48 / 10.01 | 48.42 / 88.20 | 0.1 / 0.18 | 3 / 1.62 |
| 6 | 47.00 / 9.85 | 46.90 / 86.69 | 0.1 / 0.18 | 6 / 3.28 |

As cast structure (Figure 10) shows a solidification microstructure comprising a primary phase (U,Zr)(Al,Si)₃, a eutectic two-phase component Al(Si) + (U,Zr)(Al,Si)₃ and a eutectic three-phase component Al(Si) + (U,Zr)(Al,Si)₃ + U(Al,Si)₄. In the last two components, the amount of Zr involved is less than 0.3 at % and then not detectable by EDS.

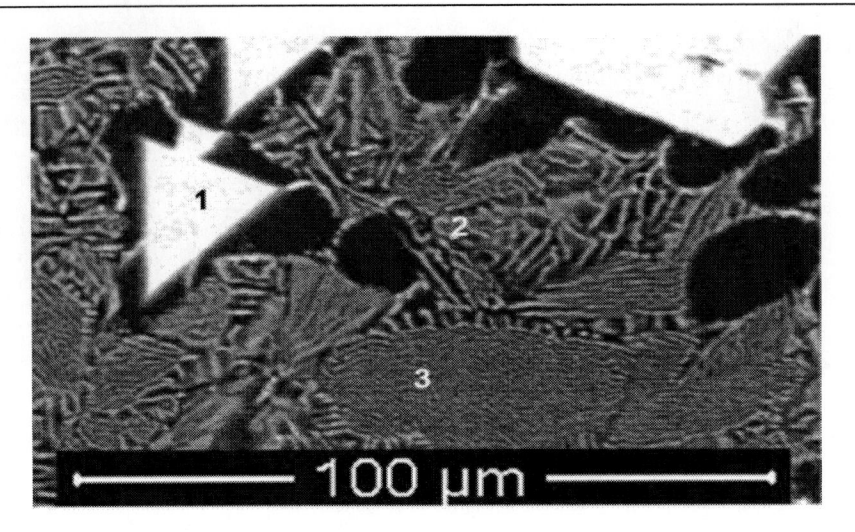

Figure 10. Secondary electron image of the eutectic-type areas into the as-cast sample 1 alloy microstructure. 1 is the $(U,Zr)(Al,Si)_3$ primary phase, 2 is the univariant eutectic-type zone $(U,Zr)(Al,Si)_3 + Al$ and 3 is the final eutectic $(U,Zr)(Al,Si)_3 + Al(Si) + U(Al,Si)_4$.

Micrographs of heat treated samples are shown in Figure 11. From composition analysis (electron-probe microanalysis measurements) two-phase and three-phase fields are established in the equilibrium phase diagram. An agreement within experimental uncertainty is found with our estimations from X Ray Diffraction spectra analysis. Measurements in the heat treated samples from alloys 1 and 3 lie within the three-phase region $Al(Si) + U(Al,Si)_4 + (U,Zr)(Al,Si)_3$, while the heat treated sample from alloy 6 belongs to the two-phase region $Al(Si) + (U,Zr)(Al,Si)_3$.

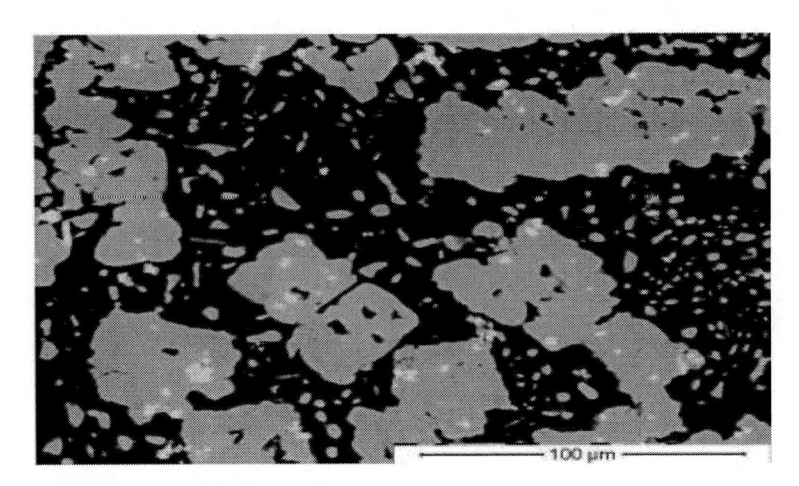

Figure 11. Secondary electron image of the annealed sample 1 alloy microstructure. The darkest region contain $Al(Si)$ and $Al(Si) + (U,Zr)(Al,Si)_3 + U(Al,Si)_4$ eutectic. The brightest contrast phases are $(U,Zr)(Al,Si)_3$. The lamellar structure of eutectic components is transforming into small spheroid precipitates.

Our analysis of samples of the quaternary U-Al-Zr-Si alloys stabilized for 1000 h at 600°C reveals that the necessary Zr and Si content to inhibit the transformation of the quaternary $(U,Zr)(Al,Si)_3$ to $UAl_4$ can be reached with the combination of 3.3 at% Zr and 0.3

at% Si. This pair of values does not comply with the relationship $C_{Si} = 4 - \frac{2}{3} C_{Zr}$ predicted for the ideal mixing of the (U,Zr)Al$_3$ and U(Al,Si)$_3$ phases, thus showing that the Zr and Si bonds play a significant role in the reduction of the enthalpy of mixing in the compound (U,Zr)(Al,Si)$_3$.

## 3.4. Kinetics of Interaction Layer Growth in an UAl$_3$/Al Diffusion Couple

In order to build the thermodynamic database of the U-Al system suitable for the Thermo-Calc code, we used all the model parameters given in reference [44] for the thermodynamic assessment of U-Al system with the exception of the molar Gibbs free energy of aluminum in the structures of UAl$_4(\alpha)$ and UAl$_4(\beta)$ which were found incorrect to describe the solubility of this compound. In this way, our optimization [84, 43] was carried out to fit the data reviewed by Kassner et al [45] with a UAl$_4$ solubility range between 81.3 and 83.8 at% Al at T=823K. The values obtained are the following:

$$^{0}G_{Al:Al}^{UAl_4(\alpha)} = {}^{0}G_{Al:Al}^{UAl_4(\beta)} = \left( {}^{0}G_{Al}^{fcc} + 1500 \right) \text{ in J/mol}$$

where the molar Gibbs free energy of pure aluminum in its stable structure is taken from the compilation by Dinsdale (SGTE database) [85]. The calculated phase diagram with this set of model parameters is shown in Figure 12.

The growth rate of UAl$_4$ phase in the UAl$_3$/Al diffusion couple is modeled with the diffusion rate of atoms across two moving interfaces; these are two planar boundaries one separating the UAl$_3$ and UAl$_4$ phases and the other separating UAl$_4$ and Al phases. Under these circumstances, the moving boundary model may be applied [86] at each interface. The local equilibrium hypothesis is used at phase interfaces.

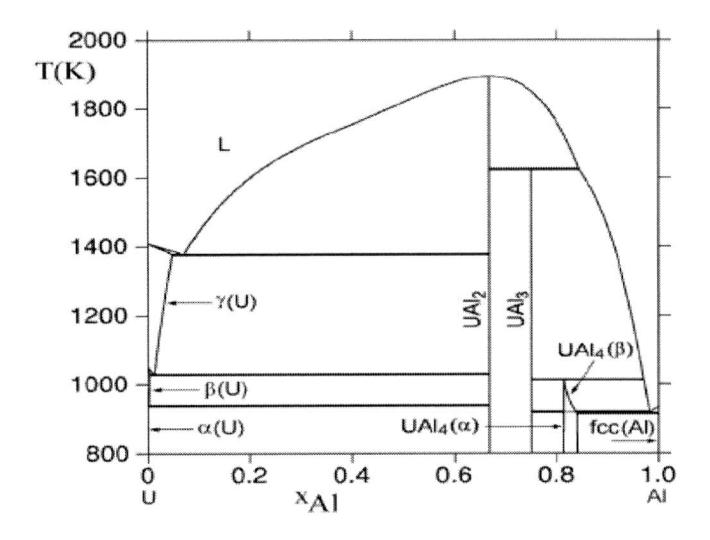

Figure 12. Calculated phase diagram of U-Al system [43].

For the U-Al binary system the interdiffusion tensor $D_{kj}$ becomes a scalar and then there is a unique interdiffusion coefficient for each phase given by:

$$\widetilde{D}^{\alpha} = \left( D_U^{*(\alpha)} \, x_{Al} + D_{Al}^{*(\alpha)} \, x_U \right) \phi^{\alpha} \quad \left( \alpha = UAl_3, UAl_4, Al \right)$$

where $x_i$ is the mole fraction of component $i$, $D_k^{*(\alpha)} = R T M_k$ is the tracer diffusion coefficient of component $k$ in the $\alpha$ phase and $\phi^{\alpha} = \dfrac{x_U}{RT} \dfrac{\partial \mu_U^{\alpha}}{\partial x_U} = \dfrac{x_{Al}}{RT} \dfrac{\partial \mu_{Al}^{\alpha}}{\partial x_{Al}}$ is the *thermodynamic factor* in that phase (the equality holds due to Gibbs-Duhem relation).

Since there are no concentration gradients for the stoichiometric compound $UAl_3$ and the terminal Al solution in the U-Al system, Fick's laws cannot be applied to them with DICTRA software. To circumvent this problem, it was proposed to introduce narrow ranges of homogeneity for these phases by using the two-sublattice model [87]. Then, the thermodynamic database of U-Al system (reference [44] and our corrections in reference [43]) was slightly modified so that a small solubility range is added in both sub-lattices for $UAl_3$, and a small solubility range of uranium is added in the sub-lattice of aluminum for $UAl_4$. A similar procedure was applied to Al phase to obtain a small uranium solubility range. We have checked that these modifications do not alter substantially the temperatures of the invariant reactions.

DICTRA simulation of the growth rate of $UAl_4$ phase in the $UAl_3$/Al diffusion couple requires the knowledge of all $D_k^{*(\alpha)}$ coefficients. However, Al self-diffusion coefficient and U tracer diffusion coefficient in Al matrix [88] were the only ones reported. There was no information about tracer diffusion coefficients in the intermediate compounds of U-Al system. Some indirect experimental information allowed us [43] to assume the tracer diffusion coefficient of U in the $UAl_4$ phase to be much smaller to that of Al in the same phase. On the other hand, we could used data from Alvares [46] who measured the thickness of the $UAl_4$ layer growing between the solid solution Al(U) and crystals of the $UAl_3$ phase in two U-Al alloys (U-80.2at%Al, U-81.7at%Al) in order to determine the growth rate of $UAl_4$ as an intermediate layer in the "diffusion couple" $UAl_3$/Al.

We used the following guess values for the simulation of $UAl_4$ growth in a $UAl_3$/Al diffusion couple using DICTRA software package:

$$\left( D_{Al}^{*(UAl_3)} / D_U^{*(UAl_3)} \right) \approx 1.5 \text{ and } D_{Al}^{*(UAl_3)} \approx k_1 \quad \text{in the } UAl_3 \text{ phase,}$$

$$\left( D_{Al}^{*(UAl_4)} / D_U^{*(UAl_4)} \right) \approx 10^4 \text{ and } D_{Al}^{*(UAl_4)} \approx k_2 \quad \text{in the } UAl_4 \text{ phase.}$$

These values were then optimized (Table 2) to reproduce the experimental data reported in reference [46].

Due to the narrow solubility range and their initial phase region size in the corresponding region of the simulation cell, both $UAl_3$ and Al(U) end members concentration profile are flat and a null concentration gradient apply. This means that the $D_k^{*(\alpha)}$ values in these regions have a little influence in the simulation of $UAl_4$ phase growth.

**Table 2. Optimized $D_{Al}^{*(UAl_4)}$ coefficients for the growth of the reaction layer UAl₄ in UAl₃/Al diffusion couples as determinate from DICTRA simulation**

|  | T=773 K | T=823 K | T=873 K |
|---|---|---|---|
| $D^*(m^2/s)$ | $1.85 \times 10^{-15}$ | $1.381 \times 10^{-14}$ | $7.743 \times 10^{-14}$ |

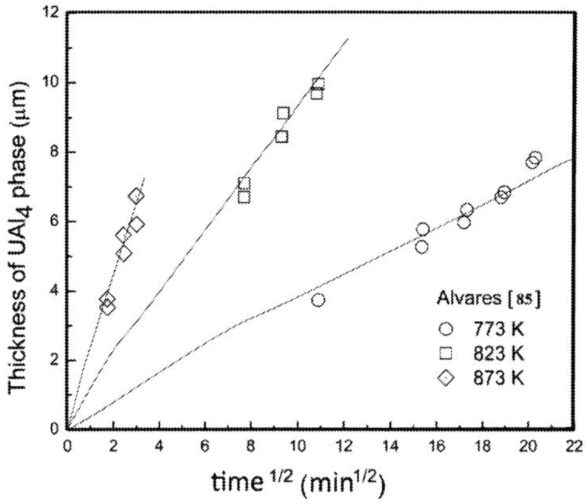

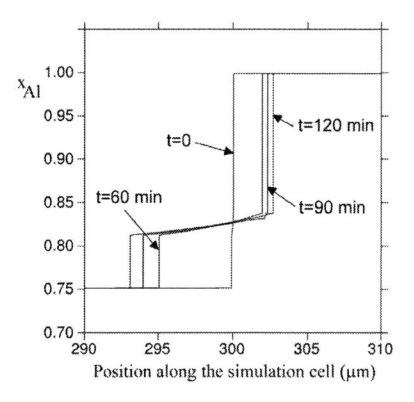

Figure 13. Model-predicted and measured [46] thicknesses of the reaction layer UAl₄ in UAl₃/Al diffusion couples vs time at different temperatures. The solid lines are calculated by DICTRA software.

Figure 13 shows the model-predicted growth rates of UAl₄ phase in the UAl₃/Al diffusion couples at 773, 823 and 873 K along with measured data [46].

The calculated concentration profiles of U and Al in each phase are shown in Figure 14. The calculated ratio between the moving interfaces shows that the interface UAl₃/UAl₄ moves faster than the interface UAl₄/Al [40].

Figure 14. Simulated Al concentration profiles in the UAl₃/Al diffusion couple at four different times and T=823K.

## 3.5. U-Al System and UAl₄ Stability

The question around UAl₄ stability and the nature of its stable structure deserved a deeper insight. As a first step, we decided to evaluate its stability in the U-Al system as a stoichiometric compound without defects. We thus presented the calculated ground state of

U-Al system by a first principles method [58]. Lattice parameters agreed within 2% with experimental values (Tables 3 and 4). An agreement was found for the bulk modulus $B$ of pure U and Al within 4% (Table 5), with

$$B = -\frac{1}{V_0}\frac{\partial^2 E}{\partial V^2}\bigg|_{V=V_0}$$

where $E$ is the total energy, $V$ the volume and $V_0$ the equilibrium volume. However, the ground state diagram at T=0K showed $UAl_2$ and $UAl_3$ as stable compounds while the $UAl_4$ compound appeared as metastable (Table 3 and 4 and Figure 15). We also show in Figure 15 experimental values for formation energies [45] measured at 298K.

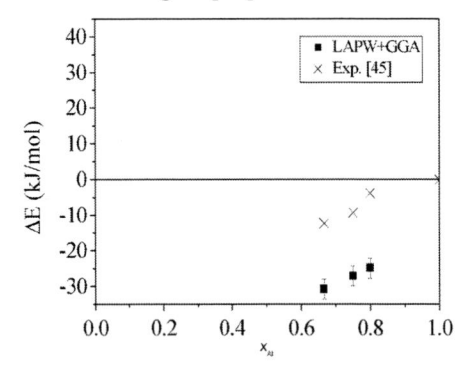

Figure 15. Calculated (T=0K) and experimental ([45], 298K) formation energies in the U-Al system.

**Table 3. Formation energies and equilibrium lattice parameters for the pure elements U and Al. Experimental values [45] are shown between brackets**

| Elem. | Structure | Pearson symbol, Spatial group | ΔE (kJ/mol) | a(Å) | b(Å) | c(Å) |
|---|---|---|---|---|---|---|
| Al | A1 | cF4, 225 | 0 | 4,054 (4,0496) | 4,054 (4,0496) | 4,054 (4,0496) |
| U | A20 | oC4, 63 | 0 | 2,814 (2,8537) | 5,788 (5,8695) | 4,886 (4,9548) |

**Table 4. Formation energies and equilibrium lattice parameters for the intermetallic compounds in the U-Al system. Experimental values [45] are shown between brackets**

| Comp. | Structure | Pearson symbol, Spatial group | ΔE (kJ/mol) | a(Å) | b(Å) | c(Å) |
|---|---|---|---|---|---|---|
| $UAl_2$ | C15 | cF24, 227 | -12,3 (-30,8±2,8) | 7,635 (7,76) | 7,635 (7,76) | 7,635 (7,76) |
| $UAl_3$ | L12 | cP4, 221 | -9,3 (-27,1±2,8) | 4,238 (4,26) | 4,238 (4,26) | 4,238 (4,26) |
| $UAl_4$ | D1b | oI20, 74 | -3,8 (-24,9±2,8) | 4,356 (4,41) | 6,197 (6,27) | 13,671 (13,71) |

**Table 5. Calculated bulk modulus (B) for the pure elements ad for the intermetallic compounds in the U-Al system. Experimental values are shown between brackets**

| Compound | Struktur. | B (GPa) |
|---|---|---|
| Al | A1 | 78 |
|  |  | (75,2) [89] |
| $UAl_2$ | C15 | 111 |
| $UAl_3$ | L12 | 94 |
| $UAl_4$ | D1b | 94 |
| U | A20 | 136 |
|  |  | (133) [90] |
|  |  | (135,5) [90] |

The disagreement between experimental stability of $UAl_4$ and calculations at T=0K lead us to search for structural defects that could stabilize the compound. Based on the discussion present in the literature about the role of vacancies in the stabilization of $UAl_4$, we reported later [59] a new investigation of stoichiometry of $UAl_4$ through a calculation method. Supercells were made up containing vacancies so as to compare the stability with the stoichiometric compound. We calculated formation energies of oI20 superstructures with atomic ratios U:Al of 3:16 and 7:32 [59], and 11:48 [this chapter] at T=0K, placing a vacancy in the unoccupied U site. All uranium sites are equivalent in the $UAl_4$ unit cell, so the 3:16 superstructure was constructed by replacing an uranium atom for a vacancy in the primitive cell. In constructing the 7:32 superstructure, two cells were considered, with 40 atoms and one uranium was replaced for a vacancy. It had to be taken into account that in this case adding another cell along each axis lead to different distances between vacancies and thus to different structures. However, calculations adding a cell along the z axis and along the x axis led to formation energies within 0.0001 Ry/at. The same comment holds for the three cells superstructure. Formation energies (Figure 16) are positive for all structures with vacancies, and thus we can assert that $UAl_4$ intermetallic compound is not stabilize at T=0K by vacancies.

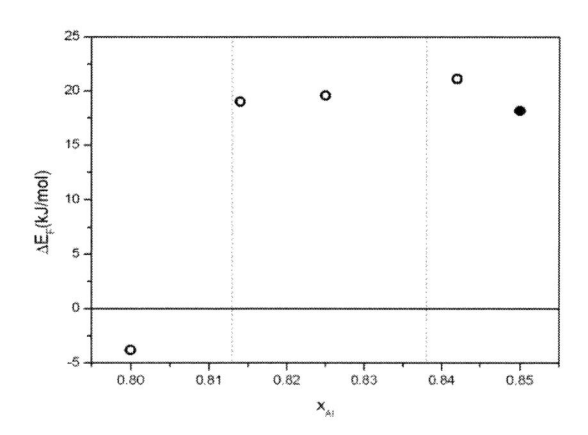

Figure 16. Formation energies for $UAl_4$ with vacancies (open circles) and Al antisites (dark circle). Experimental stability range for $UAl_4$ [45] is marked as the region in between vertical dashed lines.

As the results could still not explain experimental off stoichiometric composition of UAl$_4$, we are facing new calculations under the hypothesis of the existence of antisite defects, this is the occupation of U sites by Al atoms. Preliminary results of the calculation of formation energy of U$_3$Al$_{17}$ are encouraging since it presents lower formation energy than the structure with vacancies at similar Al content (Figure 16). Further calculations are under development to obtain values for antisite structures with several Al contents approaching the stoichiometric composition of UAl$_4$.

## 3.6. Elastic Constants of γU(Mo) Alloys

The cluster expansion of the internal energy of a multicomponent crystalline structure is given by equation (4) in section 2.2, we give emphasis here to the dependence with the parent Bravais lattice writing the equation (4) as:

$$E_G(\mathbf{z}) = \sum_{i=0}^{N} \zeta_{G,i} \; e_i(\mathbf{z}) \tag{17}$$

where $z$ is the matrix of the base of vectors that describe the parent Bravais lattice; $G$ is a configuration of that crystalline structure and is characterized by site occupation ("decoration").

By inversion of the system of equation (17) the ($N$+1) ECIs, $e_i(\mathbf{z})$, can be written in terms of the internal energies of the chosen ordered compounds [91]:

$$e_i(\mathbf{z}) = \sum_{G'=0}^{M} \xi_{i,G'} E_{G'}(\mathbf{z}) \tag{18}$$

where the $\xi_{i,G'}$ coefficients are established as a function of the MCFs in the following usual way. If we consider ($M$+1) ordered configurations to fit the ($N$+1) ECIs, with $M \geq N$, the $\xi_{i,G'}$ coefficients ( $0 \leq i \leq N$ , $0 \leq G' \leq M$ ) are found by means of a least squares fit, and the matrix expression of the solution is:

$$\xi = \left( \zeta^T . \zeta \right)^{-1} . \zeta^T \tag{19}$$

In case that $M = N$ we get $\xi = \zeta^{-1}$. Thus, using equations (18) and (19), the internal energy of the configuration $G$ is written as a superposition of known internal energies of ordered configurations $G'$:

$$E_G(\mathbf{z}) = \sum_{G'=0}^{M} f_{G,G'} \; E_{G'}(\mathbf{z}), \tag{20}$$

where the weights of that superposition are

$$f_{G,G'} = \sum_{i=0}^{N} \zeta_{G,i}\, \xi_{i,G'} \tag{21}$$

The set of clusters $\alpha$ we use for elastic energy description for the bcc $\gamma$U(Mo) alloys includes the empty cluster, the point and one pair of first neighbours. Then we must calculate at least three ordered structures to get a minimal cluster expansion, we have choose the pure bcc U and Mo (A2, cI2, W type) and the intermetallic compound with B2 structure (cP2, CsCl type).

The MCFs corresponding to U (A2), Mo (A2) and UMo (B2) are the following:

$$\zeta^{U,A2} = \begin{bmatrix} 1 & 2 & 1 \end{bmatrix} \; ; \quad \zeta^{Mo,A2} = \begin{bmatrix} 1 & -2 & 1 \end{bmatrix} \; ; \quad \zeta^{UMo,B2} = \begin{bmatrix} 1 & 0 & -1 \end{bmatrix} \tag{22}$$

and the MCF corresponding to U(Mo) A2 solid solution is:

$$\zeta^{A2}(x) = \begin{bmatrix} 1 & 2(2x-1) & (2x-1)^2 \end{bmatrix} \tag{23}$$

where $x$ stands for Mo atomic fraction.

Then, by using equations (20) up to (23), we can write the internal energy of the bcc $\gamma$U(Mo) alloys in terms of the internal energies of the chosen ordered structures $E_{A2,U}(z)$, $E_{A2,Mo}(z)$ and $E_{B2,UMo}(z)$:

$$E_{A2,UMo}(x,z) = 2x(1-x)\,E_{B2,UMo}(z) + x^2\,E_{A2,U}(z) + (1-x)^2\,E_{A2,Mo}(z) \tag{24}$$

If the dependence of structures $E_{A2,U}(z)$, $E_{A2,Mo}(z)$ and $E_{B2,UMo}(z)$ with the $z$ base vectors is known then, we have a description of variation of internal energy with orientation and magnitude of those vectors, this is to say, we can describe the variation of internal energy with lattice deformation. After the deformation, a primitive vector describing the lattice at equilibrium $z_0^G$ is transformed to $z = (1+\varepsilon)\cdot z_0^G$. Then the functional dependence of the internal energy with deformations can be written by explicitly indicating the base vectors at equilibrium and their deformation as:

$$E_G(\mathbf{z}) = E_G(\boldsymbol{\varepsilon}, z_0^G). \tag{25}$$

From elasticity theory it is known that change in energy due a small deformation $\underline{\underline{\varepsilon}}$ in the harmonic approximation can be written as a function of the elastic constants tensor $\underline{\underline{C}}$. For cubic symmetry only three constants are needed:

$$\Delta E_{el} = \frac{1}{2}V_0 \sum_{i,j,k,l=1}^{3} C_{ijkl}\varepsilon_{ij}\varepsilon_{kl} =$$

$$= \frac{1}{2}V_0\left[\left(\varepsilon_{11}^2 + \varepsilon_{22}^2 + \varepsilon_{33}^2\right)C_{11} + 2\left(\varepsilon_{11}\varepsilon_{22} + \varepsilon_{11}\varepsilon_{33} + \varepsilon_{22}\varepsilon_{33}\right)C_{12} + 4\left(\varepsilon_{12}^2 + \varepsilon_{13}^2 + \varepsilon_{23}^2\right)C_{44}\right] \quad (26)$$

Finally, if we consider an equivalence between variation in elastic energy due to macroscopic deformation (Eq (26)) and variation in internal energy due to deformation in structure base vectors (Eqs (24) and (25)), we have a tool to calculate elastic constants for the bcc $\gamma$U(Mo) alloys from first principles thermodynamics. This hypothesis had been previously used by other authors and applied to pure elements and intermetallic compounds [92,93]. We want to emphasize that it had not been previously used for solid solutions.

We have employed an hydrostatic deformation to the crystal to obtain the Bulk modulus ($B_0$) value and the first relation between $C_{11}$ and $C_{12}$:

$$B_0 = V_0\left[\frac{\partial^2 E}{\partial V^2}\right]_{V_0} = (C_{11} + 2\,C_{12})/3 \quad (27)$$

The remaining elastic moduli were obtained by applying tetragonal and trigonal distorsions:

$$\frac{1}{3}\left[\frac{\partial^2 U}{\partial \delta^2}\right]_{tetra}\Bigg|_{\delta=0} = \frac{1}{2}(C_{11} - C_{12}) = \frac{1}{2}C', \qquad C_{44} = \frac{1}{12}\left[\frac{\partial^2 U}{\partial \delta^2}\right]_{trig}\Bigg|_{\delta=0} \quad - \quad (28)$$

The procedure continues in the following way: deformations in the three modes are applied to the three base structures (U, Mo and UMo-B2) and thus, we obtain the dependence of their internal energies with volume, for hydrostatic deformation, and with deformation parameter $\delta$ in the other two cases. Then the equation (24) is written for each mode, thus obtaining the dependence of internal energy of the disordered bcc phase for the same modes of deformation. The function of the internal energy with the volume for each structure was then derived and minimize to obtain equilibrium values of lattice parameter and finally, through equations (27) and (28), the bulk modulus, $C'$ and $C_{44}$ were obtained. Table 6 resumes the results for the ordered structure.

**Table 6. Results for equilibrium lattice parameter a₀, Bulk modulus B₀, C′, and elastic constants C₁₁, C₁₂ and C₄₄**

| Structure | $a_0$ (Å) | $B_0$ (GPa) | C' (GPa) | $C_{11}$ (GPa) | $C_{12}$ (GPa) | $C_{44}$ (GPa) |
|---|---|---|---|---|---|---|
| Mo(A2) exp. [93, 94] | 3.15 | 268 | 302 | 469 | 167 | 107 |
| Mo(A2) | 3.1655 | 262 | 299 | 461 | 162 | 112 |
| U(A2) | 3.4520 | 93 | -85 | 37 | 122 | 71 |
| UMo(B2) | 3.3980 | 147 | 68 | 192 | 124 | 3 |

Predictions for the U(Mo) disordered bcc phase as a function of composition are shown in Figures 17 y 18. Reported lattice parameter values [95] were measured at room temperature in samples quenched from high temperature where U(Mo) disordered bcc phase is stable. The good agreement between measured and calculated values of lattice parameter encourages us to rely on bulk modulus calculations.

Our first principles calculations predict a negative value for C′ of U cubic phase. This result agrees with experimental knowledge of equilibrium orthorrombic phase at low temperatures, reflecting the instability of γ phase with respect to tetragonal distortions. This instability is extended in the binary U-Mo system to γ U(Mo) phase, giving also place to a negative value of C′ for high values of U content in the disordered solution. For smaller values, closer to Mo phase, bcc phase is predicted as stable and positive values are obtained. The result implies that a deformation of a retained γ U(Mo) phase at low temperature can produce a phase transformation to a more stable structure.

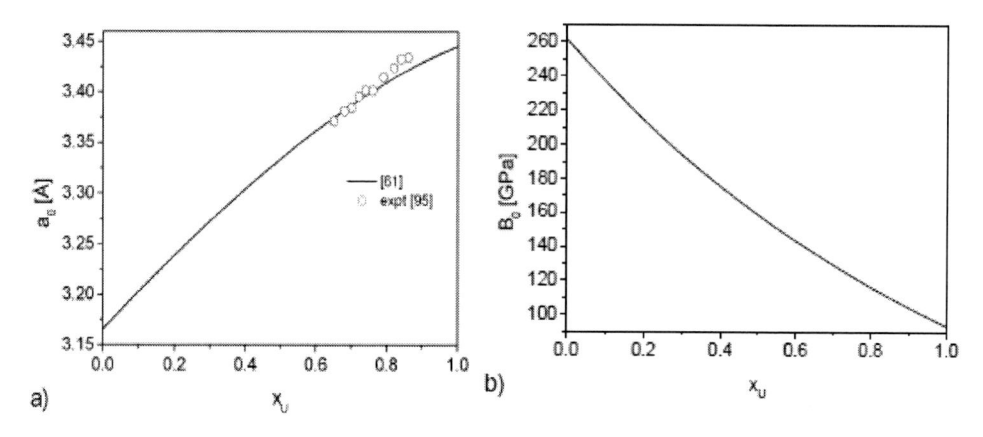

Figure 17. Lattice parameter (a) and bulk modulus (b) as a function of U content $x_U$ for U(Mo) phase.

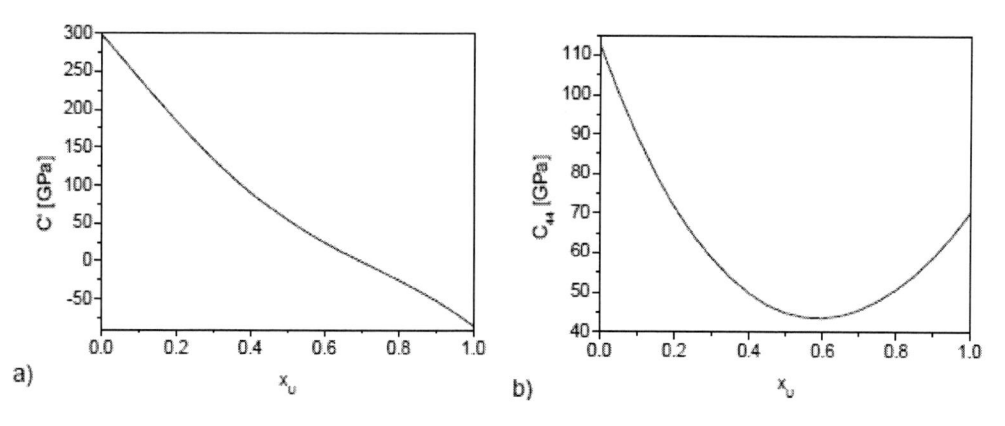

Figure 18. Calculated values of a) C′, and b) $C_{44}$, for the bcc disordered phase as a function of U content.

# CONCLUSION

Several features concerning the development of a γU based nuclear fuel have been investigated successfully applying first principles theory and methods.

1.  It was demonstrated that the stability of $\gamma U(Mo)$ phase is dominated by a three-body multisite interaction consisting in two pairs of first neighbors and one pair of third neighbors instead of pairs interactions as had been previously proposed by others authors. In consequence, it should be noted that the ultimate effectiveness of third element substitution on $\gamma$ stabilization depends on how much this element modifies the afore mentioned multisite interaction.

2.  The problem of the stability of the interdiffusion layer in dispersed $\gamma U$ based fuels in Al based matrix has been faced:

    *   Stability of $U(Al,Si)_3$ was demonstrated in the pseudo binary system $U(Al)_3$-$U(Si)_3$. It was also shown that a miscibility gap appears at low temperatures.
    *   Our experimental studies in U-Al-Zr-Si system establish that a 3.3 at% Zr content is necessary to inhibit $UAl_4$ formation in the quaternary system with 0.3 at% Si content; and confirm that Zr and Si bonds play a significant role in the reduction of the enthalpy of mixing in the compound $(U,Zr)(Al,Si)_3$
    *   The successful description for the diffusion growth of $UAl_4$ compound in the $UAl_3/Al$ diffusion couple shows that the present approach is of general validity and applicable to establish the kinetic database of technically important U-Al alloys.
    *   $UAl_4$ is an undesired compound in the interdiffusion layer, however its structure is not well established. We carried out ab-initio to study the stability of the three observed intermetallic compounds in the U-Al system. Ab-initio results show that $UAl_2$ and $UAl_3$ are stable structures while $UAl_4$ is metastable at very low temperatures. We have also studied constitutional uranium vacancies in the latter compound as an additional element of stability, as suggested in the literature, with negative results. Further calculations are in progress in order to investigate the possibility of a stabilization of the $UAl_4$ compound by antisite occupation of Al in U sites.

3.  Monolithic fuel postulates another challenge, since internal stresses will arise, during fabrication and operating conditions, in the interface between fuel and cladding materials due to the difference in thermal expansion coefficients and elastic constants of the solid solution. We have implemented a method to accurately calculate elastic constants for the solid solution U(Mo) bcc from first principles methods that had not been previously used for solid solutions. The method can be used to predict mechanical properties even in unstable phases, such as U(Mo) bcc phase at low temperature, which is predicted to transform to a more stable structure by deformation.

## ACKNOWLEDGMENTS

The work in this chapter was partially supported by the Secretaría de Ciencia y Tecnología del Gobierno Argentino under grant BID 1201/OC-AR, PICT N° 12-11186 (program 2004-2006) and grant BID 1728/OC-AR, PICT N° 38240 (2007-2009 program); by the Universidad Nacional de San Martín under grant C033 (program 2004-2006), grant C046 (2007-2008 program) and grant C054 (2009-2010 program); by CONICET under PIP 00965 (2010-2012 program); and by Departamento Materiales, CAC – CNEA.

# REFERENCES

[1]    Van den Berghe, S.; Van Renterghem, W.; Leenaers, A. *J. Nucl. Mat.* 2008, 375, 340–346.

[2]    Nazaré, S.; Ondracek, G.; Thümmler, F. J. Nucl. Mat. 1975 56 251-259; Nazaré, S. *J. Nucl. Mat.* 1984 124 14-24.

[3]    Hofman, G. L.; Snelgrove, J. L.; Hayes, S. L.; Meyer, M. K. Transactions 6th *International Topical Meeting RRFM Ghent*, Belgium 2002 17-20 50-58.

[4]    Hayes, S.L.; Meyer, M.K.; Hofman, G.L.; Strain, R.V. 1998 *Proceedings 21$^{st}$ International Meeting RERT*R Sao Paulo, Brazil.

[5]    van Thyne, R. J.; McPherson, D. *J. Trans. ASM* 1957 49 598-617.

[6]    Donzé, G.; Cabane, G.; *J. Nucl. Mat.* 1959 1 4 364-373.

[7]    Hoffman, G. L.; Meyer, M. K. *Proceedings 21$^{th}$ International Meeting RERTR* Sao Paulo, Brazil, 1998 S22 18-23.

[8]    Miedema, A. R.; de Chatel, P. F.; de Boer, F. R. Cohesion in Alloys: Fundamentals of a Semi-Empirical Model; North Holland Publishing Co.: Amsterdam, The Netherlands, 1980; pp 1-28.

[9]    Parida, S. C.; Dash, S.; Singh, Z.; Prasad, R.; Venugopal, V. *J. Phys. Chem. Sol.* 2001 62 585-597.

[10]    Brewer, L.*; In Phase Stability in Metals and Alloys*; Rudman, P; Stringer, J.; Jaeffe, R. I.; McGraw-Hill; New York, 1967; p 39.

[11]    Inden, G.; Pitsch, W.; *In Phase Transformations in Materials;* Haasen, P.; vol 5, Weinheim VCH Press: 1991; p 499.

[12]    Alonso, P. R.; Rubiolo, G. H.; Anales Jornadas SAM-CONAMET – Simposio MATERIA 2003 (San Carlos de Bariloche, Argentina) 2003 pp 987-990; Alonso, P. R.; Rubiolo, G. H.; *Materia* 2004, 9, 203-207.

[13]    Alonso, P. R.; Rubiolo, G. H. *Modelling Simul. Mater. Sci. Eng.* 2007, 15, 263-273.

[14]    Leenaers, A.; Van den Berghe, S.; Koonen, E.; Jarousse, C.; Huet, F.; Trotabas, M.; Boyard, M.; Guillot, S.; Sannen, L.; Verwerft, M. *J. Nucl. Mat.* 2004, 335, 39–47..

[15]    Mirandou, M.I.; Balart, S.N.; Ortiz, M.; Granovsky, M.S. *J. Nucl. Mater.* 2003 323 299.

[16]    Ryu, H.J.; Kim, Y.S.; Hofman, G.L.; Keiser, D.D. *Proceedings International Meeting RERTR* 2006, Cape Town, South Africa.

[17]    Nazarg, S.; Ondracek, G.; Thtimmler, F. *J. Nuclear Mater.* 1975, 56 251.

[18]    Gan, J.; Keiser, D.; Wachs, D.; Miller, B.; Allen, T.; Kirk, M.; Rest, J. *31th International Meeting RERTR* 2009, Beijing, China, S14-P1.

[19]    Boucher, R. *J. Nucl. Mater.* 1959, 1 13-27.

[20]    Thurber, W.C.; Beaver, R.J. Oak Ridge (USA) *Report* 1959, ORNL-2602.

[21]    Picklesimer, M.L.; Thurber, W.C.*; US Patent* 2950188, USPO 1960.

[22]    Chakraborty, A.K.; Crouse, R.S.; Martin, W.R. *J. Nucl. Mater.* 1971, 38 93-104

[23] Hofman, G.L.; Snelgrove, J.L.; Hayes, S.L.; Meyer, M.K. *Transactionse 6th International Topical Meeting RRFM* 2002, pp 50-58.

[24] Chae, H.T.; Kim, H.; Lee, C.S.; Jun, B.J.; Park, J.M.; Kim, C.K.; Sohn, D.S. *J. Nucl. Mat.* 2008 373 9–15.

[25] Böning, K.; Petry, W. *J. Nucl. Mat.* 2009 383 254–263.

[26] Kim, Y.S.; Hofman, G.L.; Ryu, H.J.; Rest, J. *Proceedings International Meeting RERTR* 2005, Boston, USA, November 6–10, 2005.

[27] Mirandou, M.; Arico, S.; Gribaudo, L.; Balart, S. *Proceedings International Meeting RERTR* 2005, Boston, USA.

[28] Park, J.M.; Ryu, H.J.; Lee, G.G.; Kim, H.S.; Lee, Y.S.; Kim, C.K.; Kim, Y.S.; Hofman, G.L. *Proceedings International Meeting RERTR* 2005, Boston, USA.

[29] Kim, Y.S.; Hofman, G.L.; Ryu, H.J.; Finlay, M.R.; Wachs, D. *Proceedings International Meeting RERTR* 2006, Cape Town, South Africa.

[30] Ripert, M.; Dubois, S.; Boulcourt, P.; Naury, S.; Lemoine, P. *Transactions 10th International Topical Meeting RRFM* 2006, ENS, Sofia, Bulgaria.

[31] Leenaers, A.; Van den Berghe, S.; Koonen, E.; Jacquet, P.; Jarousse, C.; Guigon, B.; Ballagny, A.; Sannen, L. *J. Nucl. Mat.* 2004 327 121–129.

[32] Mirandou M.; Granovsky M.; Ortiz M.; Balart S.; Aricó S.; Gribaudo L. *Proceedings International Meeting RERTR* 2004, Vienna, Austria.

[33] Dwight, A.E. *Argonne National Laboratory* 1982. ANL-82-14.

[34] Park, J.M.; Ryu, H.J.; Oh, S.J.; Lee, D.B.; Kim, C.K.; Kim, Y.S.; Hofman, G.L. *J. Nucl. Mater.* 2008 374 422-430.

[35] Pizarro, L.M.; Alonso, P.R.; Rubiolo, G.H. *J. Nucl. Mat.* 2009, 392, 1, 70-77.

[36] Mazaudier, F.; Proye, C.; Hodaj, F. *J. Nucl. Mater.* 2008, 377, 476–485.

[37] Palancher, H.; Martin, P.; Nassif, V.; Tucoulou, R.; Proux, O.; Hazemann, J.L.; Tougait, O.; Lahéra, E.; Mazaudier, F.; Valot, C.; Dubois, S. *J. Appl. Crystallogr.* 2007, 40 1064–1075.

[38] DeLuca, L.S.; Sumsion, H.T. Unclassified Document KAPL-1747, UC-25, *Metallurgy and Ceramics,* 1957, USAEC Report.

[39] Castleman, L. S. *J. Nucl. Mater* 1961, 3, 1, 1-15.

[40] Soba, A.; Denis, A. *J. Nucl. Mater* 2007, 360, 231-241.

[41] Olander, D. *J. Nucl. Mater.* 2009, 383, 201-208.

[42] Ryu, H. J.; Han, Y. S.; Park, J. M.; Park, S. D.; Kim, C. K. *J. Nucl. Mater* 2003, 321, 210-220.

[43] Kniznik, L. ; Alonso, P.R.; Gargano, P.H.; Rubiolo, G.H. submitted to *J. Nucl. Mat.* october 2010.

[44] Wang, J.; Liu, X.J.; Wang, C.P. *J. Nucl. Mater.* 2008 374, 79-86.

[45] Kassner, M.E.; Adler, P.H.; Adamson, M.G.; Peterson, D.E. *J. Nucl. Mater.* 1989, 167 160-168.

[46] Alvares da Cunha, C. Etudo da cinética da transformação de fase no estado sólido UAl3+Al→UAl4 1986, *Master of Science Thesis, Instituto de Pesquisas Energéticas e Nucleares*, Universidade de São Paulo (Brasil).

[47] Gordon, P. ; Kaufmann, A. R. *Trans. Amer. Inst. Min. Met. Eng.* 1950, 188, 182.

[48] Hills, R. F. *J. Inst. Met.* 1957-58, 86, 438.

[49] Chiotti, P.; Kateley, J.A. *J. Nucl. Mater.* 1969, 32, 135-145.

[50] Borie, B.S. Trans. *AIME* 1951, 191, 800.

[51]  Boucher, R. *J. Nucl. Mater.* 1959, 1, 13.

[52]  Runnals, O.J.C. ; Boucher, R.R. Trans. *AIME* 1965, 233, 1726.

[53]  Jesse, A.; Ondracek, G.; Thummler, F. *Powder Metall.* 1971, 14, 289.

[54]  Kassner, M.E.; Adler, P.H.; Adamson, M.G. *J. Nucl. Mater.* 1989, 167, 160-168.

[55]  Zenou, V.Y.; Kimmel, G.; Cotler, C.; Aizenshtein, M. *J. Alloy Comp.* 2001, 329, 189-194.

[56]  Tougait; Noël, H. *Intermetallics* 2004, 12, 219-223

[57]  Dienst,, W.; Nazare, S.; Thummler, F. *J. Nucl. Mater.* 1977, 64, 1

[58]  Pizarro, L.M.; Gargano, P.; Alonso, P. R.; Fernández, J. R. Anales 93° RNF – XI Reunión SUF 2008, 20, Buenos Aires, Argentina, 1850 – 1158.

[59]  P. R. Alonso, J. R. Fernández, P. H. Gargano, G. H. Rubiolo, U-Al system, *Physica B: Condensed Matter, Volume* 404, Issue 18, 1 October 2009, Pages 2851-2853.

[60]  Clark, C.R.; Knighton, G.C.; Meyer, M.K.; Hofman, G.L. *25th International Meeting RERTR* 2003, Chicago, IL, U.S.

[61]  Gargano, P.; Alonso, P.R.; Mosca, H.O.; Ríos, J.C.; Ortiz Albuixech, M.; Rubiolo, G.H. *27th International Meeting RERTR* 2005, Boston, Massachusetts, USA, pp S14-4

[62]  Dreizler, R.M.; Gross, E.K.U. Density Functional Theory*: an Approach to the Quantum Many-Body Problem*; Springer-Verlag: Berlin; 1990.

[63]  Perdew, J.P.; Burke, K.; Ernzerhof, M. *Phys. Rev. Lett.* 1996, 77, 3865.

[64]  Perdew, J.P.; Burke, K.; Ernzerhof, M. *Phys. Rev. Lett.* 1997, 78, 1396.

[65]  Singh, D.J. Planewaves,*Pseudopotentials and the LAPW Method*; Kluwer Academic Publishers: London, 1994.

[66]  Andersen, O.K. *Phys. Rev. B* 1975 , B12, 3060.

[67]  Blaha, P.; Schwarz, K.; Luitz, J. Computer Code WIEN97, Vienna University of Technology: 1997. (Improved and updated Unix version of the original copyrighted WIEN-code which was published by Blaha, P.; Schwarz, K.; Sorantin, P.; Trickey, S.B. *Comp. Phys. Commun.* 1990, 59 399).

[68]  Blaha, P.; Schwarz, K.; Madsen, G.K.H.; Kvasnicka, D.; Luitz, J. WIEN2k, An Augmented Plane Wave + Local Orbitals Program for Calculating Crystal Properties; Schearz, K.; Ed.; *Tech. Universität Wien*: Austria, 2001.

[69]  Sedmidubsky, D.; Konings, R.J.M.; Novák, P. *J. Nucl. Mater.* 2005; 344, 40.

[70]  Sanchez, J. M.; Ducastelle, F.; Gratias, D. *Physica A* 1984, 128, 334-350

[71]  de Fontaine, D. Solid State Physics, 47; Ehrenreich, H.; Turnbull, D.; Ed.; New York Academic: 1994, pp 33-176.

[72]  Lu, Z.W.; Wei, S.H.; Zunger. A.; Frota-Pessoa., S.; Ferreira, L.G. *Phys Rev B*, 1994, 44, 2, 512-544.

[73]  van de Walle, A.; Ceder, G. *J. Phase Equilibria* 2002, 23, 348-359.

[74]  Zhong, Y.; Wolverton, C.; Austin Chang, Y.,; Liu, Z. K. *Acta Materialia*, 2004, 52, 2739-2754.

[75]  Connolly, J.W.D.; Williams, A.R. *Physical Review B*, 1983, 27, 5169-5172.

[76]  van de Walle, A.; Asta, M.; Ceder, G. *CALPHAD* 2002, 26, 539.

[77]  Saunders, N.; Miodownik, A. P. CALPHAD; *Pergamon Materials Series;* Vol. 1; Cahn, R.W.; Ed.;, Elsevier Science: Oxford, 1998.

[78]  Redlich, O.; Kister, A. T. *Industrial and Engineering Chemistry, Vol.* 4, 1948, pp 345-348.

[79]  Sundman, B. ; Jansson, B.; Andersson, *J-O. CALPHAD* 1985; 9; 153-190.

[80] Andersson, J-O.; Ågren, J. *Journal Applied Physics* 1992; 72; 4; 1992; 1350-1355.

[81] J. O. Andersson, L. Höglund, B. Jönsson, J. Ågren. In: Prudy G. R., editor. *Fundamentals and applications of ternary diffusion.* New York: Pergamon Press; 1990. p. 153-163.

[82] Massalski, T. B.; Okamoto, H.; Subramanian, P.; Kacprzak, L. Binary Alloy Phase Diagrams; 2nd ed; vol 3; *ASM International: Materials Park OH USA*, 1990.

[83] Wolverton, C.; Asta, M.; Dreyssé, H.; de Fontaine, D. *Phys. Rev. B* 1991, 44, 10, 4914-4924.

[84] Kniznik, L.; Alonso, P.R. ; Gargano, P.H.; Rubiolo, G.H. Actas Congreso SAM/CONAMET 2009, CNEA, 2009, pp 1287- 1292.

[85] Dinsdale, A.T. *CALPHAD*, 1991, 15, 317-425.

[86] Borgenstam, A.; Engström, A.; Höglund, L.; Ågren, J. *Journal Phase Equilibria*, 2000, 21, 3, 269-280.

[87] Du, Y.; Schuster, J. C. *Metallurgical and Materials Transactions A*, 2001, 32A, 2396-2400 (2001), p. 2396-2400.

[88] Diffusion in Solid Metals and Alloys; Meher, H.; Ed., Landolt-Börnstein, New Series, Group III, 26; Springer, Berlin and Heidelberg, 1990.

[89] Smithells Metals Reference Book, Gale, W. F.; Totemeir, T. C.; Ed., 8th Edition, 2004, Elsevier ASM, The Netherlands.

[90] Söderlind, P.; *Phys. Rev. B* 2002, 66, 085.

[91] Lu, Z.W.; Wei, S.H.; Zunger, A.; Frota-Pessoa, S.; Ferreira, L.G. *Phys. Rev. B* 1991; 44; 512:544.

[92] Alonso, P. R.; Rubiolo. G. H. Anales Jornadas SAM-CONAMET, 2003, 987; Alonso, P.R.; Rubiolo. G. H. Anales Jornadas *CONAMET-SAM*, 2004, 127.

[93] Simmons, G.; Wang, H. *Single Crystal Elastic Constants and Calculated Aggregate Properties: A Handbook*, 1971, MIT Press, Cambridge, MA, 2nd. ed.

[94] Donohue, J. *The Structure of the Elements*, 1974, Wiley, New York.

[95] Dwight, E. *J Nucl. Mat.*, 1960, 2, 81.

In: Nuclear Materials
Editor: Michael P. Hemsworth, pp. 81-109

ISBN 978-1-61324-010-6

*Chapter 3*

# RADIATION DAMAGE AND RECOVERY OF CRYSTALS: FRENKEL VS. SCHOTTKY DEFECT PRODUCTION

## *V. I. Dubinko*

NSC Kharkov Institute of Physics and Technology NAS of Ukraine,
Kharkov, Ukraine

## ABSTRACT

A majority of radiation effects studies are connected with creation of radiation-induced defects in the crystal bulk, which causes the observed degradation of material properties. The main objective of this chapter is to describe the mechanisms of recovery from radiation damage, which operate during irradiation but are usually obscured by the concurrent process of defect creation. Accordingly, the conventional rate theory is modified with account of radiation-induced Schottky defect formation at extended defects, which often acts against the mechanisms based on the Frenkel pair production in the crystal bulk. The theory is applied for the description of technologically and fundamentally important phenomena such as irradiation creep, radiation-induced void "annealing" and the void ordering.

## 1. INTRODUCTION

Radiation damage in crystalline solids originates from the formation of Frenkel pairs of vacancies and self-interstitial atoms (SIAs) and their clusters. The difference in the ability to absorb vacancies and SIAs by extended defects is thought to be the main driving force of microstructural evolution under irradiation. A recovery from radiation damage is usually observed under sufficiently high temperatures and it is driven by thermal fluctuations resulting in the evaporation of vacancies (Schottky defects) from voids and dislocations and in the fluctuation-driven overcoming of obstacles by gliding dislocations. These recovery mechanisms can be efficient only at sufficiently high temperatures. However, there is increasing evidence that the so-called "thermally activated" reactions may be modified under irradiation. Thus, in refs. [1-3] it was demonstrated that Schottky defects could be produced not only by thermal fluctuations but also by interactions between extended defects and *excitons* in di-atomic ionic crystals. Excitons are the excitations in the electron sub-system, which can be formed under irradiation in *insulators* and move by diffusion mechanism towards extended defects thus providing a mechanism of Schottky defect formation at the defect surfaces. Well established excitations in ionic system include unstable Frenkel pairs (UFPs) [4, 5] and focusing collisions ("*focusons*") [6-9]. More recently, essentially non-linear lattice excitations, called *discreet breathers* (DBs) or *quodons* have been considered [10-14]. Accordingly, the rate theory of microstructure evolution in solids has been modified with account of the production of Schottky defects at voids and dislocations due to their interaction with the radiation-induced lattice excitations [15-21]. In this chapter, recent developments and applications of the new theory are presented.

The chapter is organized as follows. In the section 2, the competition between the Frenkel and Schottky defect production is illustrated at such technologically important phenomenon as irradiation creep.

In the sections 3 and 4, original experimental data on the radiation-induced void "annealing" are presented and analyzed in the framework of the rate theory modified with account of the radiation-induced Schottky defect formation.

In the section 5, we consider one of the most spectacular phenomena in physics of radiation effects, namely, the void lattice formation [22-28]. It is often accompanied by a saturation of the void swelling with increasing irradiation dose, which makes an understanding of the underlying mechanisms of both scientific significance and practical importance for nuclear engineering. We compare the popular mechanisms of void ordering based on anisotropic *interstitial transport* [22-27] with the original mechanism based on the anisotropic *energy transfer* provided by quodons [28].

## 2. IRRADIATION CREEP

Irradiation creep is one of the most outstanding puzzles of the theory of radiation damage, which has been studied rather extensively due to its technological importance [29-36]. It is known that under typical neutron fluxes, irradiation creep dominates over thermal creep below about 0.5 $T_m$ and shows rather *weak or no dependence on irradiation temperature*. However, for a long time, even this general trend was not well understood.

Conventional creep models are based on the so called *stress-induced preferential absorption* (SIPA) of radiation-produced point defects by dislocations [29, 30] or other extended defects [33] differently oriented with respect to the external stress. In these models it is assumed that under irradiation Frenkel pairs of vacancies and SIAs are created in the bulk and annihilate at extended defects (EDs) that are thus considered as *sinks* for freely migrating point defects and their small mobile clusters. Consequently, these models can yield a temperature independent irradiation creep only when the bulk recombination of point defects is negligible, which is not the case under sufficiently low irradiation temperature.

An alternative approach has been proposed in ref. [17], which takes into account that EDs can act as *sources* for the production of *radiation-induced* Schottky defects that do not exist in the bulk. A Schottky defect is a single vacancy or SIA (or a small defect cluster), which can be emitted from the ED surface and which does not require a counterpart of opposite sign in contrast to the bulk production of Frenkel pairs, in which the total numbers of vacancies and SIAs *must be equal*. Results of molecular dynamics (MD) simulations [37] have shown that more vacancies than SIAs can be produced in the vicinity of the dislocation cores, and the required energy, $E_v$, is lower than the threshold energy for Frenkel pair production in the bulk.

In this chapter we develop further a mechanism of irradiation creep [17], which is based on the *radiation and stress induced preference in emission* (RSIPE) of vacancies from dislocations differently oriented with respect to the external stress. It is essentially temperature independent in contrast to the SIPA mechanism. Let us make a clear distinction between the creep mechanisms based on the absorption of Frenkel defects and emission of Schottky defects.

## 2.1. SIPA and SIPE Mechanisms of Creep

For the sake of simplicity, we shall consider the creep rate along the tensile stress axis in a crystal containing two families of straight edge dislocations perpendicular to the external load axis, namely, dislocations with a Burgers vector parallel (type A) or perpendicular (type N) to this axis and presenting densities of $\rho_d^A$ and $\rho_d^N$, respectively. Then the creep rate is simply related to a dislocation climb velocity, $V_d$, which is determined by the difference in absorption of SIAs and vacancies and by emission of vacancies[1] [17]:

$$\dot{\varepsilon} = \rho_d^A b V_d^A = \rho_d^N b V_d^N \tag{1}$$

$$b V_d^Y = D_i \bar{c}_i Z_i^Y - D_v \bar{c}_v Z_v^Y + D_v c_v^{eq,Y} Z_v^Y, \ Y = A, N, \tag{2}$$

where $b$ is the magnitude of the Burger's vector, $D_i$ and $D_v$ are the diffusion coefficients of SIAs and vacancies, respectively, $\bar{c}_i$, $\bar{c}_v$ are their mean concentrations, $Z_{i,v}^Y$ is the capture efficiency of dislocations for vacancies (subscript "v") and SIAs (subscript "i"), and $c_v^{eq,Y}$ is the equilibrium concentration of vacancies in a crystal containing Y-type dislocations. The point defect concentrations obey the following rate equations

$$\frac{d\bar{c}_i}{dt} = K_{FP}(1 - \varepsilon_i) - k_i^2 D_i \bar{c}_i - \beta_r (D_i + D_v)\bar{c}_i \bar{c}_v, \ K_{FP} = k_{eff} K, \tag{3}$$

$$\frac{d\bar{c}_v}{dt} = K_{FP}(1 - \varepsilon_v) - k_v^2 D_v (\bar{c}_v - \bar{c}_v^{eq}) - \beta_r (D_i + D_v)\bar{c}_i \bar{c}_v, \tag{4}$$

---

[1] Emission of SIAs from dislocations is not considered due to relatively high activation energies of SIA formation in metals.

$$\bar{c}_v^{eq} = \frac{Z_v^A \rho_d^A c_v^{eq,A} + Z_v^N \rho_d^N c_v^{eq,N}}{Z_v^A \rho_d^A + Z_v^N \rho_d^N}, \; k_{i,v}^2 = Z_{i,v}^A \rho_d^A + Z_{i,v}^N \rho_d^N, \tag{5}$$

where $K$ is the displacement rate, measured in displacements per atom (dpa) per second, $k_{eff}$ is the cascade efficiency for production of stable Frenkel pairs in the bulk, which determines the strength of the source of freely migrating point defects, $K_{FP}$; $\varepsilon_{i,v}$ is the fraction of point defects formed in the in-cascade clusters, $\beta_r$ is a bulk recombination constant and $k_{i,v}^2$ is the dislocation sink strength for SIAs (subscript "$i$") or vacancies (subscript "$v$").

Usually, in the technologically relevant temperature range, steady-state conditions are attained at very early irradiation stages, in which the point defect production is balanced by their loss to sinks, so that $d\bar{c}_{i,v}/dt = 0$, and one obtains from (3) and (4) the relation

$$D_v \bar{c}_v = \frac{k_i^2}{k_v^2} D_i \bar{c}_i + D_v \bar{c}_v^{eq} \tag{6}$$

Thus, the creep rate (1) can be written as

$$\dot{\varepsilon} = \rho_d^A D_i \bar{c}_i \left( Z_i^A - \frac{k_i^2}{k_v^2} Z_v^A \right) + \rho_d^A D_v Z_v^A \left( c_v^{eq,A} - \bar{c}_v^{eq} \right) \tag{7}$$

and, accounting for (2) - (6), one obtains:

$$\dot{\varepsilon} = \frac{\rho_d^A \rho_d^N}{k_v^2} D_i \bar{c}_i \left( Z_i^A Z_v^N - Z_i^N Z_v^A \right) + \frac{\rho_d^A \rho_d^N Z_v^A Z_v^N}{k_v^2} D_v \left( c_v^{eq,A} - c_v^{eq,N} \right) = \dot{\varepsilon}_{SIPA} + \dot{\varepsilon}_{SIDE} \tag{8}$$

The first term in (8) is determined by the stress-induced preference in *absorption* of SIAs and vacancies by dislocations of different types (SIPA mechanism). It is proportional to the difference in absorption (or capture) efficiencies by dislocations of different types, $Z_i^A Z_v^N - Z_i^N Z_v^A$, which is a function of applied stress and material parameters, and to the mean steady-state flux of SIAs, $D_i \bar{c}_i$, that depends on the irradiation dose rate and temperature as follows:

$$D_i \bar{c}_i = -\frac{1}{2} \frac{D_v}{\beta_r} \frac{k_v^2}{k_i^2} \left( k_i^2 + \beta_r \bar{c}_v^{eq} \right) + \left( \frac{1}{4} \left( \frac{D_v}{\beta_r} \frac{k_v^2}{k_i^2} \right)^2 \left( k_i^2 + \beta_r \bar{c}_v^{eq} \right)^2 + K_{FP} \frac{D_v}{\beta_r} \frac{k_v^2}{k_i^2} \right) \tag{9}$$

The product $D_i \bar{c}_i$ is essentially temperature independent and proportional to the dose rate when the bulk recombination is neglected, and it rapidly decreases with decreasing temperature in the recombination dominant region:

$$D_i \bar{c}_i \approx \begin{cases} \dfrac{K_{FP}}{k_i^2}, \dfrac{K_{FP}\beta_r}{D_v k_i^2 k_v^2} \to 0, & \text{sink region} \\[4mm] \left(\dfrac{K_{FP} D_v}{\beta_r}\right)^{1/2}, \dfrac{K_{FP}\beta_r}{D_v k_i^2 k_v^2} >> 1, & \text{recombination region} \end{cases} \tag{10}$$

The capture efficiencies of dislocations under tensile stress, $\sigma$, are given in the conventional SIPA models [29,30, 36] by the following expressions

$$Z_n^Y = \frac{2\pi}{\ln\left(R_d / R^Y\right)}, \quad R_d = \sqrt{\frac{1}{\pi\rho_d}}, \quad \rho_d = \rho_d^A + \rho_d^N \tag{11}$$

$$R_n^Y = \frac{1}{2} L_n\left(1 + A_n^Y \frac{\sigma}{\mu}\right), \quad L_n = \frac{\mu b(1+\nu)}{3\pi k_B T(1-\nu)}|\Omega_n|, \quad n = i, v, \quad Y = A, N \tag{12}$$

where $R_d$ is the radius of a cylindrical "region of influence" around a straight dislocation, which is determined by the dislocation density, $R_n^Y$ is the effective "capture radius" of a dislocation determined by its elastic interaction with the point defects, $\mu$ is the shear modulus of the matrix, $\nu$ is the Poisson ratio, $\Omega$ is the point defect relaxation volume, and $A_n^Y$ is a function of the dislocation orientation, $Y$, and on the differences in shear modulus, $\mu$, and bulk modulus, $\kappa$, between the matrix and point defects. The maximum SIPA effect is obtained in the extreme case of $\mu_i \to 0$, $\kappa_i \to \infty$ for SIAs and $\mu_v \to \mu$, $\kappa_v \to 0$ for vacancies: $A_v^Y = 0$, $A_i^A \approx 0.1$, $A_i^N \approx -0.1$. In this case, the external stress does not influence the dislocation capture efficiency for vacancies ($Z_v^A = Z_v^N \equiv Z_v$), while it increases the SIA capture efficiency of A type dislocations and decreases the SIA capture efficiency of N type dislocations. In the first order of approximation for $\sigma/\mu << 1$ one obtains from (8) that the SIPA creep rate is proportional to the applied load:

$$\dot{\varepsilon}_{SIPA} \approx \frac{\rho_d^A \rho_d^N}{k_v^2} D_i \bar{c}_i Z_v \left(A_i^A - A_i^N\right)\frac{\sigma}{\mu} \le 0.2 \frac{\rho_d^A \rho_d^N}{k_v^2} D_i \bar{c}_i Z_v \frac{\sigma}{\mu} \tag{13}$$

The second term in (8) is determined by the stress-induced preference in *emission* (SIPE) of vacancies by dislocations of different types. It is proportional to the difference between equilibrium vacancy concentrations corresponding to dislocations of different types. Without irradiation, the equilibrium concentrations can be obtained from thermodynamics by minimizing the free energy of a crystal containing identical dislocations [17]:

$$c_v^{th,N} = \exp\left(-\frac{E_v^{f,d}}{k_B T}\right) \equiv c_v^{th}, \quad c_v^{th,A} = \exp\left(-\frac{E_v^{f,d} - \sigma\omega}{k_B T}\right) = c_v^{th} \exp\left(\frac{\sigma\omega}{k_B T}\right) \tag{14}$$

where $E_v^{f,d}$ is the vacancy formation energy at a free dislocation, and $\omega$ the atomic volume. Substituting (14) in (8) one obtains the creep rate due to the *temperature and stress induced preference in emission* of vacancies (TSIPE) (known also as classical Nabarro creep mechanism) that is a special example of a more general SIPE mechanism. In the first order of approximation for $\sigma\omega/k_B T \ll 1$ one obtains that the TSIPE creep rate is proportional to the applied load and to the mean steady-state flux of *thermally-produced* vacancies, $D_v c_v^{th}$, that depends exponentially on temperature:

$$\dot{\varepsilon}_{TSIDE} \approx \frac{\rho_d^A \rho_d^N Z_v^A Z_v^N}{k_v^2} D_v c_v^{th} \frac{\sigma\omega}{k_B T} \tag{15}$$

The temperature dependence of the net creep rate due to both SIPA and TSIPE mechanisms given by eqs. (8) - (15) is shown in Figure 1 versus experimentally measured irradiation creep compliance (the net part of creep per 1 MPa stress and irradiation dose of 1 dpa, which is independent from the void swelling), which is known to be about $10^{-6} MPa^{-1}dpa^{-1}$ for a large range of austenitic steels irradiated in nuclear reactors in a wide temperature interval [32] .

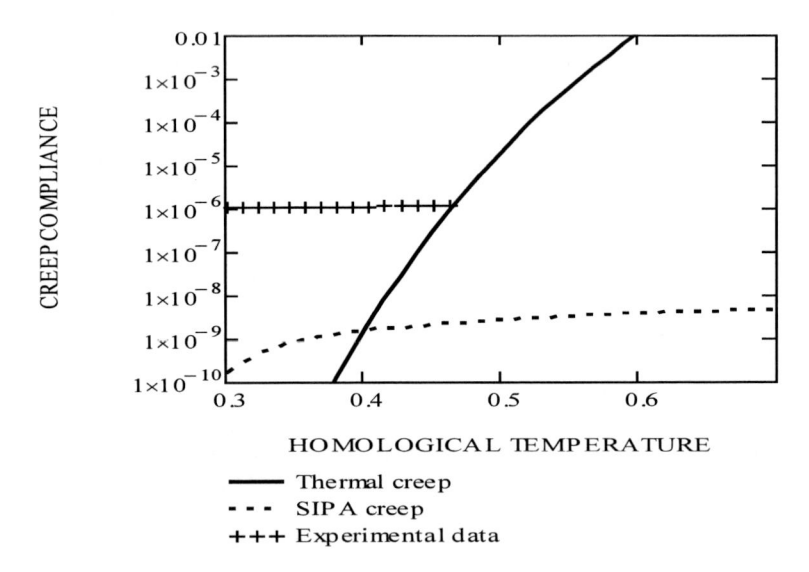

Figure 1. SIPA and TSIPE creep compliance versus experimentally observed irradiation creep compliance [32]. The dislocation density is $\rho_d^A = \rho_d^N = 10^{14} m^{-2}$. Other material parameters are given in Table 1.

It can be seen that the SIPA mechanism, even at the best choice of material parameters, predicts much lower creep rates than experimentally observed. Besides, the SIPA creep rate is proportional to the mean SIA concentration, which decreases with decreasing temperature (see eq. (10)). On the other hand, the TSIPE mechanism based on the *thermal emission* of vacancies does not depend on irradiation, but it depends exponentially on temperature and becomes inefficient at T < 0.5 Tm (Tm is the melting temperature). However, when we consider how irradiation can enhance the rate of vacancy emission from dislocations and how this enhancement depends on the dislocation type, then we will obtain an alternative mechanism of irradiation creep based on the *radiation and stress induced preference in emission* of vacancies (RSIPE). In the following section we describe possible mechanisms of radiation-induced emission of vacancies from dislocations.

**Table 1. Material parameters used in calculations**

| Parameter | Value |
|---|---|
| Atomic spacing, b, m | $3.23 \times 10^{-10}$ |
| Atomic volume of the host lattice, $\omega$, m$^{-3}$ | $2.36 \times 10^{-29}$ |
| Matrix shear modulus, $\mu$, GPa | 35 |
| Interstitial dilatation volume, $\Omega_i$, | $1.2\omega$, $0.6\omega$, |
| Vacancy dilatation volume, $\Omega_v$, | $-0.6\omega$ |
| Bulk recombination rate constant, $\beta_r$, m$^{-2}$ | $8 \times 10^{20}$ |
| Displacement energy, $E_d$, eV | 30 |
| Cascade efficiency for the stable defect production, $k_{eff}$ | 0.1 |
| Fraction of point defects in the in-cascade clusters, $\varepsilon_{i,v}$ | 0.9 |
| Maximum focuson energy, $E_F$, eV | 60 |
| Vacancy formation energy at a free surface, $E_v^f$, eV | 1.8 (Ni), 1.3 (Cu) |
| Vacancy formation energy at a void surface, $E_v^V$, eV | $E_v^f - 2\gamma\omega/R$ |
| Surface energy, $\gamma$, J/m$^2$ | 2 |
| Stacking fault energy, J/m$^2$ | 0.13 (Ni), 0.03 (Cu) |
| Irradiation-induced vacancy formation energy at dislocations, $E_v^d$ | 2.215 (Ni), 1.562 (Cu) |
| Migration energy of vacancies, $E_v^m$, eV | 1.1 (Ni), 0.98 (Cu) |
| Pre-exponent factor, $D_v^0$ | $10^{-5}$ |

## 2.2. Radiation-Induced Emission of Vacancies from Extended Defects

It is known that not all the energy of the primary knock-on atom (PKA) is spent for the production of stable defects. A considerable part of the PKA energy is spent for production of unstable Frenkel pairs (UFPs) [4, 5] and mobile lattice vibrations, such as focusons [6-9] and quodons [10-14], which can interact with dislocations and other extended defects and produce Schottky defects in their vicinity [20]. Consider the mechanisms of such interaction.

### 2.2.1. Unstable Frenkel Pairs

According to MD simulations [37], vacancies can be produced in the vicinity of dislocation cores, and the required energy, $E_v$, is lower than the threshold energy for Frenkel pair production in the bulk. The underlying mechanism can be understood as follows. The dislocation core is surrounded by a region of a radius $R_{cap}$, in which a point defect is unstable since it is captured by the dislocation athermally [38]:

$$R^n_{cap} = \left( \frac{\mu(1+\nu)}{3\pi(1-\nu)E^m_n} |\Omega_n| \right)^{1/2} b, \; R^i_{cap} \approx 3b, \; R^\nu_{cap} \approx b \tag{16}$$

where $E^m_n$ is the migration energy of point defects. The dislocation capture radius for SIAs is larger than the one for vacancies. If a regular atom in the region $R^\nu_{cap} < r < R^i_{cap}$ gets an energy $E > E_\nu$ it may move to an interstitial position, where it can be athermally captured by the dislocation, leaving behind it a stable vacancy. The capture time is about $10^{-11} - 10^{-12} s$ so that the process can be described as an effective emission of a vacancy by the dislocation due to its interaction with an unstable Frenkel pair. Since the energy of the system is increased as a result of vacancy formation and corresponding dislocation climb due to the SIA capture, the minimum transferred energy, $E_\nu$, should exceed the energy of vacancy formation *accounting for the work due to dislocation climb in the stress field*. N-type dislocations do not interact with a tensile stress, and so one has $E^N_\nu = E^f_\nu$, while for A-type dislocations one has $E^A_\nu = E^f_\nu - \sigma\omega$, as in the case of the conventional Nabarro climb.

Let us estimate the rate of radiation-induced emission of vacancies from dislocations by UFPs created under electron irradiation. The rate of radiation-induced emission of vacancies from a dislocation is given by the product of the UFP production rate per atom, $K_{UFP}(E^Y_\nu)$, and the number of atomic sites in the region $R^\nu_{cap} < r < R^i_{cap}$, from where a SIA can be captured by the dislocation, leaving behind it a vacancy. Then, the rate of vacancy emission per unit dislocation length is given by

$$J^Y_\nu \approx 2\pi R^\nu_{cap} R_{i\nu} \frac{K_{UFP}(E^Y_\nu)}{\omega}, \; R_{i\nu} \equiv R^i_{cap} - R^\nu_{cap} \tag{17}$$

$$K_{UFP}(E^Y_\nu) = j_e \int_{E^Y_\nu}^{E_d} \frac{d\sigma}{dE} dE \tag{18}$$

where $j_e$ is the electron flux, and $d\sigma/dE$ is the differential cross section for producing a PKA of energy $E$, which is written in the McKinley - Feshbach approximation as [0]

$$\frac{d\sigma}{dE} = \pi \left( \frac{Z\varepsilon^2}{m_e c^2} \right)^2 \frac{1-\beta^2}{\beta^4} \frac{E_m}{E^2} \left[ 1 - \beta^2 \frac{E}{E_m} + \frac{\pi\beta Z}{137} \left\{ \left( \frac{E}{E_m} \right)^{\frac{1}{2}} - \frac{E}{E_m} \right\} \right]$$

$$\tag{19}$$

$$\beta = \left[ 1 - \left( 1 + \frac{E_e}{m_e c^2} \right)^{-2} \right]^{1/2}, \; E_m = \frac{2E_e(E_e + 2m_e c^2)}{Mc^2} \tag{20}$$

where $E_e$ is the electron energy, $Z$ is the atomic number, $\varepsilon$ and $m_e$ are the electron charge and mass, respectively, $M$ is the target atom mass, $c$ is the light velocity, $E_d$ is the displacement energy and $E_m$ is the maximum energy that can be transferred to a PKA.

The rate of production of stable Frenkel pairs in the bulk is given by

$$K_{FP} = j_e \int_{E_d}^{E_m(E_e)} \frac{d\sigma}{dE} dE \tag{21}$$

Under low-energy sub-threshold irradiation, i.e. $E_m < E_d$, the Frenkel pair production in the bulk is suppressed ($K_{FP} = 0$) but the unstable Frenkel pairs can still be produced until $E_m > E_v$.

Now we will relate the rate of vacancy emission per unit dislocation length (17) to the radiation-induced vacancy *out-flux* density across the surface surrounding the dislocation vacancy capture region:

$$\left(j_v^Y\right)_{out}^{irr} = \frac{J_v^Y}{2\pi R_{cap}^v} = \frac{R_{iv}}{\omega} K_{UFP}\left(E_v^Y\right), \tag{22}$$

Then, the boundary condition of general type at the dislocation capture surface can be written as the difference between the influx and out-flux of vacancies:

$$\left(j_v^Y\right)_{r=R_{cap}^v} = \left(j_v^Y\right)_{in} - \left(j_v^Y\right)_{out} = \frac{v_v}{\omega} c\left(R_{cap}^v\right) - \left(j_v^Y\right)_{out} = \frac{v_v}{\omega} {}_v\left[c\left(R_{cap}^v\right) - \frac{\omega\left(j_v^Y\right)_{out}}{v_v}\right] \tag{23}$$

$$\left(j_v^Y\right)_{out} = \left(j_v^Y\right)_{out}^{irr} + \left(j_v^Y\right)_{out}^{th} = \left(j_v^Y\right)_{out}^{irr} + \frac{v_v}{\omega} c_v^{th,Y} = \frac{v_v}{\omega}\left[c_v^{irr,Y} + c_v^{th,Y}\right] \tag{24}$$

where $v_v \approx D_v/b$ is the vacancy transfer velocity from the matrix to the capture region, and $c\left(R_{cap}^v\right)$ is the actual vacancy concentration at the dislocation core surface, which should be obtained from the solution of the diffusion problem accounting for boundary conditions (23) at all dislocations in the system. In equilibrium conditions, which can be realized *without irradiation* or under *sub-threshold* irradiation ($K_{FP} = 0$) in a system with *identical* EDs, e.g. dislocations of one type, there would be *no flux* across their capture surfaces. Then $c\left(R_{cap}^v\right)$ will be equal to the equilibrium vacancy concentration, $c_v^{eq,Y}$:

$$c_v^{eq,Y} = c_v^{irr,Y} + c_v^{th,Y}, \tag{25}$$

$$c_v^{irr,Y} = \frac{\omega\left(j_v^Y\right)_{out}^{irr}}{v_v} \approx \frac{\omega\left(j_v^Y\right)_{out}^{irr} b}{D_v} = \frac{bR_{iv}}{D_v} K_{UFP}\left(E_v^Y\right) \tag{26}$$

*Without irradiation*, one has simply $c\left(R^v_{cap}\right)=c^{th,Y}_v$ given by eq. (14). Under *sub-threshold* irradiation, $c\left(R^v_{cap}\right)$ is given by the sum $c^{eq,Y}_v = c^{irr,Y}_v + c^{th,Y}_v$, where $c^{irr,Y}_v$ is due to the *radiation-induced emission* of vacancies and is given by eq. (26). In both cases, the mean steady-state concentration in the system coincides with the equilibrium concentration (no gradients and defect fluxes between the EDs).

Finally, under over-threshold irradiation ( $K_{FP} > 0$ ), the flux across the capture surface is proportional to the difference between the mean steady-state and equilibrium concentrations: $\left(j^Y_v\right)_{r=R^v_{cap}} \propto \bar{c}_v - c^{eq}_v$ .

The calculated vacancy concentrations for electron irradiations of different energies are presented in Figure 2. It can be seen that the radiation-induced equilibrium concentration dominates completely over the thermal one at temperatures below 0.5 $T_m$. At low temperature end, the thermal equilibrium concentration is practically zero, whereas the radiation-induced equilibrium concentration increases with decreasing temperature due to decreasing diffusion coefficient, and may exceed the melting point level. It saturates at very low temperatures, where the radiation-induced diffusion coefficient, $D^{irr}_v \approx K_D b^2$ , dominates over the thermal one, $D^{th}_v = D^0_v \exp\left(-\dfrac{E^m_v}{k_B T}\right)$, where $E^m_v$ is the vacancy migration activation energy, and a frequency of radiation-induced "jumps" , $K_D$ , is given by

$$K_D = j_e \int_{E^m_v}^{E_m(E_e)} \frac{d\sigma}{dE} dE , \qquad (27)$$

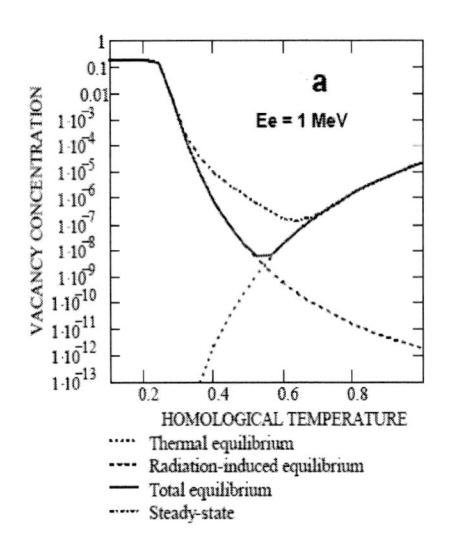
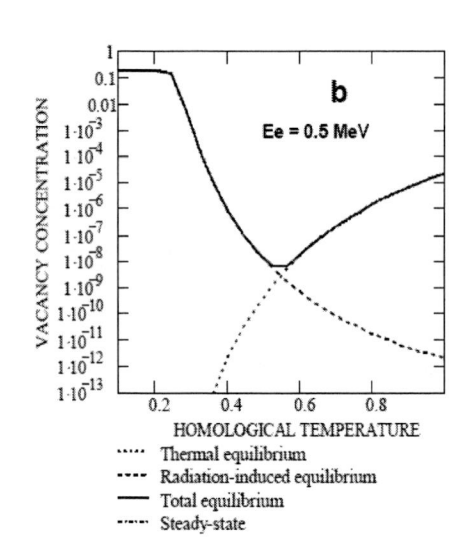

Figure 2. Equilibrium and steady-state vacancy concentrations in Ni (a) under over-threshold ( $E_e = 1MeV$ , $j_e = 10^{19} e/cm^2$, which corresponds to $K_{FP} = 2 \times 10^{-4} s^{-1}$ ) and (b) sub-threshold ( $E_e = 0.5MeV$ , $K_{FP} = 0$) electron irradiations. In the second case, the steady-state and equilibrium concentrations coincide. Dislocation density, $\rho^A_d = 10^{14} m^{-2}$ . Other material parameters are given in Table 1.

The mean steady-state concentration is larger than the equilibrium concentration between 0.3 and 0.6 $T_m$ (swelling region) but both concentrations are exactly the same under sub-threshold irradiation in a system with identical EDs (Figure 2b: true equilibrium state).

### 2.2.2. Focusons

Focusons [6-9] are produced in the recoil events and *transfer energy* along close packed directions of the lattice, but there is *no interstitial transport* by a focusing collision, which enlarges their range considerably as compared to that of a *crowdion*. The energy range in which focusons can occur has an upper limit $E_F$, which has been estimated to be about 60 eV for Cu, 80 eV for Ag and 300 eV for Au [9]. In an ideal lattice a focuson travels in a close packed direction and loses its energy continuously by small portions, $\varepsilon_F$, in each lattice spacing, which determines its propagation range, $l_0(E, E_v)$, as a function of initial energy, $E$, and the final energy, $E_v^Y$, required for the vacancy ejection [9]:

$$l_0\left(E, E_v^V\right) = l_F^0 \ln\left(E/E_v^V\right), \ l_F^0 = b/\varepsilon_F, \ \varepsilon_F \approx 0.1 \div 0.01 \tag{28}$$

The initial focuson energy is completely transformed into lattice vibrations or heat. However, if a focuson has to cross a region of lattice disorder, a defect may be produced in its surroundings [8]. The presented derivation of the radiation-induced equilibrium concentration can be extended to include the focuson mechanisms of Schottky defect formation. To do so let's estimate the rate of vacancy emission from a cylinder of the radius $R_{cap}^v$ and unit length surrounding a dislocation core due to its interaction with incoming focusons, $J_v^Y$. It is equal to the number of energetic focusons ($E > E_v^Y$) absorbed by the dislocation core from the bulk per unit time, which is proportional to the focuson production rate per atom and the number of atomic sites from which a focuson can reach the dislocation core.

If a dislocation line is oriented along one of the $n_c$ close-packed directions, focusons can reach a dislocation core from $n_c - 2$ cylinders of a radius $R_{cap}^v$ and length $l_0(E, E_v)$ protruded along the closed packed directions. Accordingly, the rate of the vacancy emission due to absorption of focusons is given by

$$J_v^Y = \frac{2 R_{cap}^v \left(n_c - 2\right)}{\omega n_c} \int_{E_v^Y}^{E_F} l_0\left(E/E_v^Y\right) z\left(E_p, E\right) dE \tag{29}$$

where $E_F$ is the focuson maximum energy and $z(E_p, E) dE$ is the number of focusons produced by a primary recoil atom (PKA) with energy, $E_p > E_F$, in the energy range $(E, dE)$:

$$z\left(E_p, E\right) = \frac{2 E_p}{E_F^2} \ln \frac{E_F}{E}, \ E \leq E_F, \tag{30}$$

The rate of production of focusons can be expressed via the displacement rate per atom, $K$:

$$K_F = K_{PKA} \int_{E_v^Y}^{E_F} z(E_p, E) dE = 4K \frac{E_d}{E_F} \int_1^{E_v^Y/E_F} \ln x\, dx \tag{31}$$

where $K_{PKA} = K(2E_d/E_p)$ is the rate of PKA production and $E_d$ is the displacement energy.

In the equilibrium state, the vacancy flux out of the dislocation core (29) is balanced by the vacancy flux into the core, which is given by the product of the local vacancy concentration, $c_v^{irr,Y}$, the cylinder surface, $2\pi R_{cap}^v$, and the rate, at which a vacancy cross the capture surface, $v_v$:

$$J_v^Y = \frac{2\pi}{\omega} R_{cap}^v c_v^{irr,Y} v_v \tag{32}$$

Equalizing eqs. (29) and (32) one obtains the following expression for the focuson-induced equilibrium concentration of vacancies at a dislocation core:

$$D_v c_v^{irr,Y} = b l_F^0 K_F^d(E_v^Y), \quad K_F^d(E_v^Y) \equiv K \frac{4(n_c - 2)}{\pi n_c} \frac{E_d}{E_F} \int_1^{E_v^Y/E_F} \ln x \ln\left(x \frac{E_F}{E_v^Y}\right) dx \approx \frac{1}{2} K_F \tag{33}$$

which is very similar to the expression (26), but the effective production rate of focusons, $K_F^d(E_v^Y)$, stands here for the UFP production rate, and the focuson propagation range, $l_F^0$ stands for $R_{iv} \approx b$. It can be shown that the focuson mechanism of vacancy production is more efficient than the UFP one if $l_F^0 > 5 R_{iv}$, which is likely to be the case at sufficiently low temperatures. Focusons are unstable against thermal motion because they depend on the alignment of atoms. Typically, at elevated temperatures, the focuson range is limited to several unit cells and their lifetime is measured in picoseconds [9]. However, there exists much more powerful, essentially non-linear, mechanism of the lattice excitations having large lifetimes and propagation distances, which is called *discreet breathers* (DBs) or *quodons* [10-14].

### 2.2.3. Quodons

According to molecular dynamic (MD) simulations, DBs are highly anharmonic vibrations, being sharply localized on just a few sites, which have frequencies above or below the phonon band, and so they practically don't interact with phonons [10]. Breathers can be sessile or mobile depending on the initial conditions. The relative atomic motions in a breather can be of transverse or longitudinal type. Quodons, by definition, are high energy mobile longitudinal optical mode discreet breathers. As the incident focuson energy is dispersed, on-site potentials and long range co-operative interactions between atoms can influence the subsequent dispersal of energy in the lattice by the creation of quodons that don't interact with phonons and thus are thermally stable. Russell and Eilbeck [11] have presented evidence for the existence of energetic, mobile, highly localized quodons that propagate great distances in atomic-chain directions in crystals of muscovite, an insulating solid with a layered crystal structure. Specifically, when a crystal of muscovite was bombarded with alpha-particles at a given point at 300 K, atoms were ejected from remote points on *another face of the crystal*, lying in atomic chain directions at more than $10^7$ unit cells distance from the site of bombardment.

Although these results relate to layered crystals there is evidence that quodons can occur in non-layered crystals, but with shorter path lengths of order $10^4$ unit cells. This was reported in connection with radiation damage studies in silicon [12] and with diffusion of interstitial ions in austenitic stainless steel [13]. This points out to the possibility of vacancy emission from any extended defect (void, edge dislocation or grain boundary) in a process of quodon-induced energy deposition.

Let us modify the model with account of quodon-induced production of Schottky defects. Assuming that quodons and focusons can eject vacancies from EDs independently by a similar mechanism, one can write the focuson and quodon -induced equilibrium concentration of vacancies as a sum:

$$D_v c_v^{irr,Y} = b l_F^0 K_F^d \left(E_v^Y\right) + b l_Q^0 K_Q^d \left(E_v^Y\right) \tag{34}$$

where $K_Q$ and $l_Q^0 \gg l_F^0$ are the effective production rate and the propagation range of quodons in a perfect lattice, respectively. Determination of the quodon production rates and ranges requires a detailed model of their formation and propagation in real crystals, which is lacking. So we will use the quodon propagation range as a free parameter and assume below that the quodon production rate is equal to the focuson production rate given by eq. (33). Its dependence on the focuson energy is shown in Figure 3. It can be seen that in the energy range required for the Schottky defect formation, the rate of focuson production is close to the displacement rate.

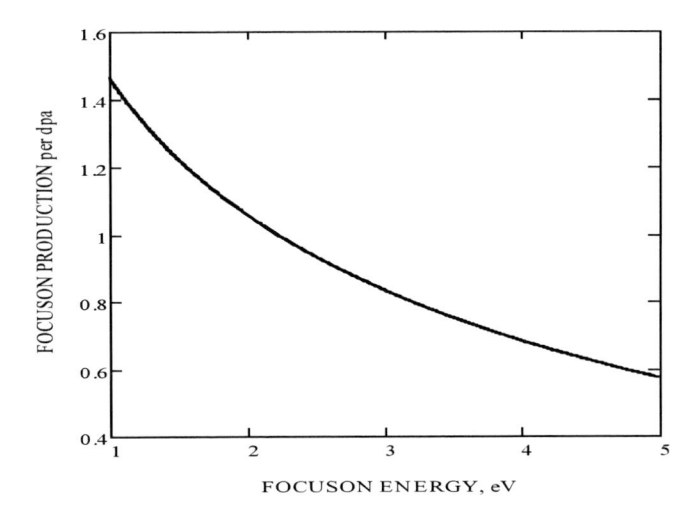

Figure 3. The focuson production rate per dpa as a function of the focuson energy given by eq. (33).

## 2.3. Creep Due to Radiation and Stress Induced Preference in Emission (RSIPE)

In a system with different EDs, such as A-type and N-type dislocations, the true equilibrium state is impossible to reach even without irradiation, due to the difference in equilibrium concentrations at A-type and N-type dislocations, which is a driving force for thermal creep as pointed out in the previous section. Under irradiation, this difference starts to depend on the irradiation dose rate, resulting in a new mechanism of creep, based on *radiation and stress induced preference in emission* of vacancies (RSIPE). Substituting eqs. (26) and (34) into the expression for SIPE creep rate (8) one obtains:

$$\dot{\varepsilon}_{SIPE} = \dot{\varepsilon}_{TSPE} + \dot{\varepsilon}_{RSIPE} \tag{35}$$

$$\dot{\varepsilon}_{RSIPE} \approx \frac{\rho_d^A \rho_d^N Z_v^A Z_v^N}{k_v^2} D_v c_v^{irr,N} \frac{\sigma\omega}{E_v^{f,d}} \tag{36}$$

$$D_v c_v^{irr,N} = bR_{iv} K_F^d\left(E_v^{f,d}\right) + bl_F^0 K_F^d\left(E_v^{f,d}\right) + bl_Q^0 K_Q^d\left(E_v^{f,d}\right) \tag{37}$$

where the terms in eq. (37) correspond to the UFP, focuson, and quodon mechanisms of vacancy emission, respectively. In this derivation, we have taken into account that each constituent in the radiation-induced vacancy emission is inversely proportional to the vacancy formation energy: $D_v c_v^{irr,N} \propto 1/E_v^{f,d}$, $D_v c_v^{irr,A} \propto 1/\left(E_v^{f,d} - \sigma\omega\right)$.

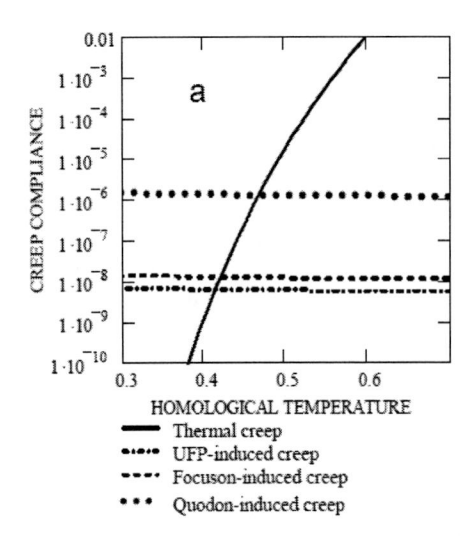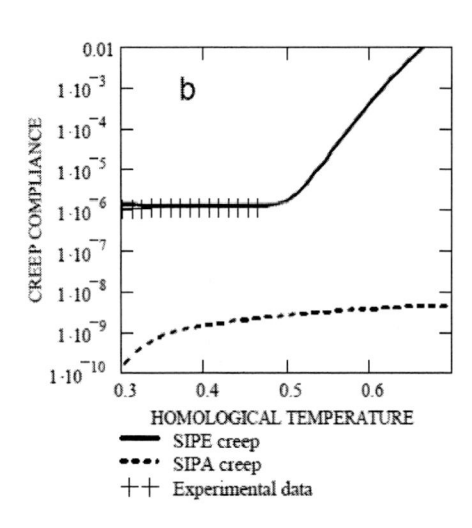

Figure 4. Comparison of different creep mechanisms. (a) Creep rates due to different SIPE mechanisms: thermal; UFP ( $R_{iv} = b$ ); focuson ( $l_F^0 = 10b$ ); quodon ( $l_Q^0 = 1000b$ ). (b) SIPA and SIPE (total) creep compliance versus experimentally observed irradiation creep compliance [32]. The dislocation density is $\rho_d^A = \rho_d^N = 10^{14} m^{-2}$ . Other material parameters are given in Table 1.

Considering the first order of approximation for $\sigma\omega/E_v^f << 1$ one has $D_v c_v^{irr,A} \approx D_v c_v^{irr,N}\left(1+\sigma\omega/E_v^f\right)$, which results in a creep rate (36) proportional to the applied load, which is very similar to the expression for thermal creep (15) but $c_v^{irr,N}$ stands here for $c_v^{th,N}$ and $E_v^f$ stands for $k_B T$.

Comparison of different SIPE mechanisms (Figure 4a) shows that an efficiency of each one is proportional to the range of the corresponding lattice excitation, which is expected to be much higher for quodons than for focusons or UFPs. The RSIPE creep rate agrees perfectly with experimental data (Figure 4b) assuming the quodon propagation range to be about $10^3$ unit cells, which is well within the range previously reported for metals [13]. In contrast to SIPA, the RSIPE creep rate does not depend on production and recombination of Frenkel pairs, and so it can be temperature independent as demonstrated in Figure 4. For over-threshold irradiation it is proportional to the displacement rate:

$$\dot{\varepsilon}_{RSIDE}\left(E_m > E_d\right) \approx \rho_d b l_Q^0 \frac{E_d}{E_F}\frac{\sigma\omega}{E_v^{f,d}}K, \;\; \rho_d^A = \rho_d^N = \rho_d \tag{38}$$

which explains why the experimentally measured creep compliance is about $10^{-6} MPa^{-1}dpa^{-1}$ for a large range of austenitic steels irradiated in nuclear reactors in a wide temperature interval [32] .

## 2.4. SIPE vs. SIPA Summary

It is interesting to note that in the first models of irradiation creep, (see e.g. ref. [36]) it was suggested that below 0.5 $T_m$ the thermal self-diffusion coefficient in the expression for *thermal creep* rate could be replaced by a larger value, namely, by the product of the diffusion coefficient with a steady-state concentration of vacancies increased by the irradiation. This model was criticized because irradiation was thought to enhance the vacancy concentration *in only one way*, i.e. by producing Frenkel pairs in the bulk, which plays no role in *emission driven mechanisms*. We have considered another way of increasing the vacancy concentration by irradiation, which is based on their enhanced emission from extended defects. This radiation-induced emission can be different for dislocations of different orientations in the stress field due to variations in the vacancy formation energy, which are proportional to the external stress. Hence, there are some similarities between RSIPE and TSIPE creep models, which differ qualitatively from SIPA models based on the *long-range* interaction of point defects with dislocations.

The second difference is that the bulk recombination of point defects is efficient in suppressing SIPA creep (Figure 4b), while the SIPE creep is not affected by the recombination.

The third difference between the RSIPE and SIPA models is their different dependence on the electron beam energy as has been pointed out in [17]. RSIPE creep should be observed under sub-threshold irradiation, when SIPA creep is zero, which could be used for an unambiguous experimental verification of the RSIPE mechanism.

# 3. IRRADIATION SWELLING

A schematic representation of the radiation effect on perfect and real crystals is shown in Figure 5.

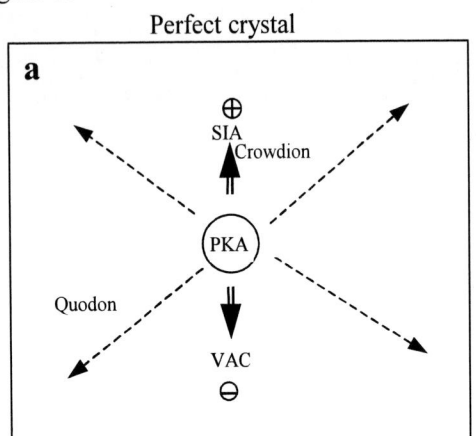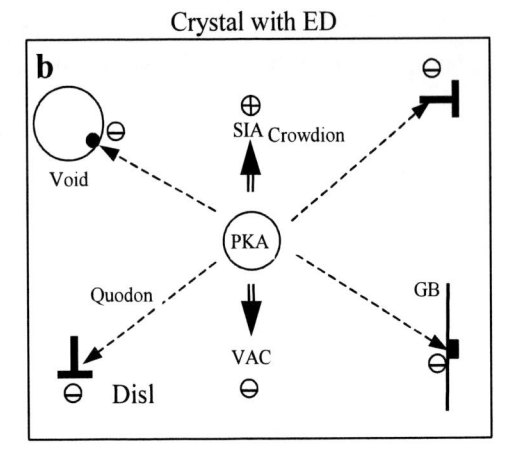

Figure 5. Illustration of Frenkel and Schottky defect production in perfect and real crystals. (a) Frenkel pair formation in the bulk by short-ranged crowdions: SIA is the self-interstitial atom, VAC is the vacancy, arrows show the propagation directions of quodons that do not produce defects in ideal lattice; (b) vacancy formation at extended defects due to interaction of long-ranged quodons with voids, dislocations (⊥) and grain boundaries (GB).

Irradiation produces stable Frenkel pairs by a crowdion mechanism, which is a major driving force of the radiation damage. Even in a perfect crystal (Figure 5b), a considerable part of energy is spent on production of quodons[2] and only subsequently dissipates to heat. In a real crystal (Figure 5b), if a quodon has to cross a region of lattice disorder, a Schottky defect may be produced in its vicinity, which is another driving force of microstructural evolution. In the previous section, we have considered one example of its action, i.e. irradiation creep due to the interaction of quodons with dislocations of different orientations in respect to the stress field. A similar effect can be shown to arise due to the interaction of quodons with grain boundaries (GBs) of different orientations in respect to the stress field, which may be important for description of irradiation creep of ultra fine grain materials. In this section, we will consider another prominent example, namely, irradiation swelling due to the void formation, and will show some important consequences of the quodon interaction with voids. We will start with a description of experimental observations of the radiation-induce void "annealing", which requires special irradiation conditions, since under typical irradiation, this phenomenon is obscured by the void growth due to absorption of vacancies produced in the bulk. Our approach is based on the suppression of radiation-induced defect formation in the bulk. This can be achieved if the irradiation temperature is decreased following the void formation. The bulk recombination of Frenkel pairs increases with decreasing temperature suppressing the production of freely migrating vacancies (the driving force of void growth). On the other hand, the rate of radiation-induced vacancy emission from voids remains essentially unchanged, which can result in the void dissolution. Two types of

---

[2] Quodons are defined here as quasi-one-dimensional lattice excitations that transfers energy along close-packed directions, which applies also to a classic focuson [4-9].

irradiating particles (heavy ions and protons) and irradiated materials (Ni and Cu) will be described in this chapter.

## 3.1. Experimental Observations of the Radiation-Induced Void Annealing

### 3.1.1. Irradiation of Nickel with Cr Ions

Nickel foils of 100 micron thickness have been irradiated with 1.2 MeV Cr ions at 873 K up to the total ion fluence of $10^{21}$ m$^{-2}$, which corresponded approximately to the irradiation dose of 25 displacements per atom (dpa) at the dose rate, $K = 7 \times 10^{-3} s^{-1}$ [19]. Examination of control samples in transmission electron microscope (TEM) has revealed formation of a high number density (~$10^{21}$ m$^{-3}$) of voids of 40-50 nm in diameter. The remaining foils have been irradiated subsequently up to the ion fluence of $10^{21}$ m$^{-2}$ at two different temperatures, 798 K and 723 K, respectively. The resulting microstructure is shown in Figure 6a, from which it is evident that the irradiation at lower temperatures has made the voids to decrease in size. The quantitative analysis of the void swelling confirms this conclusion: the void swelling has decreased by a factor of ~ 5 (Figure 6b).

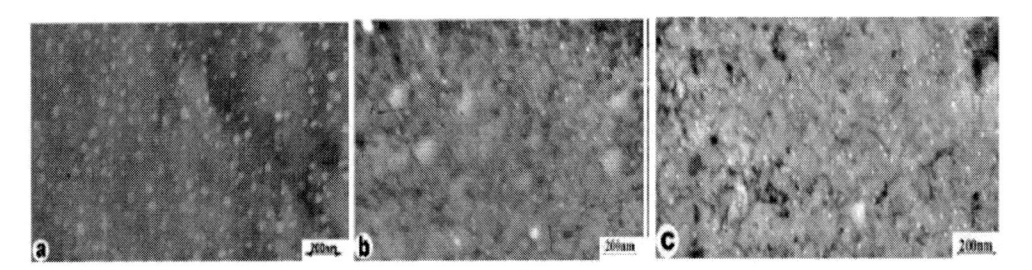

Figure 6a. TEM micrographs showing voids in (a) Ni irradiated with Cr ions up to 25 dpa at 873 K; (b) 25 dpa at 873 K + 25 dpa at 798 K ; (c) 25 dpa at 873 K + 25 dpa at 723 K [19].

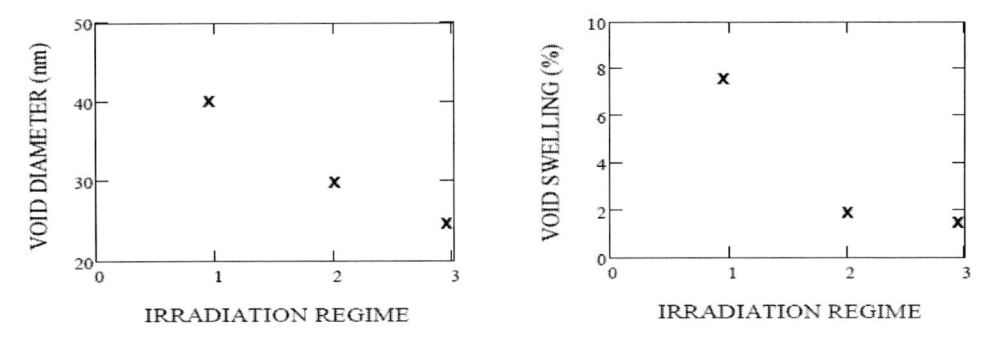

Figure 6b. Mean void diameter and swelling in Ni irradiated with Cr ions up to 25 dpa at 873 K (regime 1); 25 dpa at 873 K + 25 dpa at 798 K (regime 2) and 25 dpa at 873 K + 25 dpa at 723 K (regime 3). The size of symbols "x" corresponds to the mean error in void measurements [19].

### 3.1.2. Irradiation of Nickel with Protons

In the second type experiment, nickel films of 100 nm thickness have been irradiated with 30 keV protons at 873 K up to the total proton fluence of $10^{22}$ m$^{-2}$, which corresponded to the irradiation dose of 6 dpa at the dose rate of $K = 6 \times 10^{-4} s^{-1}$ and to the concentration of implanted hydrogen of 14 at.%. TEM examination of the control samples has revealed formation of a bimodal distribution of cavities (Figure 7) consisting of a high number density (~$10^{22}$ m$^{-3}$) of compact hydrogen bubbles of 10-15 nm in diameter and a relatively low density population of vacancy voids, which have irregular shapes and sizes ranging from 20 to 25 nm, which is consistent with the conventional view on the bubble-void transition effects in metals irradiated with gas ions [39]. Some of the films have been irradiated subsequently up to a maximum proton fluence of $2 \times 10^{22}$ m$^{-2}$ at 723 K. Figure 7b shows that the low temperature irradiation resulted in an increase of the number density of small bubbles and disappearance of large vacancy voids. As a result, the net swelling has decreased by a factor of 3 (Figure 8).

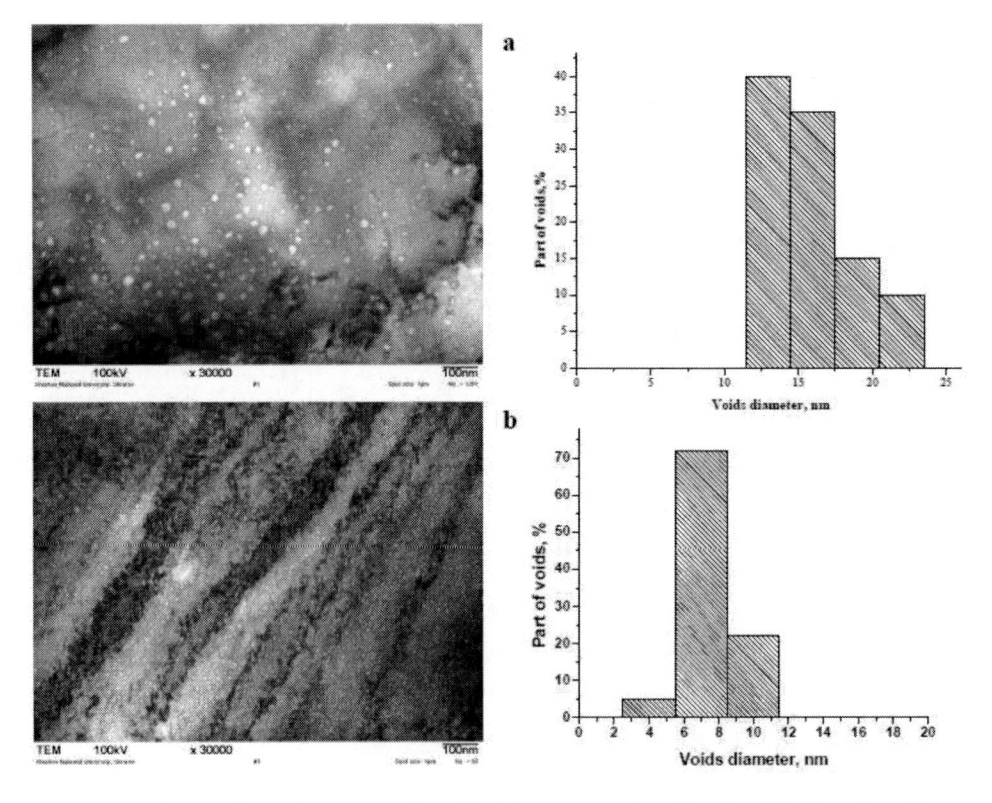

Figure 7. TEM micrographs and corresponding size histograms of cavities in nickel irradiated by 30 keV protons at different temperatures. Picture (a) shows a mixed population of voids and hydrogen bubbles after initial irradiation up to 3 dpa at 873 K. Picture (b) shows a high density population of hydrogen bubbles after subsequent irradiation up to 6 dpa at 723 K.

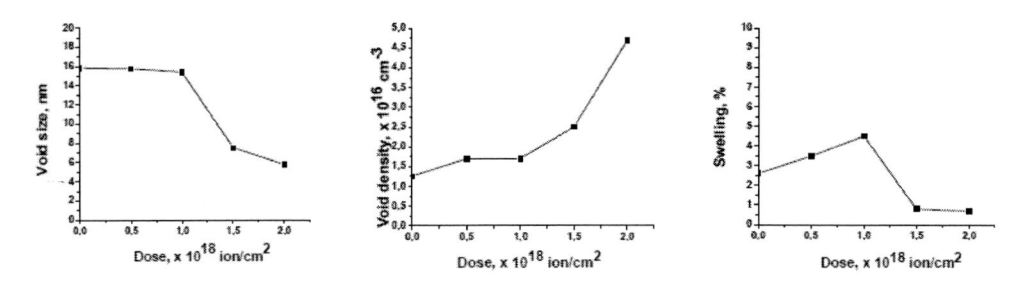

Figure 8. Void size, number density and swelling in nickel pre-irradiated by 30 keV $H^+$ ions at 873 K as a function of irradiation dose at 723 K.

### 3.1.3. Irradiation of Copper with Protons

In the third type experiment, copper films of 100 nm thickness have been irradiated with 30 keV protons at 773 K up to the total proton fluence of $10^{22}$ $m^{-2}$, which corresponded to the irradiation dose of 10 dpa at the displacement rate, $K = 10^{-3} s^{-1}$, and to the concentration of implanted hydrogen of 14 at.%. TEM examination of the control samples has revealed formation of a bimodal distribution of cavities (Figure 9) consisting of a high number density ($\sim 10^{21}$ $m^{-3}$) of hydrogen bubbles of 20-40 nm in diameter and a relatively low density population of vacancy voids in the size range of 60-80 nm. Some of the films have been irradiated subsequently up to a maximum proton fluence of $6 \times 10^{21}$ $m^{-2}$ at 573 K. Figure 9b shows that the low temperature irradiation resulted in a gradual increase of the number density of small bubbles and shrinkage of large vacancy voids. As a result, the net swelling has decreased by a factor of 3 (Figure 10).

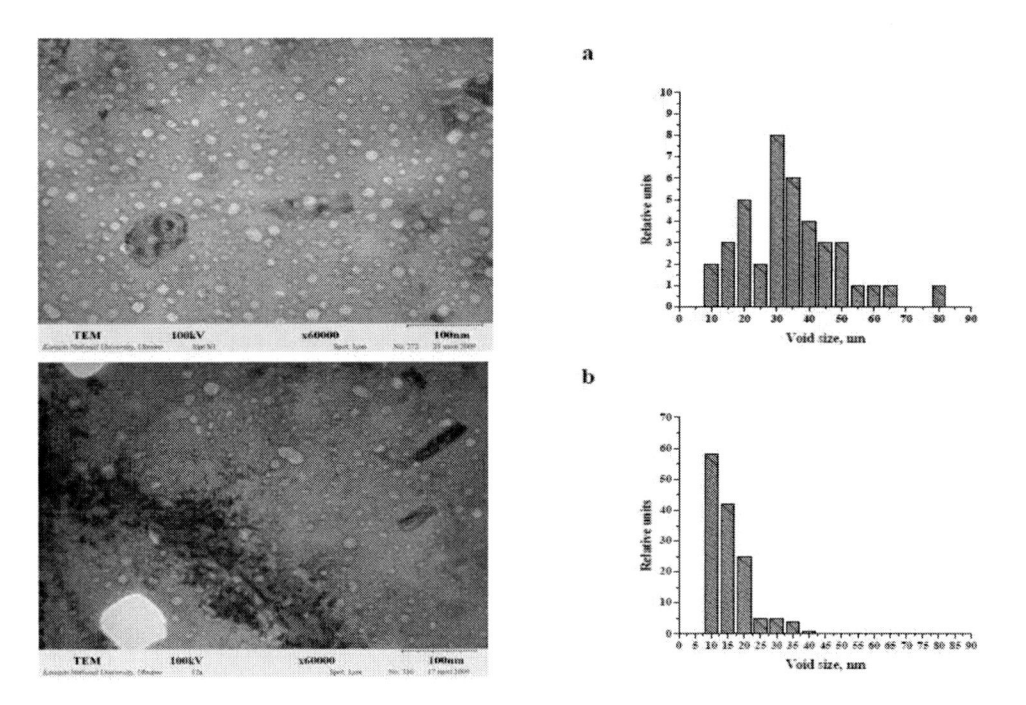

Figure 9. TEM micrographs and corresponding size histograms of cavities in copper irradiated by 30 keV protons at different temperatures. Picture (a) shows a mixed population of voids and hydrogen bubbles after initial irradiation up to 10 dpa at 773 K. Picture (b) shows the microstructure after subsequent irradiation up to 6 dpa at 573 K.

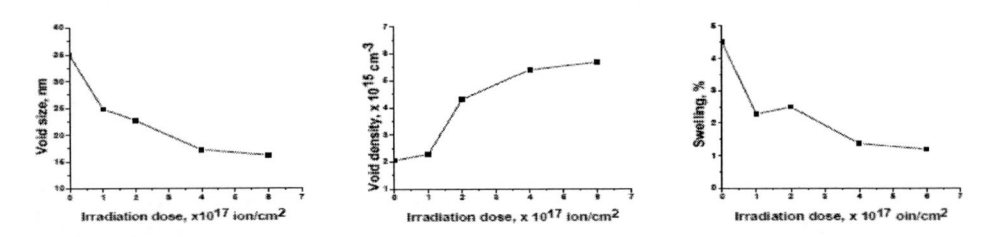

Figure 10. Void size, number density and swelling in copper pre-irradiated by 30 keV $H^+$ ions at 773 K as a function of irradiation dose at 573 K.

A few previous experiments have indicated that voids can shrink with decreasing temperature after they have been formed under more favorable conditions [41, 42]. According to Steel and Potter [42], voids formed during 180 keV $Ni^+$ ion bombardment of Ni at 923 K shrink rapidly when subjected to further bombardment at temperatures between 298 and 823 K. The authors have attempted to explain the observations using the rate theory modified to include the additional metal interstitial atoms injected by the ion beam. However, this effect was shown to be negligible due to a very low "production bias" introduced by injected SIAs (about 0.1%) [19]. What is more, in the experiment [19] described above, the energy of Cr ions was an order of magnitude higher than that in Ref. [42]. Accordingly, the Cr ions came to rest at a distance of about $10^4$ nm, which exceeded the depth, at which voids have been produced, by orders of magnitude. In the case of the proton irradiation of Ni and Cu, there were no additional metal interstitial atoms, but a high concentration of hydrogen bubbles was formed, which were stable under low temperature irradiation in contrast to voids. The underlying physical mechanisms for the observed radiation-induced annealing of voids are considered in the following section.

## 3.2. Quodon Model of the Radiation-Induced Void Annealing

The energy of vacancy formation at the void surface is lower than that at a flat stress-free surface because of the surface tension, which „tries" to make a void shrink. On the other hand, the vacancy formation energy at the bubble surface is higher than that at the flat stress-free surface because of the gas pressure, which opposes the surface tension [40]. The same is true for the vacancy formation at the dislocation core due to the stalking fault energy. It means that the energy delivered by quodons to the void and bubble surfaces and dislocation cores ejects more vacancies from voids than from bubbles or dislocations. The fate of a void or a bubble depends on the balance between the absorption and emission of vacancies, which may become negative for voids when the absorption of vacancies formed in the bulk decreases. This is what happens at lower irradiation temperature due to enhancement of the bulk recombination of vacancies with SIAs.

In order to incorporate this qualitative picture in the modified rate theory, let us evaluate the *local equilibrium concentration of vacancies* at the surface of a cavity (void or gas bubble), $c_v^{eq,C}$, which is determined by the rates of vacancy emission due to thermal or radiation-induced fluctuations of energy states of atoms surrounding the void. Generally, $c_v^{eq,C}$ is given by the sum of the thermal and the radiation-induced constituents:

$$c_v^{eq,C} = c_v^{irr,C} + c_v^{th,C}, \quad c_v^{th,V} = \exp\left(-\frac{E_v^f + P_g\omega - 2\gamma\omega/R}{k_B T}\right) \tag{39}$$

where $E_v^f$ is the thermal vacancy formation energy at a free surface, $\gamma$ is the surface energy, $P_g$ is the gas pressure inside a gas bubble. In order to evaluate the radiation-induced constituent, $c_v^{irr,C}$, let's estimate the number of vacancies ejected from a void of the radius $R$ due to its interaction with incoming quodons, $dN_v/dt$. Similar to the dislocation case (29), it is equal to the number of energetic focusons ($E > E_v^C$) absorbed by the cavity from the bulk per unit time, which is proportional to the quodon production rate per atom and the number of atomic sites from which a quodon can reach the cavity:

$$J_v^C = \frac{4\pi R}{\omega} \int_{E_v^V}^{E_F} l_0\left(E/E_v^C\right) z\left(E_p, E\right) dE \tag{40}$$

In the equilibrium state the vacancy flux out of the cavity (40) is balanced with the vacancy flux into the cavity, which is given by the product of the local vacancy concentration, $c_v^{irr,C}$, the cavity surface, $4\pi R$, and the rate, at which a vacancy cross the cavity surface, $v_v$:

$$J_v^C = \frac{4\pi}{\omega} R c_v^{irr,C} v_v \tag{41}$$

Equalizing eqs. (40 and (41) one obtains, similar to eq. (33), the following expression for the quodon-induced equilibrium concentration of vacancies at a cavity surface:

$$D_v c_v^{irr,C} = b l_Q^0 K_Q^C\left(E_v^C\right), \quad K_Q^C\left(E_v^C\right) \equiv K \frac{E_d}{E_F} \int_1^{E_v^C/E_F} \ln x \ln\left(x \frac{E_F}{E_v^C}\right) dx \tag{42}$$

where $l_Q^0$, is the quodon propagation range in ideal crystal, $K_Q^C(E_v^C)$, is the effective production rate of quodons. The quodon-induced equilibrium concentrations of vacancies at voids, gas bubbles and dislocations are different due to the difference in the vacancy formation energies (see Table 1), which a physical origin of the driving "force" that tries to minimize the total energy of the system and recover it from the damage being produced by Frenkel defects.

Now the cavity growth (or shrinkage) rate is given by the usual expression [18, 40]

$$\frac{dR_C}{dt} = \frac{1}{R_C}\left(D_v \bar{c}_v Z_v^C - D_i \bar{c}_i Z_i^C - D_v c_v^{eq,C} Z_v^C\right), \quad C = V, B \tag{43}$$

where $z_{i,v}^C$ is the cavity capture efficiency for vacancies [40] (superscript "V" stands for voids and "B" for bubbles), $\bar{c}_{i,v}$ are the mean concentrations of point defects determined by the rate equations (3) and (4), in which $k_{i,v}^2$ is the sink strength of all microstructure components that determine mean concentrations of Frenkel and Schottky defects:

$$\bar{c}_v^{eq} = \frac{Z_v^d \rho_d^d c_v^{eq,d} + Z_v^C N_C \bar{R}_C \bar{c}_v^{eq,C}}{k_v^2}, \ k_{i,v}^2 = Z_v^d \rho_d^d + Z_v^C N_C \bar{R}_C \tag{44}$$

where $N_C$, $\bar{R}_C$ are the cavity concentration and the mean radius, respectively. The product $D_v c_v^{eq,C} Z_v^C$ in eq. (43) determines the rate of the void shrinkage due to emission of thermal and radiation-induced Schottky defects. The latter does not depend on temperature in contrast to the difference $D_v \bar{c}_v Z_v^C - D_i \bar{c}_i Z_i^C$, which is due to the biased absorption of Frenkel defects that decreases with decreasing irradiation temperature due to enhancement of the point defect recombination in the bulk. As a result, the net growth rate may become negative below some temperature (or above some dose rate [18]).

Figure 11 shows the temperature dependence of the void growth/shrinkage rate in Ni and Cu irradiated under the present irradiation conditions calculated for quodon and focuson ranges assuming the quodon production rate to be equal to the focuson production rate given by eq. (42). Other material parameters are presented in Table 1. The focuson range is limited to about ten unit cells, which would result in negligible deviance from the conventional theory shown as the "classical" limit. The observed void shrinkage with decreasing irradiation temperature can be explained by the quodon model assuming the quodon range to be about $10^3$ unit cells and the quodon-induced vacancy formation energy at dislocations to be 2.215 eV and 1.562 eV in Ni and Cu, respectively. This difference may be due to different stacking fault energies (see Table 1), which makes it more difficult to eject vacancies from dislocations in Ni than in Cu.

## 4. VOID LATTICE FORMATION

The radiation-induced void dissolution phenomenon demonstrated in the previous sections is intrinsically connected with a void ordering in a super-lattice, which copies the host lattice [22-27]. The void ordering is often accompanied by a saturation of the void swelling, which makes an understanding of the underlying mechanisms to be both of scientific significance and of practical importance for nuclear engineering. Currently the most popular void ordering concept is based on the mechanisms of anisotropic interstitial transport along the close packed planes [23] or directions [24-27]. In these models, anisotropic diffusion of SIAs [23, 24] or propagation of small SIA loops along the close packed directions [25 -27] makes the competition between growing voids to be dependent on spatial void arrangements. A common difficulty inside the concept of anisotropic SIA transport is the explanation of saturation and even reduction in the void swelling accompanying the void lattice formation [41].

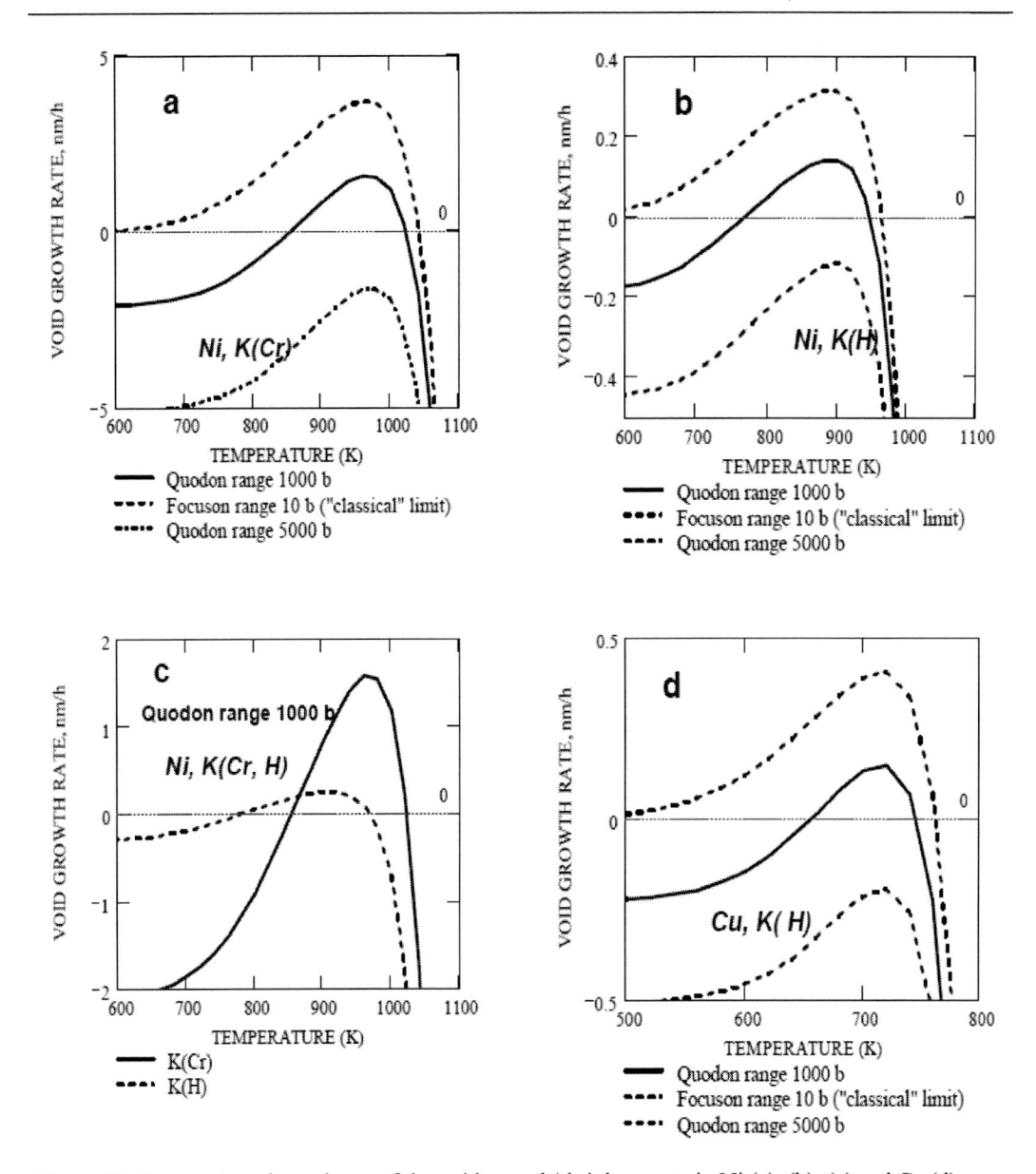

Figure 11. Temperature dependence of the void growth/shrinkage rate in Ni (a), (b), (c) and Cu (d) calculated for different displacement rates: $K(Cr) = 7 \times 10^{-3} s^{-1}$ (a), $K(H) = 6 \times 10^{-4} s^{-1}$ (b), $K(H) = 10^{-3} s^{-1}$ (c). The focuson range is limited to about ten unit cells, which would result in negligible deviance from the conventional theory shown as the "classical" limit.

As was suggested in ref. [28], the ordering phenomenon may be considered as a consequence of the energy transfer along the close packed directions provided by long propagating DBs or quodons. If the quodon range is larger than the void spacing, the voids shield each other from the quodon fluxes along the close packed directions, which provide a driving force for the void ordering, as illustrated in Figure 12.

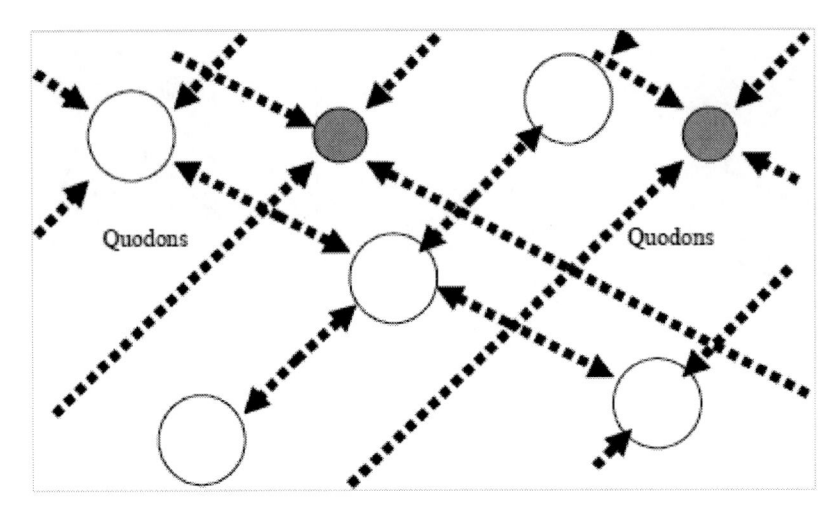

Figure 12. Illustration of the radiation annealing of voids in the "interstitial" positions due to the absorption of quodons coming from larger distances as compared to locally ordered voids that shield each other from the "quodon wind" along the close packed directions.

The emission rate for "locally ordered" voids (which have more immediate neighbors along the close packed directions) is smaller than that for the "interstitial" ones, and so they have some advantage in growth. In terms of the rate theory, the void solubility limit starts to depend on the void position. For the "locally ordered" voids it is proportional to the mean distance from the nearest neighbors along the close packed directions, $l_V$, while for the "interstitial" voids it is proportional to the quodon path length, $l_Q > l_V$ or to the mean distance from neighbors from next coordination spheres, $\langle l_V \rangle > l_V$. Accordingly, the growth rate of the "interstitial" voids is lower than that of the "ordered" ones and so they become smaller. In this way, quodons point out the "interstitial" voids, while their eventual shrinkage is driven by the mechanism of the radiation-induced competition (coarsening) of the voids of different sizes [43], which makes the smaller ("interstitial") voids shrink away, as has been demonstrated in details by Dubinko et al [25-27]. The mechanism of the radiation-induced coarsening [43] is similar to the classical Ostwald ripening of the voids but it is based on the void bias in absorption of SIAs rather than on the evaporation of vacancies. Due to this mechanism the maximum void number density, $N_V$, under irradiation is limited by the value, $N_V^{\max}$, proportional to the dislocation density and the dislocation to void bias factor ratio. At the "nucleation stage" ($N_V \ll N_V^{\max}$) the void growth rate due to the dislocation bias exceeds the void shrinkage rate due to the vacancy emission, and all voids grow. But eventually, at the "ripening stage" ($N_V \to N_V^{\max}$), the net vacancy gain rate is reduced down to the vacancy loss rate (for the "ordered" voids) or below that (for the "interstitial" ones). Consequently, the "interstitial" voids shrink away while the "ordered" ones form a stable void lattice, in which voids neither grow nor shrink.

Let us estimate the void lattice parameters given by the present mechanism. As the void sink strength increases with time, the rate of the quodon-induced vacancy emission from voids, $(d\bar{c}_v/dt)_Q$ equals the rate of the net vacancy absorption by voids due to the dislocation bias, $(d\bar{c}_v/dt)_B$, thus providing the condition for swelling saturation:

$$(d\bar{c}_v/dt)_Q \approx k_V^2 D_v c_v^{irr} \tag{45}$$

where $k_V^2 = 4\pi N_V \bar{R}_V$ is the void sink strength. The bias-induced constituent to the void growth is given by the usual expression (see e.g. [25-27]) :

$$(d\bar{c}_v/dt)_B \approx \Delta^* k_V^2 k_D^2 \delta_D / (k_V^2 + k_D^2), \quad \Delta^* \equiv D_v (\bar{c}_v - \bar{c}_v^{eq}) \approx K_{FP} / (k_V^2 + k_D^2), \tag{46}$$

where $k_D^2$, $\delta_D$ are the dislocation sink strength and the bias factor, respectively, and $\Delta^*$ is the vacancy supersaturation, which is determined by the ratio of the production rate of freely migrating Frenkel pairs, $K_{FP}$, to the microstructure sink strength if the latter dominates over the bulk recombination of Frenkel pairs (see eq. (10)). In this case, equalizing the right sides of eqs. (45) and (46) with account of (42) one obtains a relation between the void and dislocation sink strengths corresponding to the saturation of swelling:

$$(k_V^2 + k_D^2) \approx (K_{FP} k_D^2 \delta_D / K_Q l_Q^0 b)^{1/2} \tag{47}$$

The second relation is determined by the radiation-induced coarsening mechanism, which becomes efficient at $N_V = N_V^{max}$ making the smaller ("interstitial") voids shrink away [25-27]:

$$N_V = N_V^{max} = \Delta^* \delta_D k_D^2 / 4\pi \alpha_{iv}^* \tag{48}$$

$$\alpha_{iv}^* \equiv \alpha_{iv} + D^* \alpha_\gamma / \Delta^*, \quad D^* \cong D_v c_v^{th} \tag{49}$$

where $\alpha_{iv}^*$ is the coarsening parameter, $\alpha_{iv} \equiv \alpha^{im} + \alpha^d (2\gamma/\mu b)$ is the constant of the void bias due to elastic interaction with point defects, $\alpha^{im}$ and $\alpha^d$ are the constants of image and elasto-diffusion interaction, $\gamma$ is the surface energy, $\mu$ is the shear modulus.

Substituting (48) into (47) one obtains both the stable void radius and the number density expressions, from which the ratio of the void lattice parameter (VLP) to the void radius is estimated as follows:

$$a_{VL}/\bar{R} \approx (\alpha_{iv}^*)^{-2/3} (k_D^2 \delta_D)^{1/6} (K_Q l_Q^0 b / K_{FP})^{1/2} \tag{50}$$

Figure 13a shows the dependence of the VLP/R ratio on the DB path length given by eq. (50) assuming that $k_D^2 \delta_D = 10^{14} m^{-2}$, and the coarsening parameter, $\alpha_{iv}^* \approx \alpha_{iv} \approx 5b$, is determined by material constants, which is true for sufficiently low irradiation temperatures or high dose rates corresponding to the void lattice formation. The tendency of the VLP/R ratio decrease with increasing temperature may be expected from eq. (50) due to the increase of the coarsening parameter, $\alpha_{iv}^*$, with increasing temperature according to eq. (49).

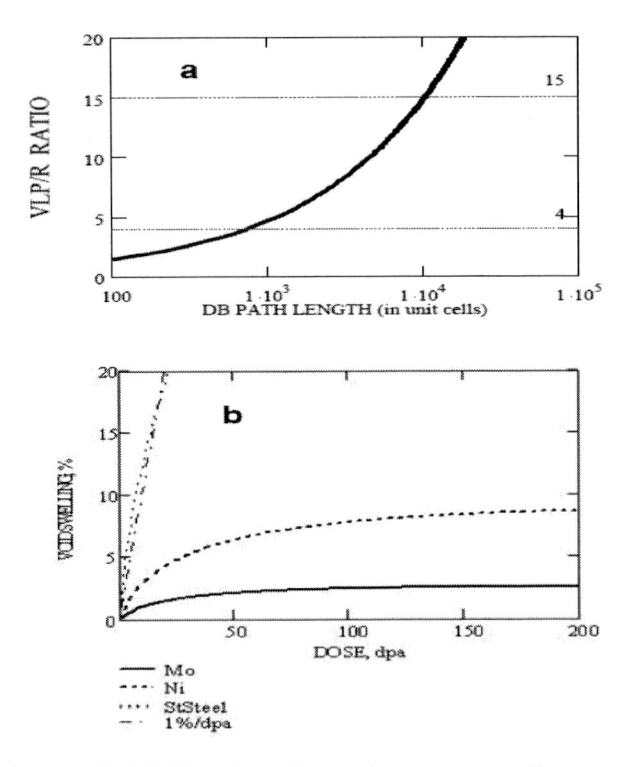

Figure 13. (a) Dependence of the VLP/R ratio on the quodon range according to eq. (50). The markers show the experimentally observed range of VLP/R ratio values in different materials. (b) Dose dependence of swelling in different metals calculated for typical reactor irradiation conditions (displacement rate, $K = 10^{-6} s^{-1}$, T = 673 K) assuming different void number densities, $N_V$ deduced from experimental data: Mo ( $N_V = 2 \times 10^{22} m^{-3}$), Ni ( $N_V = 8 \times 10^{21} m^{-3}$), St steel ( $N_V = 2 \times 10^{21} m^{-3}$).

Experimentally observed values of the VLP/R ratio range from about 4 to 15 and have a tendency of decreasing with increasing temperature [25-27]. It can be seen from Figure 13a that the present estimate of the quodon range is well within the range suggested in ref. [13] and it agrees with the quodon range of $10^3$ unit cells assumed in this chapter to describe experimental observations of the radiation-induced void annealing and irradiation creep.

Finally, the swelling saturation accompanying the void ordering can be achieved due to the competition between the absorption of Frenkel defects and emission of Schottky defects provided that the void concentration is high enough, as it is illustrated in Figure 13b for metals with different void concentrations.

# 5. CONCLUSION

In this chapter, rate theory modified to take account of quodon-induced reactions has been applied to describe irradiation creep, radiation-induced annealing of voids and the void ordering. The underlying mechanisms seem to have a common nature based on the vacancy emission from voids and dislocations due to their interaction with irradiation-induced quodons. The quodon propagation distance in cubic metals deduced from the comparison between the theory and experiment quodon range is about $10^3$ unit cells, which is consistent with data on the radiation-induced diffusion in stainless steel [13].

In order to forecast the behavior of nuclear materials in real radiation environment one has to know the generation rate of quodons and their propagation range in different crystal structures as the functions of impurity atom concentration and type. The latter factor seems to be of a primary technological importance since it offers a new insight on design of radiation-resistant materials.

## ACKNOWLEDGMENTS

This study has been supported by the STCU grant # 4962.

## REFERENCES

[1] V. I. Dubinko, D. I. Vainshtein, H. W. den Hartog, Effect of the radiation-induced emission of Schottky defects on the formation of colloids in alkali halides, *Radiat. Eff. and Defects in Solids*, 158 (2003) 705.

[2] V. I. Dubinko, D. I. Vainshtein, H. W. den Hartog, Mechanism of void growth in irradiated NaCl based on exiton-induced formation of divacancies at dislocations, *Nuclear Inst. and Methods in Physics Research*, B228 (2005) 304.

[3] V. I. Dubinko, A.N. Dovbnya, D. I. Vainshtein, H. W. den Hartog, Simulation of radiation damage in rocks considered for safe storage of nuclear waste, *Proc. Int. Conf. Waste Management and the Environment* IV, 2 - 4 June, 2008, Granada, Spain, P. 891.

[4] Y. R. Zabrodsky and V. M. Koshkin, Unstable pairs – new point defect type in solids, DAN SSSR (in Russian) 227 (1974) 1323.

[5] V. L. Indenbom, New hypothesis on the mechanism of radiation-induced processes, Pisma v ZhTF (in Russian), 5 (1979) 489.

[6] R. H. Silsbee, Focusing in Collision Problems in Solids, *J. Appl. Phys.* 28 (1957) 1246.

[7] R. S. Nelson, M. W. Tompson, Atomic collision sequences in crystals of copper, silver and gold revealed by sputtering in energetic ion beams, *Proc. Roy. Soc.*, 259 (1960) 458.

[8] G. Leibfried, Defects in dislocations by focusing collisions in f.c.c. lattices, *J. Appl. Phys.* 31 (1960) 117.

[9] M. W. Tompson, Defects and Radiation Damage in Metals, Cambridge, 1969.

[10] S. Flach, A.V. Gorbach, Discrete breathers — Advances in theory and applications, *Phys. Rep.* 467 (2008) 1.

[11] F.M. Russell, J.C. Eilbeck, Evidence for moving breathers in a layered crystal insulator at 300 K, *Europhys. Lett.*, 78 (2007) 10004.

[12] P. Sen, J. Akhtar, and F. M. Russell. MeV ion-induced movement of lattice disorder in single crystal silicon. *Europhys. Lett*, 51 (2000) 401.

[13] G. Abrasonis, W. Moller, X.X. Ma, Anomalous ion accelerated bulk diffusion of interstitial nitrogen, *Phys. Rev. Lett.*, 96 (2006) 065901.

[14] J. F. R. Archilla, J. Cuevas, M. D. Alba, M. Naranjo, J. M. Trillo, Discrete breathers for understanding reconstructive mineral processes at low temperatures, *J. Phys. Chem.* B110 (2006) 24112.

[15] N.P. Lazarev, V.I. Dubinko, Molecular dynamics simulation of defects production in the vicinity of voids, Radiat. *Eff. Def. Solids*, 158 (2003) 803.

[16] V.I. Dubinko, N. P. Lazarev, Effect of the radiation-induced vacancy emission from voids on the void evolution, *Nucl. Instrum. Meth.* B 228 (2005) 187.

[17] V.I. Dubinko, New mechanism of irradiation creep based on the radiation-induced vacancy emission from dislocations, *Radiat. Eff. Def. Solids* 160 (2005) 85.

[18] V.I. Dubinko, V.F. Klepikov, The influence of non-equilibrium fluctuations on radiation damage and recovery of metals under irradiation, *J. Nuclear Materials*, 362 (2007) 146H.

[19] V.I. Dubinko, A.G. Guglya, E. Melnichenko, R. Vasilenko, Radiation-induced reduction in the void swelling, *J. Nuclear Materials* 385 (2009) 228.

[20] V. I. Dubinko, *Mechanisms of recovery of radiation damage based on the interaction of quodons with crystal defects*, Proc. Int. Conf. *LENCOS* 09, Sevilla (Spain) July 14-17, 2009.

[21] V.I. Dubinko, A.G. Guglya, S.E. Donnelly, Radiation-induced formation, annealing and ordering of voids in crystals: theory and experiment, COSIRES 2010, *Nuclear Inst. and Methods in Physics Research*, B, to be published.

[22] W. Jäger, H. Trinkaus, Defect ordering in metals under irradiation, *J. Nucl. Mater.* 205 (1993) 394.

[23] J. H. Evans, A computer simulation of the two-dimensional SIA diffusion model for void lattice formation, *J. Nucl. Mater.* 132 (1985) 147.

[24] C. H. Woo, W. Frank, A theory of void-lattice formation, *J. Nucl. Mater.* 137 (1985) 7.

[25] V. I. Dubinko, A. A. Turkin, A.V. Tur, V.V. Yanovsky, A mechanism of formation and properties of the void lattice in metals under irradiation, *J. Nucl. Mater.* 161 (1989) 50.

[26] V. I. Dubinko, A. A. Turkin, Self-organization of cavities under irradiation, *Appl. Phys.* A58 (1994) 21.

[27] V. I. Dubinko, Temporal and spatial evolution of dislocation and void structures under cascade damage production, *Nucl. Instrum. Meth.* B153 (1999) 116.

[28] V.I. Dubinko, Breather mechanism of the void ordering in crystals under irradiation, *Nucl. Instrum. Meth.* B 267 (2009) 2976.

[29] P. T. Heald and M. V. Speight, Steady-state irradiation creep, *Philos. Mag.* 29 (1974) 1075.

[30] P. T. Heald and M. V. Speight, *Point* defect behavior in irradiated materials, *Acta Met.* 23 (1975) 1389.

[31] K. Ehrlich, Irradiation creep and interrelation with swelling in austenitic stainless steels, *J. Nucl. Mater.* 100, 149 (1981).

[32] F. A. Garner, Recent insights on the swelling and creep of irradiated austenitic alloys, *J. Nucl. Mater.* 122and123, 459 (1984).

[33] C. H. Woo, Irradiation creep due to elastodiffusion, *J. Nucl Mater.* 120 (1984) 55.

[34] A. I. Ryazanov and V. A. Borodin, The effect of diffusion anisotropy on dislocation bias and irradiation creep in cubic lattice materials, *J. Nucl. Mater.* 210 (1994) 258.

[35] W. A. Coghlan, Review of recent irradiation creep results, *Int. Met. Rev.* 31, 245 (1986).

[36] J. R. Matthews and M. W. Finnis, Irradiation creep models – an overview, *J. Nucl. Mater.* 159 (1988) 257.

[37] V.V. Gann, I.G. Marchenko, Modeling of the collision replacement chains in the vicinity of the dislocation core [in Russian], VANT, FRM I RM, 1/24 (1983) 99.

[38] V. I. Dubinko, A. S. Abyzov, A. A. Turkin, Numerical evaluation of the dislocation loop bias, *J. Nucl. Mater.*, 336 (2005) 11.

[39] V. I. Dubinko, New insight on bubble-void transition effects in irradiated metals, *J. Nucl. Mater.* 206 (1993) 1.

[40] V. I. Dubinko, Theory of irradiation swelling in materials with elastic and diffusional anisotropy, *J. Nucl. Mater.* 225 (1995) 26.

[41] B. A. Loomis, S. B. Gerber, Effect of irradiation temperature change on void growth and shrinkage in ion-irradiated Nb, *J. Nucl. Mater.* 102 (1981) 154.

[42] L.K. Steele, D.I. Potter, The disappearance of voids during 180 keV NiP+P bombardment of nickel, *J. Nucl. Mater.* 218 (1995) 95.

[43] V. I. Dubinko, P.N. Ostapchuk and V. V. Slezov, Theory of radiation induced and thermal coarsening of the void ensemble in metals under irradiation, *J. Nucl. Mater.* 161(1989)239.

In: Nuclear Materials
Editor: Michael P. Hemsworth, pp. 111-133

ISBN 978-1-61324-010-6
© 2011 Nova Science Publishers, Inc.

*Chapter 4*

# FUEL RESTRUCTURING AND ACTINIDE RADIAL REDISTRIBUTIONS IN AMERICIUM-CONTAINING URANIUM-PLUTONIUM MIXED OXIDE FUELS IRRADIATED IN A FAST REACTOR

*Koji Maeda[1], Kozo Katsuyama[1],*
*Yoshihisa Ikusawa[2], and Seiichiro Maeda[2]*
[1]Oarai Research and Development Center, Japan Atomic Energy Agency,
4002 Narita, Oarai, Ibaraki 311-1393, Japan
[2]Nuclear Fuel Cycle Engineering Research Institute, Japan Atomic Energy Agency,
4-33 Muramatsu, Tokai-mura, Naka-gun, Ibaraki 319-1194, Japan

## ABSTRACT

The fuel irradiation test (B14) was carried out in the experimental fast reactor Joyo from May 8 to 11, 2007, to evaluate the thermal behaviors of low-density uranium and plutonium mixed oxide fuels containing several percent of americium (Am-MOX fuels). The test objective is to research early thermal behaviors of Am-MOX fuels such as fuel restructuring and radial redistribution of actinides. The test results are expected to reduce the margin of the thermal design for minor actinide (MA)-MOX fuels.

Pellet-cladding gap width and the oxygen-to-metal (O/M) molar ratio of oxide fuels were specified as experimental parameters. The contents of Pu and Am in the fuel pellets were 31 wt% and 2.4 wt%, respectively. Four fuel pins were irradiated step-by-step in consideration of fuel restructuring during 48 h as a preconditioning before full power reactor operation. The irradiation history, i.e. linear power, was simulated using the conventional FBR oxide fuel pins, and to simulate the transient condition, the linear power was rapidly increased to 470 W cm$^{-1}$ for 10 min.

After the irradiation, non-destructive and destructive post-irradiation examinations were conducted. The fuel restructuring (microstructure change) was examined with an optical microscope and a scanning electron microscope, and the radial redistribution of actinides was measured with an electron probe micro analyzer. Ceramography specimens were taken from the axial position of each fuel pin where the fuel centerline temperature reached the maximum during irradiation.

The influences of both pellet-cladding gap width and O/M molar ratio on the fuel restructuring were observed, but no fuel melting was seen. Fuel restructuring and radial distributions of actinides were investigated relative to those of other irradiated fuels. It seemed that fuel restructuring before the transient condition due to preconditioning was enough to prevent fuel melting. Under the transient condition, additional radial migration of lenticular voids toward the fuel center resulted in an insignificant increase of the central void diameter. The degree of fuel restructuring was found to be more dependent on the linear heating rate than on the small addition of americium and on the fuel O/M molar ratio. In the fuel pin with a large as-fabricated gap width, significant fuel relocation and off-centered fuel restructuring were observed, but no enhancement in the maximum fuel temperature was identified.

The radial profiles of Am and Pu contents indicated that these elements had similar redistribute tendencies in the columnar grain region. The extent of increase in Am and Pu contents around the central void showed a significant dependence on the O/M molar ratio. The fuel having a higher O/M molar ratio had a larger redistribution of Am and Pu towards the central void. In addition the difference in the Am and Pu contents was larger for the fuel having a higher O/M molar ratio.

# 1. INTRODUCTION

In the once-through nuclear fuel cycle, minor actinides (MAs) generated in light water reactor (LWR) fuels remain in the high-level radioactive waste. Due to their long storage time, the amount of americium will increase in the Pu raw materials or fast reactor (FR) uranium and plutonium mixed oxide (MOX) fuel pellets due to $\beta$-decay of Pu-241. This must be taken into account for utilization of the advanced fuel cycle including MAs.

One of the options for MA transmutation is to irradiate some MA-doped MOX (MA-MOX) fuels in FRs. One nuclear fuel cycle strategy proposes that the MAs such as Np and Am produced from LWRs are burnt as 5 wt% MA-MOX fuels in the FRs [1, 2]. But the degree of degradation in fuel performance by including MAs must be taken into account for the thermal design of the fuel pins.

Influences from the addition of low Am content (~3 wt%) would be rather small on the thermo-physical properties of MOX fuels, i.e. melting temperature [3], thermal conductivity [4], and oxygen potential [5]. However, several safety factors associated with the fuel irradiation behavior are still conservatively considered. So far MOX fuel containing 30 wt% Pu has been developed as the core fuel pellet of the FR Monju [6]. In the FR, the fuel centerline temperature rises above 2300 K and this is accompanied by a steep radial temperature gradient during irradiation because of the high linear heating rate. This large temperature gradient in the radial direction causes redistribution of Pu and U, and Pu content increases to about 40 wt% at the pellet center [7] and [8]. However the maximum temperature of the MOX fuel pellets during irradiation is suitably limited within the design criterion to prevent fuel melting. In contrast, information about the influence of MA content on fuel irradiation behavior is still insufficient for verification of the validity of the criteria of fuel pin thermal design and for authorization to use MA-MOX fuels in a conventional fast breeder reactor (FBR) [9 - 12]. These criteria are that oxide fuels should not melt even at transient conditions during irradiation and that the maximum fuel temperature should be less than the melting temperature [13].

To fill in these information gaps, a characteristic short-term irradiation test (B14) with four fuel pins was carried out under the expected operating conditions of a conventional FBR in the experimental FR Joyo and the early thermal behavior of MA-MOX fuels was investigated. For this irradiation test, low-density $(Am,Pu,U)O_{2-x}$ fuels as representative MA-MOX fuels were fabricated. The americium content of 2.4 wt% was used and the test parameters were the pellet-cladding gap width and oxygen-to-metal (O/M) molar ratio. After the irradiation test, post-irradiation examinations (PIEs) were carried out at a hot cell laboratory in the Oarai Research and Development Center in order to confirm fuel integrity and performance for low-density $(Am,Pu,U)O_{2-x}$ fuels.

The results of two other irradiation tests, B4M and A1M performed in Joyo were compared with the B14 test results. B4M and A1M tests were a short-term irradiation test and a steady state long-term irradiation test, respectively. The americium content of fuel pellets used in these irradiation tests was less than 1.0 wt%. Other specifications including the geometry were similar to those of the B14 test. This paper uses the test results to discuss irradiation behavior of low-density $(Am,Pu,U)O_{2-x}$ fuels from the viewpoints of fuel microstructure and actinide radial redistribution.

## 2. EXPERIMENTAL

### 2.1. Fuel Fabrication

Solid $(Am,Pu,U)O_{2-x}$ fuel pellets for the B14 test were fabricated in a conventional alpha particle-tight glove box facility at the Nuclear Fuel Cycle Engineering Research Institute. They were fabricated from raw $PuO_2$ powder, with americium content of 7.6 or 6.4 wt%, and raw $UO_2$ powder. Designated raw powders and pore former additives [14] for adjusting the fuel density were ball milled together. After pelletization, the green pellets were sintered at ~1973 K for 4 h. Pellets with two different diameters were produced by grinding the surfaces of the sintered pellets. Subsequently the O/M molar ratios of the pellet material were adjusted by controlling the $H_2/H_2O$ molar ratio in the feed gas and the temperature (1473-1773 K) in the sintering furnace [15].

The fuel specifications for irradiation were based on conventional FBR oxide fuels [16,17]. The nominal composition of the fuel pellets was $(Am_{0.024}Pu_{0.310}U_{0.666})O_{2-x}$. Three O/M molar ratios were chosen: 1.96 (low), 1.98 (standard) and 2.00 (high). The fuel pellet diameters were 5.40 or 5.35 mm and pellet height was 8 mm. The fuel relative density was 85% of the theoretical density (TD). The plutonium content was 31 wt%. Four kinds of $(Am,Pu,U)O_{2-x}$ fuel pellets were loaded into the 20% cold worked PNC-316 stainless steel cladding together with thermal insulator pellets, a reflector, a spring and a sleeve. The cladding tube outer diameter and its wall thickness were 6.5 mm and 0.47 mm, respectively. The test parameter of fuel-cladding gap width was adjusted by using the two pellet diameters.

The length of the fissile column was 400 mm which was shorter than the Joyo driver fuel length (500 mm). The atmosphere inside the fuel pins was a helium (91%) and krypton (9%) gas mixture, similar to the present composition for a conventional FBR, to allow easy determination of fuel pin failure. The four fuel pins with grid spacers were separately loaded in the irradiation rig at the Oarai Research and Development Center.

The test parameters of the four fuel pins were defined by different combinations of O/M molar ratio and gap width designated as: PTM001 (1.98, 160 μm); PTM003 (1.96, 160 μm); PTM002 (1.98, 210 μm) and PTM010 (2.00, 210 μm). PTM001 was a reference with nominal specifications for the O/M molar ratio and gap width. The small diameter of the pellets to get the large gap width of 210 μm for PTM002 and PTM010 was the upper limit of the fabrication tolerance for thermal design (gap conductance). The lowest O/M molar ratio of 1.96 in PTM003 was the lower limit of fabrication tolerance for thermal design (thermal conductivity). The highest O/M molar ratio of 2.00 and the small pellet diameter in PTM010 were determined to investigate the most significant phenomena of radial plutonium and americium redistributions.

The solid $(Pu,U)O_{2-x}$ fuel pellets for B4M and A1M tests were also fabricated in a conventional alpha particle-tight glove box facility at the Nuclear Fuel Cycle Engineering Research Institute. The B4M and A1M test fuel pellets were fabricated from combinations of powders and a pore former by conventional powder technology. The B4M powder was prepared by mixing $UO_2$ (ADU) powder and $PuO_2$ powder which were obtained by calcinating the oxalate. The A1M powder was a mixture of $UO_2$ (ADU) powder and MOX (mixed oxide) powder which was prepared by co-converting the mixed nitrate (Pu:U = 20:1) by microwave heating [18] and [19]. Designated raw powders and pore former additives [14] for adjusting the fuel density were ball-milled together.

The fuel specifications for B4M and A1M irradiation tests were similar to the one in B14 test. The O/M molar ratios of fuel pellets were 1.95 and 1.96 in B4M, and 1.98 in A1M. The americium content of fuel pellets used in both irradiation tests was less than 1.0 wt%. The fissile content in the fuel was adjusted to nearly 31wt%. Geometry of fuel pellets in the B14 test was based on that of conventional FBR oxide fuel pellets. The test parameter of gap width in B4M test was determined by large and small pellet diameters; the small pellet diameter gave the large gap width of 210 μm.

The fuel pellets of B4M and A1M were inserted into the cladding tubes of 20% cold worked PNC-316 along with insulator pellets, a reflector, a spring and a sleeve. The length of the fissile column was 550 mm. After spacer wires were spirally wrapped around the fuel pins, they were assembled and inserted into a wrapper tube.

## 2.2. Irradiation Conditions

The B14 irradiation test was done in the experimental FR Joyo from May 8 to 11, 2007. The irradiation rig (Figure 1), including the four fuel pins, was irradiated at the center of the MK-III core. The simulated irradiation history of the fuel pins was the same as the actual irradiation conditions of a conventional Japanese FBR.

Figure 2 shows the irradiation history of reactor thermal power and linear heating rate for fuel pins taking into account the preconditioning. Preconditioning was performed in two steps. In the first step, the reactor thermal power was increased to 92 MW (linear heating rate, 347 W cm$^{-1}$) at the rate of 0.4 % min$^{-1}$, and the reactor power was held for 24 h. In the second step, the reactor thermal power was increased to 102 MW (386 W cm$^{-1}$) at 0.4 % min$^{-1}$, and the reactor power was held for 24 h. By preconditioning, a reduction in fuel centerline temperature due to fuel restructuring was expected. After the preconditioning, the reactor thermal power was rapidly increased to 125 MW (470 W cm$^{-1}$) to simulate the transient condition assumed in the fuel design.

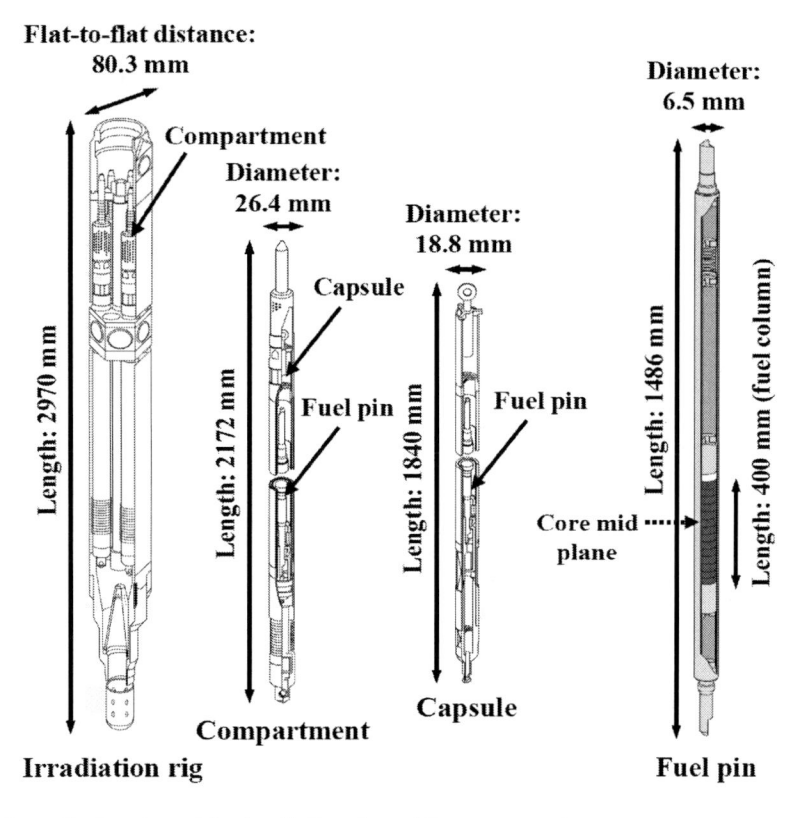

Figure 1. Schematic drawing of the irradiation rig and its components.

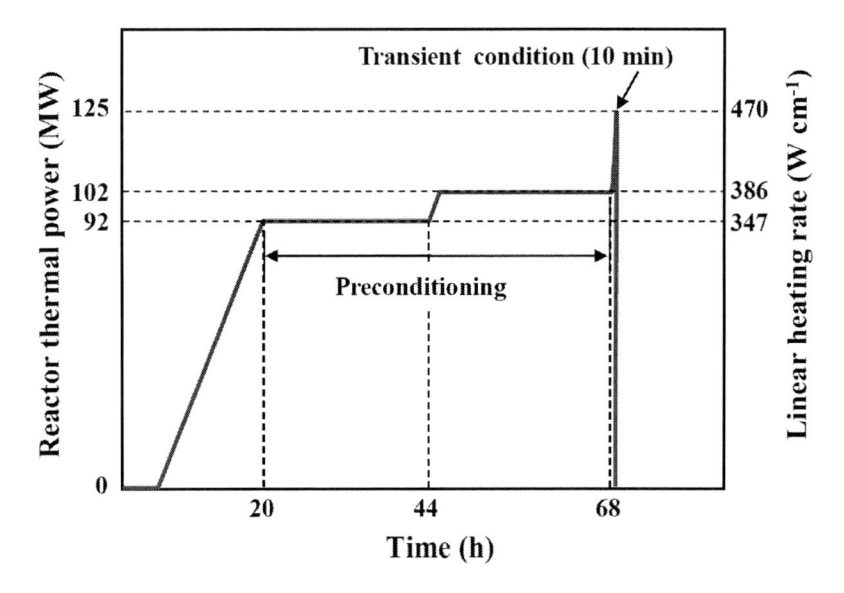

Figure 2. Schematic diagram of irradiating history of reactor thermal power and linear heating rate for fuel pins in B14 test.

At the maximum power level, the reactor power was held for 10 min, and then the reactor was shut down rapidly by a manual scram in order to preserve the fuel microstructure and to prevent additional fuel restructuring. For the transient irradiation, fuel melting was not actually expected, but due to the large safety margins derived from the limited physical property data in the thermal design of the fuel pins, the maximum linear heating rate for the test was set as 470 W cm$^{-1}$, which corresponded to the limitation of the maximum fuel melt fraction of 20%. The B14 test achieved a peak fast neutron fluence of $5.17 \times 10^{20}$ cm$^{-2}$ ($E \geq$ 0.1 MeV). In addition, the maximum linear heating rate and local burnup of the fuel achieved were approximately 470 W cm$^{-1}$ and 0.0495 at%, respectively.

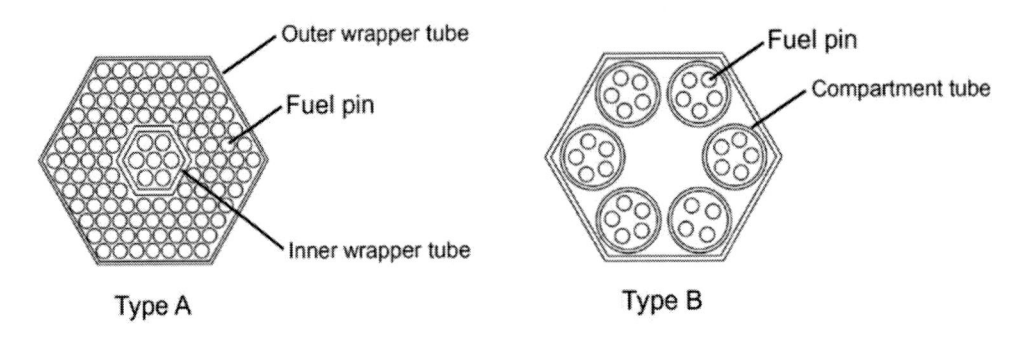

Figure 3. Cross-sectional views of Types A and B irradiation rigs.

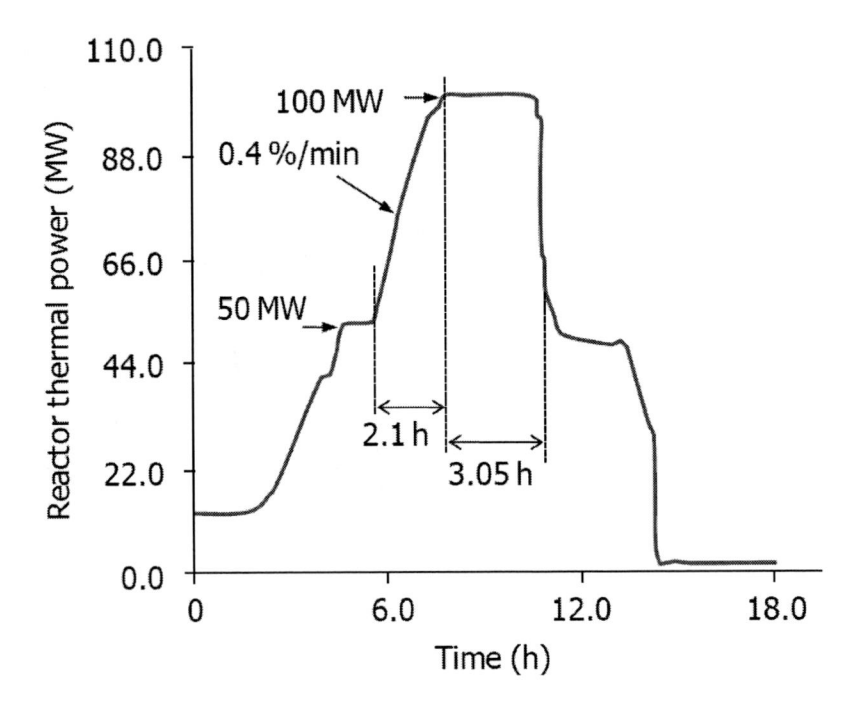

Figure 4. Schematic diagram of irradiation history of reactor thermal power in B4M test.

The B4M and A1M irradiation tests were done in the experimental FR Joyo. The irradiation rig (Figure 3), including the four fuel pins, was irradiated in the MK-II core. Irradiation rigs of Types A and B were used for A1M and B4M tests, respectively. The simulated irradiation history of the fuel pins in the B4M test was somewhat similar to the actual irradiation conditions of a conventional Japanese FBR. Figure 4 shows the irradiation history of reactor thermal power and linear heating rate for fuel pins of the B4M test. These fuel pins were briefly irradiated for 3 h to observe the process of fuel restructuring at 377 W $cm^{-1}$. The irradiation conditions of the A1M test were determined in order to demonstrate the integrity of the fuel pins at a higher linear heating rate (480 W $cm^{-1}$ in the beginning of life) in the steady state irradiation than the upper value of the linear heating rate in the thermal design of fuel pins used in the conventional FBR. Fuel pins of the A1M test reached the burnup of 42 $MWd\ kg^{-1}$.

## 2.4. Post-Irradiation Examinations (PIEs)

After the irradiation test, the irradiation rig was transferred to the hot cell laboratory for the PIEs and disassembled. Nondestructive examinations (NDEs) were made for all fuel pins. These included visual inspection, profilometry of the fuel pin diameter, and axial gamma scanning, weight measurements, X-ray radiography and X-ray computer tomography [20] (only for the B14 test). After the NDEs, all fuel pins were punctured and sectioned into specimens for optical microscopy, electron probe micro analysis (EPMA), and burnup measurements.

# 3. RESULTS AND DISCUSSION

## 3.1. NDES for B14 Test

Images on the cross section of fuel pins used in B14 test at the axial core center were obtained by X-ray computer tomography (CT) and radiography. There were no indications of fuel melting, but formation of a central void was observed. No deformation of fuel pins was observed by the visual examination or from the profilometry results. The axial distribution of gamma intensity was similar to the axial neutron fluence profile. The initial filling gas mixture of helium and krypton was detected by a puncture test. Nothing unexpected regarding the fuel behavior was recognized from the results. Integrity of all fuel pins was confirmed.

### 3.1.1. Nondestructive Observation by X-Ray CT

A non-destructive X-ray CT technique makes it possible to easily and quickly get the internal configuration of the irradiation rig. X-ray CT images were obtained along the axial direction of the irradiation rig at the same separation interval.

A typical cross-sectional image of the irradiation rig used in the B14 test was obtained at the axial center of the core fuel column and is shown in Figure 5. The fuel pellets, central voids in them, claddings, wrapping wires, and wrapper tube could be clearly distinguished from each other. Each enlarged image on the transverse cross section of a fuel pin is also shown in the figure. There were no indications of fuel melting in these images.

A number of cross-sectional images were taken at equal intervals in the axial direction of the irradiation rig. Three-dimensional images were synthesized from these cross-sectional images.

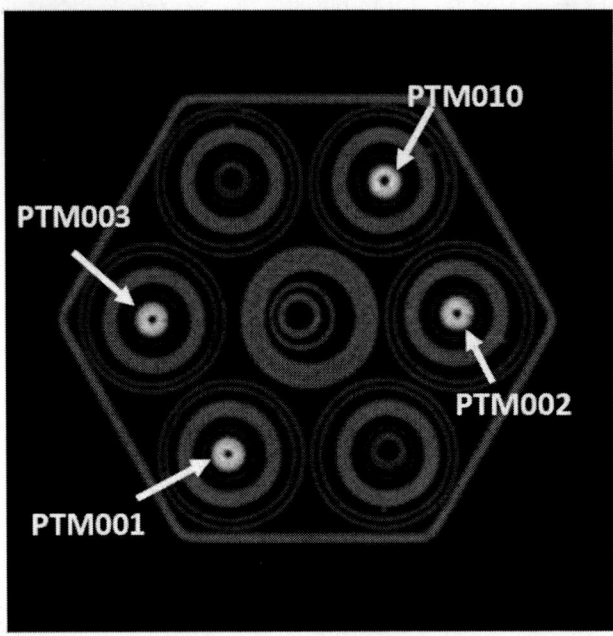

Figure 5. Cross-sectional view of B14 irradiation rig at core midplane obtained in an X-ray test.

Figure 6(a) is a three-dimensional image on the irradiation rig when the irradiation rig was axially cut behind the front wrapper tube, and the interior of the irradiation rig can be seen. Similarly, Figure 6(b) shows the distribution of the central void sizes along the axial direction, when each fuel pin in the compartment tube was axially cut into nearly equal halves. Distribution of the central voids in each fuel pin was clearly visible, and the void diameters were larger in fuel pins having the larger gap width. Generally molten fuel moves axially down towards the lower end of the fuel column, and then it plugs the downside of the axial central void [21]. In these observations, obvious features of molten fuel such as central melt cavities, the slumping molten fuel and the molten fuel plugging the central void could not be identified in the fuel column. Furthermore, gamma scanning of fuel pins did not show any unexpected features in axial distributions of the detected nuclides, supported the observations by X-ray CT.

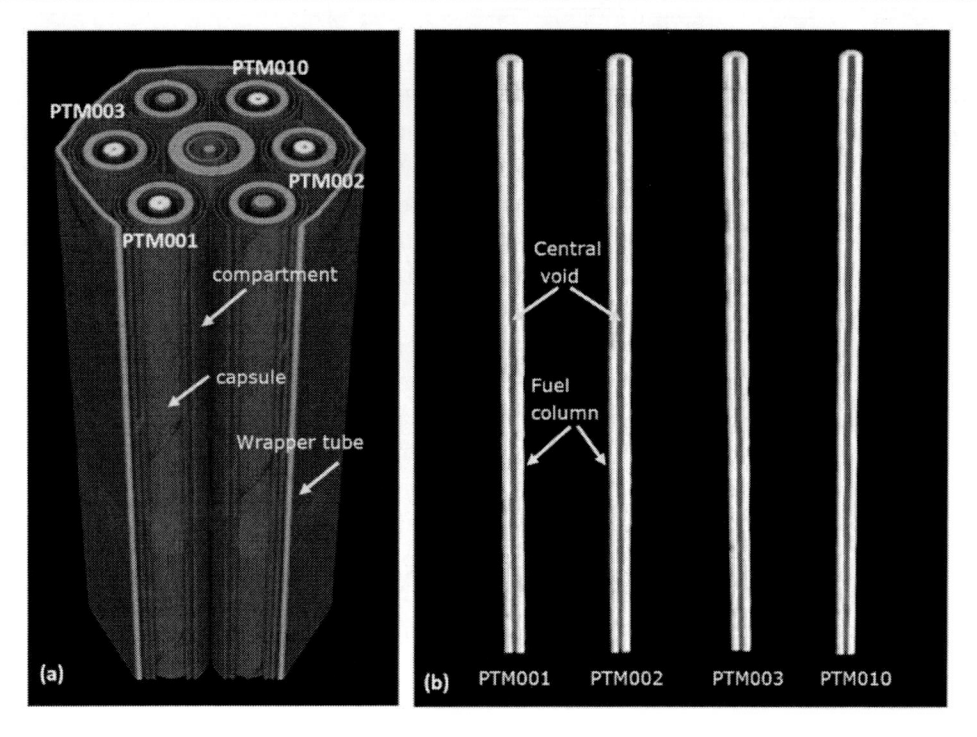

Figure 6. Three-dimensional images of B14 irradiation rig around core mid plane obtained in an X-ray test. (a) Irradiation rig and (b) axial longitudinal section of fuel column.

## 3.2. Destructive Examinations

### 3.2.1. Observation of Fuel Microstructure

After the NDEs, all fuel pins from the B14 test were sectioned for ceramography to observe the fuel microstructure. Ceramography specimens were taken from two axial positions. One specimen was obtained from the middle level (+30 mm from the core mid plane) where the fuel centerline temperature was estimated to be the maximum. The other specimen was obtained from the upper axial level (+100 mm from core mid plane) of the fissile column to investigate the dependence of linear heating rate on the irradiation behavior. Since the irradiation was terminated by a reactor scram, it helped to freeze the existing fuel microstructure after the simulated transient condition.

As-polished ceramographs from each fuel pin of the B14 test are shown in Figures 7(a)-(d). Radial sweeping of lenticular voids towards the fuel center, formation of a central void (1.1 - 1.5 mm in diameter), and radially formed columnar grain structure in the fuel were seen just as with typical MOX fuels irradiated in FRs. No indication of fuel melting was observed in any ceramographs. As shown in Figures 7(a)-(c), radial migration of lenticular voids and accumulation of lenticular voids around the central void had started within the brief 10 min irradiation at high linear heating rate. This indicated that no fuel melting occurred during the transient condition and evidently the fuel restructuring process has not yet been terminated. As shown in Figure 7(d) for PTM010, unlike for the other fuel pins, lenticular voids with typical faceted surfaces were not seen clearly. However, long columnar grains had become fairly well developed. At the outer edge of the restructured region, lenticular voids extending from pre-existing coarse sintering pores were clearly observed.

In particular, fuel restructuring in the fuel pellets of O/M molar ratio of 2.00 was the most significant and the columnar grain structure was completely developed. As for the fuel pin with pellets with O/M molar ratio below 1.98, some lenticular voids were still migrating towards the central region, and the columnar grain structure was not completely developed yet. In addition, a higher degree of fuel restructuring in the specimens of the B14 test was developed in fuel pellets of higher O/M molar ratio. It seemed that the lenticular voids had moved faster in the PTM010 specimen than in the other specimens.

Lenticular voids move at a rate which depends on both temperature and on temperature difference across the void. The temperature in particular determines the vapor pressure of fuel constituents within the void. Vapor pressure over the fuel is larger with increasing O/M molar ratio. Therefore, this suggested that the fuel O/M molar ratio of 2.00 had more influence on fuel restructuring behavior than the fuel O/M molar ratio below 1.98.

A comparison of the fuel microstructure for fuel pins of the B14 test having 160 and 210 μm as-fabricated gap widths showed significant fuel relocation which has been attributed to fuel cracking [22] and off-center fuel restructuring in the fuel having a large as-fabricated gap. Fuel pellet relocation behavior largely influences the heat transfer between cladding and fuel pellet surfaces. Typical fuel cracks were seen and lenticular void migration developed from some of the major fuel cracks. Since a fuel crack can be a source of lenticular voids, wedge shaped cracks having lenticular voids in Figures 7(a)-(c) were partly healed by the formation of lenticular voids and their transport to the center. Other large cracks connected to the central void from the fuel surface were thought to have formed by fuel fragmentation or by cooling during reactor shutdown. These cracks would form when thermal stress exceeded the fracture stress of the fuel pellet.

For the fuel pin having the smaller gap width of 160 μm, the degree of the relocation was not significant. Significant fuel relocation and partial closure of the as-fabricated gap were seen in the fuel pin having the gap width of 210 μm. The degree of the relocation was significant due to a relatively unconstrained motion of fuel fragments in the larger gap width. Fuel relocation towards the cladding decreases the gap volume and increases the volume of the central void, which would be greater with the fuel pin having the larger as-fabricated gap. It should be considered that the larger as-fabricated gap of 210 μm contributed a certain volume to the central void by fuel relocation.

### 3.2.2. Axial Distribution of Central Void Diameter

The equivalent diameters of the central void in the specimens of the B14 test were measured by X-ray CT. The decreased central region of CT counts due to formation of a central void was seen in the diametral distribution of CT counts. It was assumed that the width of this rectangle was the diameter of the artificial void. Using this method, the diameter was determined within an error of ±0.1 mm [20].

a)

b)

c)

d)

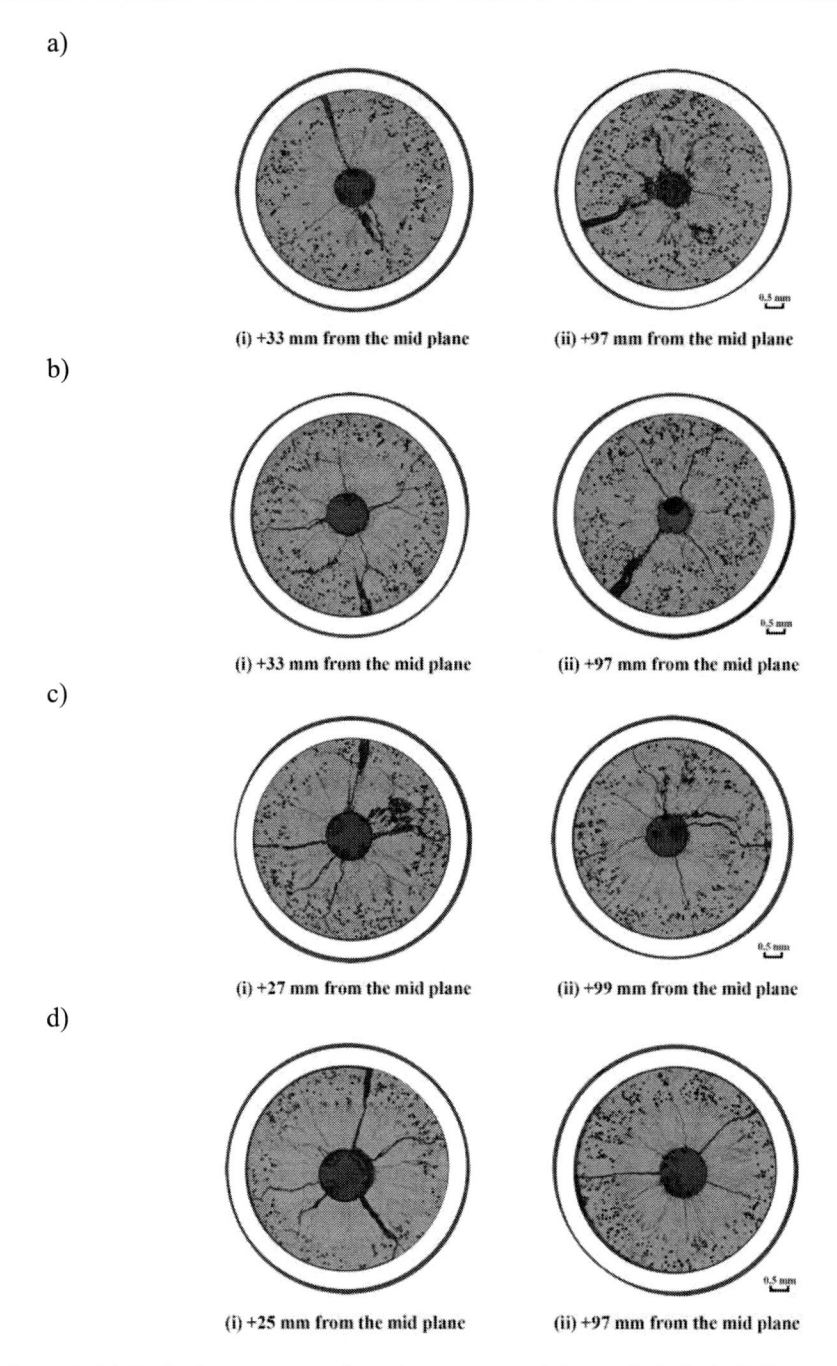

(i) +33 mm from the mid plane        (ii) +97 mm from the mid plane

(i) +33 mm from the mid plane        (ii) +97 mm from the mid plane

(i) +27 mm from the mid plane        (ii) +99 mm from the mid plane

(i) +25 mm from the mid plane        (ii) +97 mm from the mid plane

Figure 7. (a) Fuel microstructure of specimens removed from a fuel pin (PTM001: gap width, 160 μm; O/M, 1.98). (i) LHR, 469.4 W cm$^{-1}$, BU, 0.0482 at%. (ii) LHR, 436.1 W cm$^{-1}$, 0.0448 at %. (LHR, Linear heating rate; BU, local burn up). (b) Fuel microstructure of specimens removed from a fuel pin (PTM003: gap width, 160 μm; O/M, 1.96). (i) LHR, 459.7 W cm -1; BU, 0.0480 at %. (ii) LHR, 427.3 W cm -1; BU, 0.0446 at%. (LHR, liear heating rate; BU, local burn up). (c) Fuel microstructure of specimens removed from a fuel pin (PTM002, gap width: 210 μm, O/M: 1.98). (i) LHR, 458.0 W cm $^{-1}$; BU, 0.0485 at%. (ii) LHR, 422.8 W cm $^{-1}$; BU 0.0447 at %. (LHR, liear heating rate; BU, local burn up). (d) Fuel microstructure of specimens removed from a fuel pin (PTM010, gap width: 210 μm, O/M: 2.00). (i) LHR, 461.1 W cm $^{-1}$; BU, 0.0484 at%. (ii) LHR, 427.0 W cm $^{-1}$; BU 0.0448 at %. (LHR, liear heating rate; BU, local burn up).

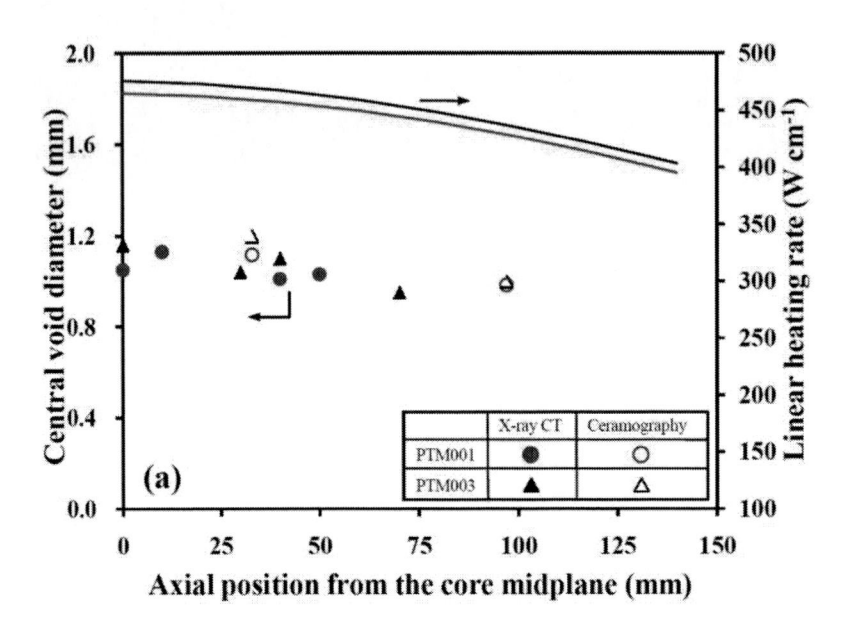

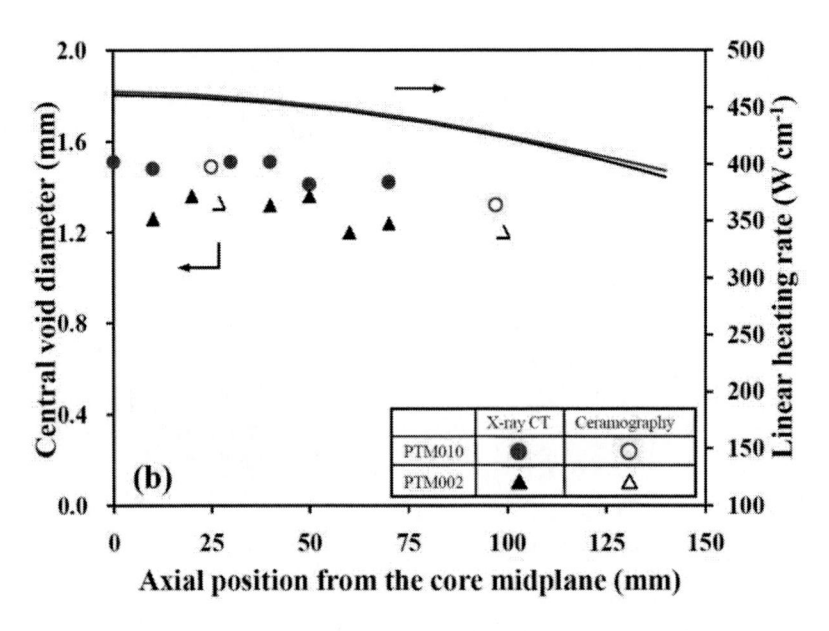

Figure 8. Axial distributions of central void diameter obtained by X-ray CT and ceramography. (a) PTM001, PTM003 (Gao width: 160 μm), (b) PTM010, PTM002 (Gap width: 210 μm).

The diameter of the central void measured by X-ray CT was compared with that observed by ceramography. The axial distributions of central void diameter and the linear heating rate are shown in Figure 8 as a function of the distance from the mid plane for the 160 and 210 μm gap width fuel pins. The axial distributions of the central void in the four fuel pins were similar for fuel pins having the same gap width. The width of the gap between the fuel and the cladding determines the gap conductance, which has a large influence on the temperature conditions. Therefore, the fuel centerline temperature increased depending on the low gap conductance due to the large gap width of 210 μm, which seemed to have resulted in significant fuel restructuring. The axial distributions of the central void diameter indicated the dependence on linear heating rate was like that of the normal steady state irradiation. Based on the axial distribution of the central void diameter, there was no apparent fuel melting in any fuel pins.

Comparison of the fuel pins of the B14 test which had fuel cladding gap widths of 160 and 210 μm showed significant differences in the central void diameters. A larger increase in central void diameter of fuel pins PTM002 and PTM010 with the larger gap width was recognized. However, it should be considered that a significant contribution to the central void volume by fuel relocation was included. Fuel relocation towards the cladding decreases in the radial gap width and it allows an increase in the radius of the central void. The relocation terminates when the fuel fragment comes into contact with the cladding. Without fuel relocation, the fuel temperature would be higher due to lower gap conductance in a larger gap width than that of a smaller gap width. Observation of specimens (at ~30 mm from mid plane) having the larger gap width (PTM002, PTM010) showed the residual gap width after the irradiation was similar to that of other specimens having the smaller gap width (PTM001, PTM003). The calculated maximum fuel temperature around the central void was 2733 - 2800K for a linear heating rate of ~470 W cm$^{-1}$. From observation of the fuel microstructure, it seemed that the maximum fuel temperature in the fuel with 210 μm gap width was lower than the value expected by the thermal design.

Comparison of the specimens of the B14 test which had fuel O/M molar ratios of 2.00 and 1.98 showed small differences in the diameters of the central void. However, the difference in the diameter of the central void was almost negligible in comparison to specimens which had fuel O/M molar ratios of 1.98 and 1.96. As shown in Figure 7(d), fuel restructuring was almost completed only with the O/M molar ratio of 2.00 (PTM010). Possibly, the high vapor pressure derived from the highest O/M molar ratio (2.00) of the fuel would affect the fuel restructuring.

### 3.2.3. Dependence of Linear Heating Rate on Fuel Restructuring

The fuel restructuring in the B14 test was compared to the experiences of other irradiation tests performed in Joyo, the short-term irradiation test B4M and the steady state long-term test A1M [16].

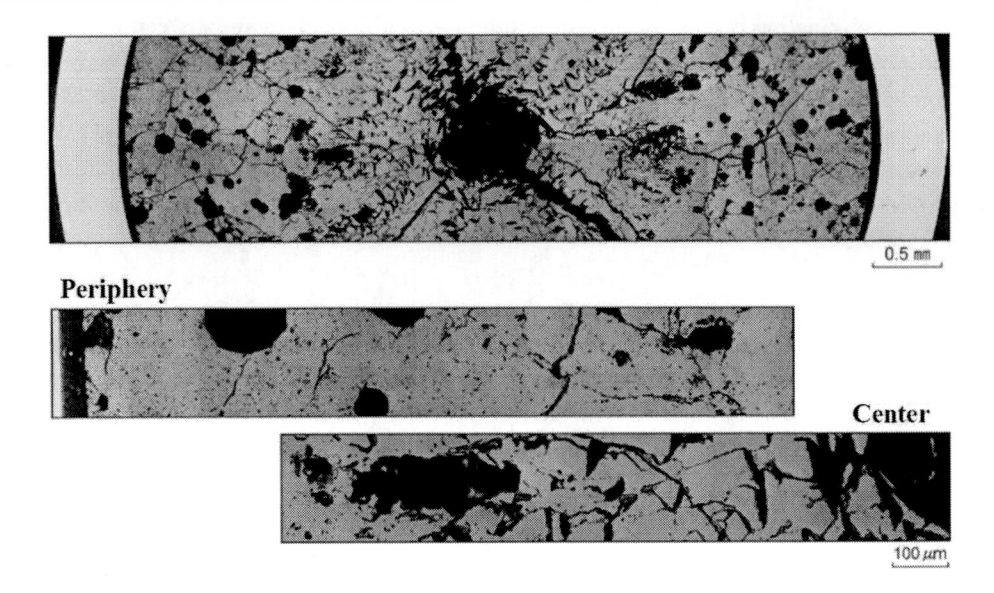

Periphery

Center

Figure 9. Fuel microstructure of B4M test specimen removed from a fuel pin. LHR, 377 W cm [-1]; BU, 0.080 MWd/kg [-1]. (LHR, linear heating rate; BU, local burn up).

Composites of as-polished ceramographs of a B4M test specimen are shown in Figure 9. There were incipient, radially formed columnar grains growing behind the lenticular voids. A small central void was formed but it was still growing due to migration of a significant number of lenticular voids towards the fuel center at the end of 3 h. Since columnar grains were in the process of being formed with many lenticular voids, fuel restructuring in the specimen had not yet been completed. This means that the linear heating rate of 377 W cm[-1] was sufficient to start fuel restructuring. It has been shown that migration of lenticular voids with subsequent formation of a central void was based on the vaporization-condensation mechanism [23].

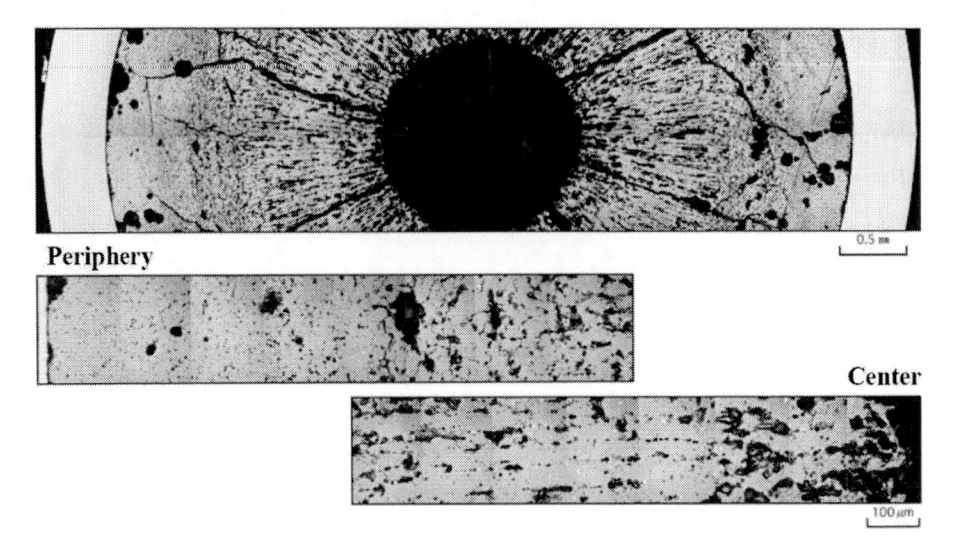

Periphery

Center

Figure 10. Fuel microstructure of A1M test specimen removed from a fuel pin. LHR, 480 W cm [-1]; BU, 42 MWd/kg [-1]. (LHR, linear heating rate; BU, local burn up).

Composites of as-polished ceramographs of an A1M test specimen are shown in Figure 10. Fuel restructuring was completed in the A1M test specimen which was irradiated under the steady state condition at 480 W cm$^{-1}$ in the beginning of life and reached 42 MWd kg$^{-1}$. A large central void and typical restructured regions such as columnar grains and equiaxed grains were seen in the specimen. The diameter of the restructured region exposed to high temperature was significantly developed. Significant pore migration at the beginning of irradiation seemed to have finally led to the columnar grain structure and central void formation. Furthermore, the wedge shaped crack was completely healed. And other cracks seemed to have closed by volume swelling of fuel. Interlinked pores along the grain boundary were seen in the columnar grain region. However, with increasing burnup, the porosity distributions in the specimen could no longer be explained only by migration of as-fabricated pores. Therefore the formation of fission gas bubbles and the transport of gap volume were thought to be included. At high linear heating rate of 480 W cm$^{-1}$, the fuel specimen showed the expected features in fuel microstructure suggesting the fuel integrity was kept.

Figure 11 compares cross-sectional ceramographs from the B4M and A1M tests and the B14 test for PTM001. The sizes of the restructured region of the specimens were compared to each other. The B4M test specimen was in the process of fuel restructuring at 377 W cm$^{-1}$. Migration of the lenticular voids occurred towards the fuel center, but a small central void was formed within 3 h. In contrast fuel restructuring was completed for the A1M test specimen which was irradiated under the steady state condition at 480 W cm$^{-1}$ in the beginning of life and reached 42 MWd kg$^{-1}$. The central void diameter of the B14 test specimen was developed due to sufficient fuel restructuring and was larger than that of the B4M test specimen but smaller than that of the A1M test specimen. As shown in Figure 12, it was obvious that the diameter of the restructured region was larger, depending on the linear heating rate and irradiation period. Also the differences in the diameter of the central void between A1M and B14 tests (except for fuel pin PTM010) were significant. The migration of lenticular voids towards the center during 10 min at high linear heating rate (~470 W cm$^{-1}$) in the B14 specimens is shown in detail in Figure 13. This accumulation of lenticular voids around the central void contributed little to the volume increase in the central void.

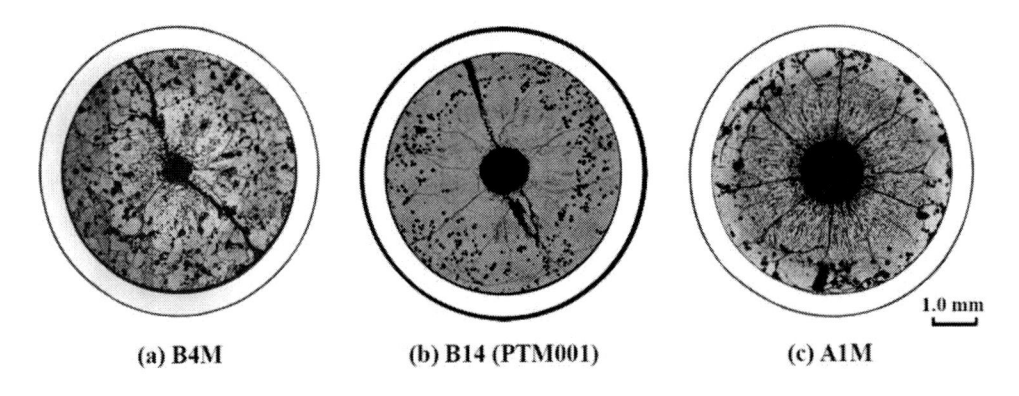

(a) B4M               (b) B14 (PTM001)               (c) A1M

Figure 11. Comparison of fuel microstructure of specimens removed from different fuel pins.

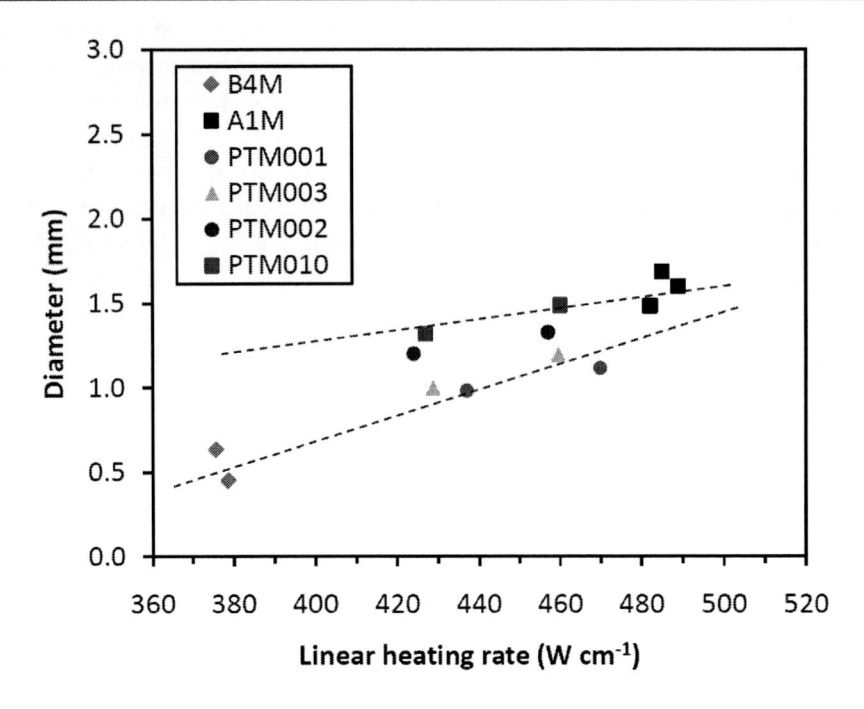

Figure 12. Central void diameter in the specimens as a function of linear heating rate.

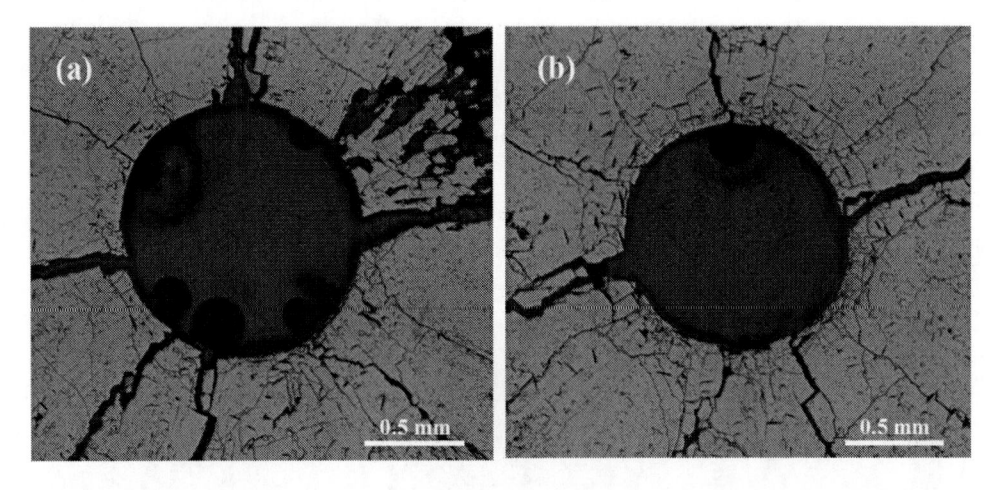

Figure 13. Appearance of the lenticular void accumulation around the central void of specimens removed at +30mm from the midplane. (a) PTM002 (gap width, 210 μm; O/M, 1.98). (b) PTM003 (gap width, 160 μm; O/M, 1.96).

### 3.2.4. Radial Redistribution of Fuel Constituents

The changes in Am, Pu and U contents of specimens in the B14 test were measured by EPMA, and were normalized to the mean values measured in the peripheral region corresponding to the unaffected region during the irradiation. The relative changes in Am, Pu and U contents in the specimens (~30 mm from the mid plane) are shown in Figures 14(a)-(d) as a function of radial distance from the fuel surface. The Am and Pu contents increased towards the central void and were redistributed in the columnar grain region. In contrast, U content decreased towards the central void. Radial redistributions of Am, Pu and U in an early stage of irradiation occur mainly by the mechanism of vapor transport, vaporization and condensation in moving lenticular voids during fuel restructuring [7,24,25]. The vaporization and condensation process is considered effective in redistributing actinides and other elements possessing any degree of solubility in the fuel. In the case of vaporization and condensation, the substance having the highest vapor pressure is transported to low temperature regions in the specimen. A higher vapor transport of uranium as $UO_3$ has been reported in some studies [26-28]. According to this, the hot face of a moving lenticular void becomes enriched in the less volatile of the fuel components of the mixed oxide (Am and Pu) by preferential evaporation of the more volatile species across the pore.

By comparison for the PTM001 and PTM002 specimens with the O/M molar ratio of 1.98, the depletions of Am and Pu contents at the outer edge of the columnar grain region were pronounced in PTM001. And Am and Pu contents increased more in PTM001 towards the central void than in PTM002. Thus redistributions of Pu and Am were not larger in PTM002 than in PTM001. It was suggested that increases in Am and Pu contents towards the central void were reduced due to the lower fuel temperature obtained from the favorable gap conductance by significant fuel relocation in PTM002. Fuel restructuring with the formation of central voids and significant relocation of fuel fragment can decrease fuel temperatures.

In the PTM003 specimen with the O/M molar ratio of 1.96 in comparison to the PTM001specimen, Pu and Am contents abruptly increased near the central void. This was attributed to a low vapor transport of uranium as $UO_3$, which is a consequence of the lower oxygen potential of the fuel with an O/M molar ratio of 1.96 than in that of the fuel with an O/M molar ratio of 1.98 as has been reported in the literature [26-28]. Under this lower oxygen potential condition, the radial profiles of Am and Pu also exhibited a similar tendency to redistribute in the columnar grain region.

In the PTM010 specimen with O/M molar ratio of 2.00 in comparison to the PTM001 specimen, Pu and Am contents were increased the most towards the central void compared to the other specimens. It was reasonable to consider that the significant increase was derived from the large vapor pressure of the fuel [26-28].

These results revealed that the radial profiles of Am and Pu contents showed a similar tendency to redistribute in the columnar grain region. In addition, a greater dependence of increase in Am and Pu contents towards the central void on the fuel O/M molar ratio than on linear heating rate was suggested. This was similar to the results reported in some relatively high-density MA-MOX fuels [9-12,29]. In addition, it was likely that the radial redistribution of Am and Pu was affected by fuel relocation due to the large as-fabricated gap width.

Figure 15 shows the relative changes in Am, Pu and U contents around the central void in the specimens as a function of O/M molar ratio. A similar dependence of increase in Am and Pu contents on the O/M molar ratio was seen in the specimens. Also the degree of increase of Am and Pu contents in specimens having the O/M molar ratio below 1.98 was similar. However the difference in the degree of increase of Am and Pu contents in specimens was significant for the O/M molar ratio above 1.98. The significant increase of Am and Pu contents was attributed to the higher vapor transport of uranium as $UO_3$ as has been reported [26-28], which was a consequence of the high oxygen potential of the fuel with an O/M molar ratio of 2.00. However, it was suggested here that the difference in vapor pressure between Pu and Am affected the redistribution at the high O/M molar ratio of 2.00.

a)

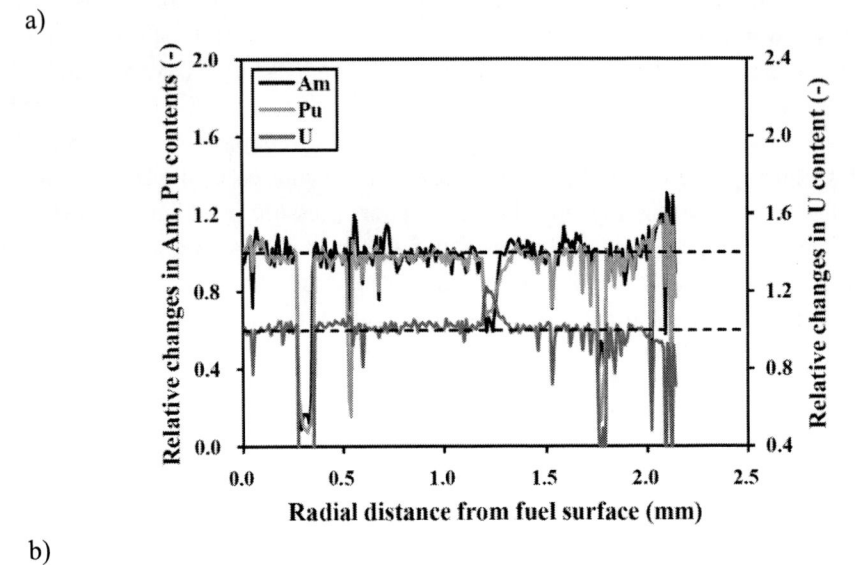

b)

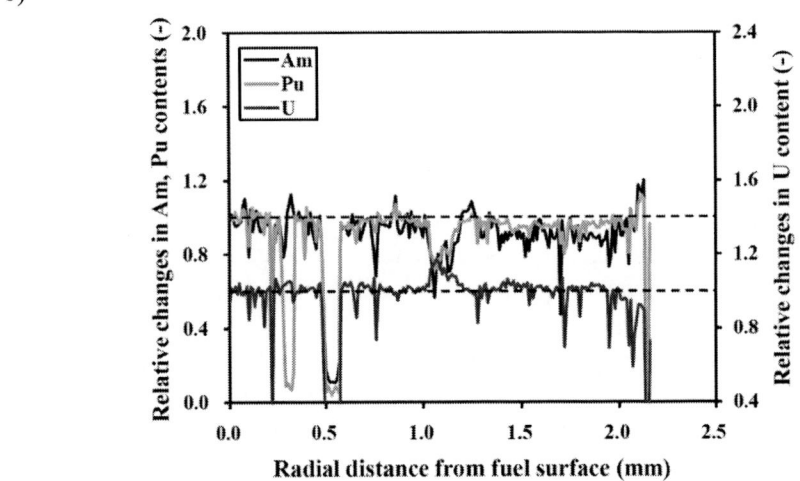

c)

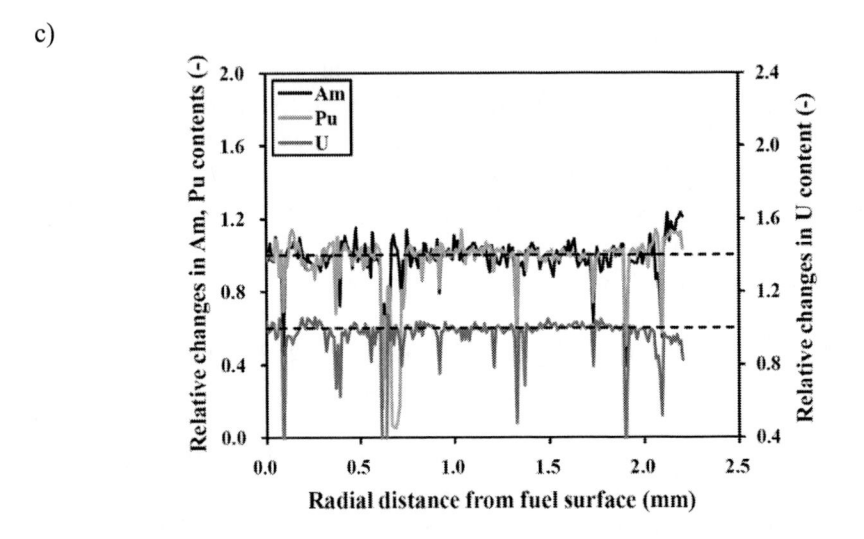

d)

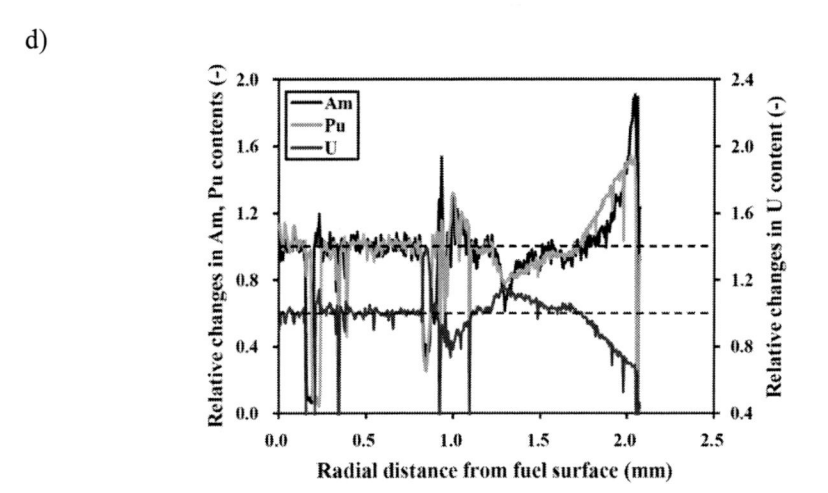

Figure 14. (a) Changes in Am, Pu and U contents of the specimen (PTM001: gap width, 160 μm; O/M, 1.98) as a function of radial distance from the fuel surface. (b) Changes in Am, Pu and U contents of the specimen (PTM003: gap width, 160 μm; O/M, 1.96) as a function of radial distance from the fuel surface. (c) Changes in Am, Pu and U contents of the specimen (PTM002: gap width, 210 μm; O/M, 1.98) as a function of radial distance from the fuel surface. . (d) Changes in Am, Pu and U contents of the specimen (PTM010: gap width, 210 μm; O/M, 2.00) as a function of radial distance from the fuel surface.

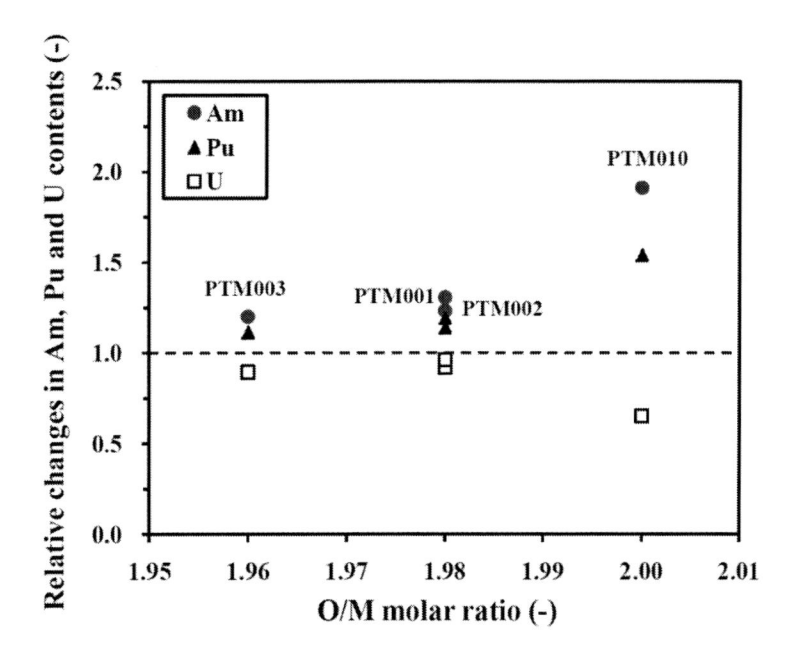

Figure 15. Dependence of O/M molar ratio on Am, Pu and U contents around the central void in the specimens.

In the case of the A1M test specimen, a large increase of Pu content towards the central void was seen. In contrast no significant redistributions of Pu and U were identified in the B4M test specimen which was in the process of fuel restructuring. Redistribution of Pu in the B14 test specimen was significant and similar to that of the A1M test specimen. And the Am and Pu content profiles were distributed similarly as a function of fuel radius as shown in Figs. 14(a)-(d). The redistribution behavior of Am and Pu from the onset through to the end of fuel restructuring seemed to be similar. In particular, the maximum contents of Pu and Am near the central void were 41.5 and 3.8 wt% in PTM010, respectively. Validity relative to the defined margin in the thermal design of fuel pins was evaluated based on the fact that no fuel melting occurred. By assuming completion of both the fuel restructuring and the radial redistribution of fuel constituents at a very early stage of irradiation, the chemical composition of the fuel at the center would conservatively be considered to $(U_{0.547}Pu_{0.415}Am_{0.038})O_{2.00}$. According to this very conservative evaluation, the maximum fuel temperature would be estimated as lower than its melting point. Kato et al. [3] represented the variation of the solidus and liquidus temperatures in the $UO_2$–$PuO_2$–$AmO_2$ ternary system by the ideal solution model. Based on the ideal solution model, the melting point of $(U_{0.547}Pu_{0.415}Am_{0.038})O_{2.00}$ was evaluated as 3023 K. This indicates that the thermal design of fuel pins in the B14 test involved more than enough margin against occurrence of actual fuel melting, because their thermal design included 20% occurrence of fuel melting. The design margin was derived mainly from uncertainties of fuel fabrication tolerance and irradiation conditions, in addition to the uncertainties in properties of MOX fuels containing americium. By evaluation of the current design based on the results of examinations, the design procedure could be modified to one with a more appropriate design margin.

## 4. SUMMARY

In the B14 short-term irradiation test, four fuel pins (PTM001, PTM002, PTM003 and PTM010) were irradiated to 470 W cm$^{-1}$ under the transient condition (10 min) for the current fuel pin thermal design after step-by-step increases of linear heating rate to 347 and 386 W cm$^{-1}$ by preconditioning. No fuel melting was observed by PIEs for any fuel pins after the B14 test. It seemed that fuel restructuring before the transient condition due to preconditioning was enough to prevent fuel melting. Under the transient condition, additional radial migration of lenticular voids toward the fuel center resulted in an insignificant increase of the central void diameter except for PTM010. Except for PTM010 (O/M molar ratio of 2.00), the degree of fuel restructuring was found to be more dependent on the linear heating rate than on the small addition of americium and on the fuel O/M molar ratio. In the fuel pin with a large as-fabricated gap width, significant fuel relocation and off-center fuel restructuring were observed, but no enhancement in the maximum fuel temperature was identified. As for PTM010, its fuel restructuring was more developed than the others and the central void diameter after the transient condition corresponded to that of the maximum linear heating rate. It might be possible that the high vapor pressure derived from the higher O/M molar ratio of the fuel affected the fuel restructuring.

The radial profiles of Am and Pu contents indicated that these elements had similar redistribution tendencies in the columnar grain region. The degree of increase in Am and Pu contents around the central void showed a significant dependence on the O/M molar ratio. The fuel having a higher O/M molar ratio showed a larger redistribution of Am and Pu towards the central void. In addition the difference in the contents between Am and Pu was larger with fuel having a higher O/M molar ratio.

Small addition of Am did not alter fuel restructuring and the redistribution characteristics of U and Pu from those of the MOX fuel without Am. The present results indicated that the margin in the current thermal design of fuel pin would be appropriate, and that it could prevent fuel melting. The results will be used to obtain a valid model of the $(Am,Pu,U)O_{2-x}$ fuels in thermal analysis codes at the early stage of irradiation, which is expected to improve defined margins in thermal design of fuel pins.

# REFERENCES

[1]   M. Konomura, M. Ichimiya, *J. Nucl. Mater.* 371 (2007) 250-269.

[2]   M. Ichimiya, T. Mizuno, M. Konomura, in: Proc. GLOBAL 2003, New Orleans, LA, USA, 16-20 November 2003 (on cd-rom), pp. 434-446.

[3]   M. Kato, K. Morimoto, H. Sugata, K. Konashi, M. Kashimura, T. Abe, *J. Nucl. Mater.* 373 (2008) 237-245.

[4]   K. Morimoto, M. Kato, M. Ogasawara, M. Kashimura, T. Abe, *J. Nucl. Mater.* 452 (2008) 54-60.

[5]   M. Kato, T. Tamura, K. Konashi, *Nucl. Mater.* 385 (2009) 419-423.

[6]   K. Asakura, T. Yamaguchi, T. Ohtani, *J. Nucl. Mater.* 357 (2006), 126-137.

[7]   M. Bober, G. Schumacher, D. Geithoff, *J. Nucl. Mater.* 47 (1973), 187-197.

[8]   T. Ishii, T. Asaga, *J. Nucl. Mater.* 294 (2001), 13-17.

[9]   C.T. Walker, G. Nicolaou, *J. Nucl. Mater.* 218 (1995) 129-138.

[10]  ] C. Prunier, F. Boussard, L. Koch, M. Coquerelle, *Nucl. Technol.*, 119 (1997) 141-148.

[11]  K. Tanaka, S. Miwa, I. Sato, T. Hirosawa, H. Obayashi, S. Koyama, H. Yoshimochi, K. Tanaka, *J. Nucl. Mater.* 385 (2009) 407-412.

[12]  K. Maeda, S. Sasaki, M. Kato, S. Kihara, *J. Nucl. Mater.* 385 (2009) 413-418.

[13]  N. Nakae, K. Tanaka, H. Nakajima, M. Matsumoto, *J. Nucl. Mater.* 188 (1992) 331-336.

[14]  K. Maeda, K. Katsuyama, T. Asaga, *J. Nucl. Mater.* 346 (2005) 244-252.

[15]  M. Kato, S. Nakamichi, T. Takano, in Proc. GLOBAL 2007, Boise, ID, USA, 9–13 September 2007 (on cd-rom), pp. 916-920.

[16]  M. Katsuragawa, H. Kashihara, M. Akebi, *J. Nucl. Mater.* 204 (1993) 14-22.

[17]  T. Asaga, S. Shikakura, S. Iwanaga, S. Nomura, I. Shibahara, M. Katsuragawa, *J. Nucl. Mater.* 188 (1992) 102-108.

[18]  M. Koizumi, K. Ohtsuka, H. Isagawa, H. Akiyama and A. Todokoro, *Nucl. Technol.* 61 (1983), 55-70.

[19]  M. Koizumi, K. Ohtsuka, H. Ohshima, H. Isagawa, H. Akiyama, A. Todokoro, K. Naruki, *J. Nucl. Sci. Technol.* 20 (1983) (7), 529-536.

[20]  K. Katsuyama, T. Nagamine, S. Matsumoto, S. Sato, *Nucl. Instrum. and Meth. B* 255(2007) 365-372.

[21]  M. Inoue, K. Yamamoto, T. Sekine, M. Osaka, N. Kushida, T. Asaga, *J. Nucl. Mater.* 323 (1993) 108-122.

[22]  M. Inoue, *J. Nucl. Mater.* 282 (2000) 186-195.

[23]  D.R. De Halas, G.R. Horn, *J. Nucl. Mater.* 8 (1963) 207-220.

[24]  M. Bober, G. Schumacher, *Adv. Nucl. Sci. Technol.* 7 (1973), 21.

[25]  C.F. Clement, M.W. Finnis, *J. Nucl. Mater.* 75 (1978), 193-200.

[26]  C. Ronchi, Hj. Matzke, in: *Proc. Fuel and Fuel Elements for Fast Reactors* 1973, IAEA Symposium, Brussels, 1974, pp. 57-70.

[27]  M.H. Rand, T.L. Markin, in: *Proc. Thermodynamics of Nuclear Materials* 1967, IAEA Symposium, Vienna, 1968, pp. 637-650.

[28]  H. Bailly, D. Menessier, C. Prunier, *The Nuclear Fuels of Pressurized and Water Reactors and Fast Reactors Design and Behaviour*, Intercept Ltd., Hampshire, U.K., 1999, p.106.

[29]  K. Maeda, S. Sasaki, M. Kato, S. Kihara, *J. Nucl. Mater.* 389 (2009) 78-84.

In: Nuclear Materials
Editor: Michael P. Hemsworth, pp. 133-156

ISBN 978-1-61324-010-6
© 2011 Nova Science Publishers, Inc.

*Chapter 5*

# MICROSTRUCTURAL CHARACTERIZATION OF STRUCTURAL MATERIALS OF PRESSURIZED HEAVY WATER REACTOR

## *P. Mukherjee, P. Barat, A. Sarkar[1],*
## *M. Bhattacharya, and N. Gayathri*

Variable Energy Cyclotron Center, 1/AF ,Bidhannagar, Kolkata – 700064, India
[1]Mechanical Metallurgy Section, Bhabha Atomic Research Centre,
Mumbai-400085, India

## ABSTRACT

Zirconium based alloys are extensively used in Pressurized Heavy Water Reactor (PHWR) as structural materials due to their low neutron absorption cross-section, acceptable corrosion resistance in hot pressurized water, adequate mechanical properties at elevated temperature and irradiation induced creep and growth resistance properties. Among them, Zircaloy-2, Zircaloy-4 and Zr-2.5% Nb are the main alloys of zirconium used in Pressurized Heavy Water Reactor (PHWR). These alloys are used for fabricating the structural materials like fuel tube, coolant tube, calandria tube, spacers, garter spring, end caps and end plates etc. Several advanced alloys like Zr-1%Nb, Zr-1%Nb-1%Sn-0.1Fe has been designed as candidate materials for fuel cladding tube to withstand higher burn up of the fuel and are found to possess desired combination of strength, higher creep and growth resistance, corrosion resistance and fabricability. The initial microstructures of the fuel cladding tubes and pressure tubes are of great concern from the point of view of their structural integrity as well as irradiation induced creep and irradiation induced growth resistance properties inside the reactor. The pre-irradiation dislocation network is thought to contribute to irradiation growth at high neutron fluencies. This growth rate increases as the degree of cold work increases and has been reported to vary with the dislocation density. The mechanical behavior of these alloys also gets altered by the irradiation induced defects like vacancy, vacancy clusters, dislocation loops etc. As a result, irradiation hardening occurs, yield strength and tensile strength of the materials increase and ductility drastically reduces. All these structure-sensitive properties are dependent on the microstructure of the materials, mobility of vacancies, the interaction of dislocations with point defects, formation of dislocation substructure or cell etc.

Therefore it is of utmost importance to characterize the microstructure of these alloys, their defect states, dislocation density, stacking and growth faults. The present article mainly deals with the characterization of several microstructural imperfections introduced by cold deformation and irradiation using light ion (proton) and heavy ions ($O^{5+}$, $Ne^{6+}$) in zirconium based alloys. The microstructural parameters like domain size, microstrain within the domains, density of dislocations, stacking faults and twin or growth fault probabilities have been characterized using X-Ray Diffraction Line Profile Analysis (XRDLPA).

## INTRODUCTION

Zirconium and its alloys are commonly used for many structural applications particularly for fuel element cans and in-core structural materials in water cooled thermal reactors. The general requirements for these structural components are: low neutron absorption cross-section, adequate strength, ductility and toughness at ambient and reactor operating temperatures, good corrosion resistance, low hydrogen absorption, long term dimensional stability in irradiation environments, compatibility with fuel and coolant under service conditions and resistance to fracture. Zirconium alloys containing very small amounts of selected alloying elements have been found to meet these requirements at reactor operating temperature. The consideration of neutron economy imposes a severe restriction on the selection of alloying elements and the permissible levels in the design of an alloy system [1,2]. Addition of alloying elements like niobium, nickel, iron, chromium, Molybdenum, copper and oxygen have helped to enhance certain properties [1,2]. Alloys like Zircaloy-4, Zr-1%Nb, Zr-2.5%Nb, Zr-2.5%Nb-0.5%Cu alloys are being used as core structural materials in Pressurized Heavy Water Reactors (PHWR). These alloys are used for fabricating the structural materials like fuel tube, coolant tube, calandria tube, spacers, garter spring, end caps and end plates etc. Several advanced alloys like Zr-1%Nb, Zr-1%Nb-1%Sn-0.1%Fe has also been designed as candidate materials for fuel cladding tube to withstand higher burn up of the fuel and are found to possess desired combination of strength, higher creep and growth resistance, corrosion resistance and fabricability. The sheets and the tubes made from these alloys are used inside the reactors in the cold-worked and heat-treated state [3]. The microstructural parameters, particularly the dislocation density and the domain size are believed to influence the deformation behavior of these alloys when used in the reactors [4]. The pre-irradiation dislocation network is thought to contribute to irradiation growth at high neutron fluences. This growth rate increases as the degree of cold work increases [4-7] and has been reported to vary with the dislocation density. The mechanical behavior of these alloys also gets altered by the irradiation induced defects like vacancy, vacancy clusters, dislocation loops etc. As a result, irradiation hardening occurs, yield strength and tensile strength of the materials increase and ductility drastically reduces. All these structure-sensitive properties are dependent on the microstructure of the materials, mobility of vacancies, the interaction of dislocations with point defects, formation of dislocation substructure or cell etc. Therefore it is of utmost importance to characterize the microstructure of these alloys, their defect states, dislocation density, stacking and growth faults. There are several techniques which can be used to study the microstructural defects or imperfections like surface and decoration method, Transmission Electron Microscopy (TEM), resistivity measurements, X-Ray Diffraction Line Profile Analysis (XRDLPA) and Positron

Annihilation Spectroscopy (PAS). The present article mainly deals with the characterization of several microstructural imperfections like domain size, microstrain within the domains, density of dislocations, stacking faults, twin or growth faults using XRDLPA on zirconium base alloys like Zircaloy-2, Zr-2.5%Nb, Zr-1%Nb, Zr-1%Nb-1%Sn-0.1%Fe. With the advent of computer based programs for microstructural characterization, this technique has become immensely popular in the field of materials science. X-ray powder diffraction, has a modest beginning in the early part of the twentieth century in various fields and has come a long way till the end of the century with the advent of most powerful methods based on Fourier line shape modeling, whole pattern fitting or profile fitting approaches, incorporating functional representation of X-ray line shapes and *ab-initio* physical modeling of material microstructure. All these methods have now given profile analysis a strong foundation to probe crystalline materials at micro- and even nano-levels to reveal microstructural parameters of 'real' crystals in a qualitative as well as quantitative manner.

X-ray diffraction peaks are broadened due to the effect of microstructural parameters, mainly the sizes of coherently diffracted domains and lattice microstrain caused by the presence of defects (dislocation) in the microstructure. Defects have been introduced by cold deformation i.e. by cold rolling and hand filing and also by irradiation using light ion (proton) and heavy ions ($O^{5+}$, $Ne^{6+}$). A detailed discussion of microstructural characterization of zirconium based alloys using different techniques of XRDLPA have been presented in the following text.

## 2. Experimental Details

### 2.1. Specimen Preparation

#### 2.1.1. Powdered and Cold Worked Samples

The samples of Zircaloy-2, Zr-2.5% Nb and Zr-1%Nb-1%Sn-0.1%Fe were collected from Nuclear Fuel complex (NFC), Hyderabad, India in the form of plates. These alloys were prepared by double vacuum arc melting technique followed by β-quenching heat-treatment and hot extrusion at a temperature of 1023K-1073K in the form of plates.

The powder samples from each alloy were then obtained by a very careful hand-filing from a portion of the plate by a jeweler's file at room temperature. Powder finer than 250 mesh was taken for preparation of the briquette to be used as the sample in the diffractometer. A portion of the cold-worked powder of each alloy was vacuum sealed in a quartz tube and annealed to a temperature of 1023K for a period of 10 hours to annul the effect of cold-work (i.e. to make it free from strains, small domain sizes etc.).

The extruded plates were hot rolled to a thickness of 10mm. The samples of these alloys were collected at this stage and then annealed in vacuum by sealing these in quartz tubes. These samples were then cold-rolled at room temperature to different deformations i.e. 30%, 50%, 60% for Zr-2.5% Nb and 30%, 50% and 70% for Zircaloy-2.

### 2.1.2. Irradiation of Zircaloy-2, Zr-1%Nb-1%Sn-0.1Fe, Zr-1Nb, using Light Ion (Proton) and Heavy Ions ($O^{5+}$ or $Ne^{6+}$)

In each case, samples were collected in the form of sheets or flat discs. Specimens of dimension 10 mm x 10 mmx0.4 mm were cut and then irradiated at different doses at Variable Energy Cyclotron (VEC), Kolkata, India. The flange used in irradiation was cooled by continuous flow of water. During irradiation, the temperature of the sample did not rise above 313 K as measured by the thermocouple connected with the closest proximity of the sample. The samples of Zircaloy-2 were irradiated with 11 MeV proton and the annealed samples of Zr-1%Nb-1%Sn-0.1Fe were irradiated by 15 MeV proton, 116 MeV $O^{5+}$ and 145 MeV $Ne^{6+}$ ions from Variable Energy Cyclotron (VEC), Kolkata, India. The irradiation doses were $9.85 \times 10^{21}$, $5.0 \times 10^{22}$ protons/m$^2$ for 11 MeV proton, $1 \times 10^{20}$, $5 \times 10^{20}$, $5 \times 10^{21}$ protons/m$^2$ for 15 MeV proton, $1 \times 10^{17}$, $5 \times 10^{17}$, $1 \times 10^{18}$ and $5 \times 10^{18}$ $O^{5+}$ ions/m$^2$ for 116 MeV $O^{5+}$ ion, $3 \times 10^{17}$, $8 \times 10^{17}$, $1 \times 10^{18}$ and $3 \times 10^{18}$ ions/m$^2$ for 145 MeV $Ne^{6+}$ ion.

## 2.2. Data Collection for XRD Analysis

For a reliable application of profile-fitting methods in line-broadening analysis such as modified Rietveld technique as well as for Warren Averbach technique, the first requirement is the collection of step intensities at known increments of $2\theta$. X-ray data from the experimental samples including the silicon standard was recorded in an X-ray Diffractometer (Model PW 1710 or Bruker D-8 Advance) in a step scan mode using Ni-filtered $CuK_\alpha$ - radiation. The $2\theta$ ranging from $32.5^0$ to $101^0$ with step scan of $0.02^0$ was used. The time per step was varied between 5-10 seconds.

## 3. METHOD OF ANALYSIS

X-ray diffraction line profile analysis (XRDLPA) has been widely applied for the evaluation of the microstructural parameters in different deformed and irradiated metals and alloy systems [8-10].

Two categories of structural imperfections can give rise to a spread of intensity around each reciprocal-lattice points and the diffraction line profiles are modified by a measurable amount. The first is the finite size of the domains, over which the diffraction is coherent, measured in the direction of the diffraction vector for a given reflection. This can be the size effects of the individual crystallites, it can also relate to a sub-domain structure, i.e. the separation of the regions bounded by the low angle grain boundaries and also the structural mistakes [11]. The second category of the structural imperfections is based on the distortion within the crystal lattice. This may arise from the microstrain due to an applied or residual stress. Dislocations contribute to both categories of broadening of the X-ray diffraction line profile. There will be a size contribution due to their mean separation, inversely proportional to the dislocation density, and microstrain arising from internal stress fields [12]. There are several model-based approaches which help to separate these two independent effects by fitting the diffraction peaks with suitable mathematical functions. In the present paper, different model-based approaches [13-18] have been adopted to assess the dose-dependent variation of domain size and microstrain. On the other hand, Warren Averbach technique is more conventional and rigorous one and it is applied on the random sample. The treatment of this technique on X-ray diffraction peak is totally unbiased and no assumption is made about the analytical form of the diffraction peak.

## 3.1. Warren Averbach Technique

The detailed Fourier analysis on the diffraction profiles has been performed for the hexagonal alloys as described in [12]. The Stokes' corrected Fourier coefficients $(A_L)$ for various values of $L(\text{Å})$ have been calculated for Zircaloy-2, Zr-1%Nb-1%Sn-0.1%Fe and Zr-2.5%Nb respectively for both fault-affected and fault-unaffected reflections. The separation of particle size coefficients $A_L^S$ and distortion coefficients $A_L^D$ was done by use of the log plot of $A_L (= A_L^S \times A_L^D)$ against $\dfrac{1}{d^2}$, where $d$ is the interplanar spacing. The intercept and slope of this plot yield $A_L^S$ and $\langle \varepsilon_L^2 \rangle$ as a function of $L$. The strain distribution is assumed to be Gaussian in nature and is given by:

$$A_L^D = \exp\left(-2\pi^2 L^2 \langle \varepsilon_L^2 \rangle / d^2\right) \tag{1}$$

For fault-unaffected reflections, i.e. $H - K = 3\text{N}$ where $H, K$ and $L_0$ are hexagonal indices, the initial slope of $A_L^S$ vs $L$ curves gives the average domain size $D_{av}(\text{Å})$

$$-\frac{dA_L^S}{dL} = \frac{1}{D_{av}} \tag{2}$$

Using the relation, $A_L (= A_L^S \times A_L^D)$ and the values of microstrain $\langle \varepsilon_L^2 \rangle^{\frac{1}{2}}$, the distortion coefficients $A_L^D$ for fault affected reflections were calculated and the size coefficients $A_L^S$ were separated out for each such fault-affected reflection. The initial slope of $A_L^S$ for $H-K=3\text{N}\pm1$ ($L_0$ odd and even reflections) is related to the effective domain size $D_e$ and stacking faults (deformation and growth or twin) by the following relations [12]:
When $H - K = 3\text{N}\pm1$ and $L_0$ even,

$$-\left(\frac{dA_L^S}{dL}\right)_0 = \frac{1}{D_e} + \frac{|L_0|d}{C^2}(3\alpha + 3\beta) \tag{3}$$

where $D_e$ is the effective domain size of the fault affected reflections.
When $H - K = 3\text{N}\pm1$ and $L_0$ odd,

$$-\left(\frac{dA_L^s}{dL}\right)_0 = \frac{1}{D_e} + \frac{|L_0|d}{C^2}(3\alpha + \beta) \tag{4}$$

These equations permit us to determine the fault probabilities $\alpha$ and $\beta$ or compound fault probability from the strain or distortion corrected $A_L^S$ coefficients.

## 3.1. Williamson-Hall Technique

According to the Williamson-Hall method [19], it is assumed that, both the size and strain broadened profiles are Lorentzian. Based on this assumption, a mathematical relation has been established between the integral breadth ($\beta$), the volume weighted average domain size ($D_v$) and the upper limit the microstrain ($\varepsilon$) as follows:

$$\frac{\beta\cos\theta}{\lambda} = \frac{1}{D_v} + 2\varepsilon\left(\frac{2\sin\theta}{\lambda}\right) \tag{5}$$

The plot of $\left(\dfrac{\beta\cos\theta}{\lambda}\right)$ versus $\left(\dfrac{2\sin\theta}{\lambda}\right)$ gives the value of $\varepsilon$ from the slope and $D_v$ from the ordinate intercept respectively.

## 3.2. Modified Rietveld Technique

X-ray data were analyzed for domain size and root mean squared strain with the help of modified Rietveld technique using the program LS1 [14]. The program LS1 includes the simultaneous refinement of crystal structure and microstructural parameters like domain size and microstrain within the domain. The method involves the Fourier analysis of the broadened peaks [14]. The initial values of the lattice parameters were obtained from a least square fit of the powder diffraction peaks. The lattice parameters, surface weighted average domain size ($Ds$) and microstrain $\langle \varepsilon_L^2 \rangle^{\frac{1}{2}}$ were used as fitting parameters simultaneously to obtain the best fit in order to calculate the average values of the fitting parameters. The effective domain size ($De$) with respect to fault-affected crystallographic planes was then refined to obtain the best fitting parameters. The X-ray diffraction peak profiles have been described by the pseudo-Voigt function [17] for the deformed, unirradiated and irradiated samples in order to fit the experimental data. The orientation parameters were also used as fitting parameters [20–22] to incorporate the corrections for the preferred orientation as Zirconium and its alloys show strong crystallographic texture along certain crystallographic direction after mechanical working [22]. The instrumental broadening correction was done by a silicon sample which has large crystallites and is free from any defect.

## 3.3. Simplified Breadth Method

In this method [23], each diffraction peak is fitted by a Voigt function if the shape factor $\phi$ of the peak (defined by $FWHM / \beta$) is in between 0.63 and 0.94, where $FWHM$ is Full Width at Half Maximum and $\beta$ is the integral breadth. It is also assumed that Cauchy component of the Voigt function arises for the small domain size and the Gaussian component arises for the microstructural strain for the individual peak. The relation between $D_v, \varepsilon, \beta_c$ and $\beta_G$ is given by [24]:

$$D_v = \frac{\lambda}{\beta_c \cos \theta} \tag{6}$$

and

$$\varepsilon = \frac{\beta_G}{\tan \theta} \tag{7}$$

where $\beta_c$ and $\beta_G$ represent the Cauchy and Gauss component of the integral breadth.

## 3.4. Double Voigt Technique

Cauchy or Gaussian function exclusively cannot model the peak broadening. Therefore, the size and strain effects are approximated by a Voigt function [25], which is basically a convolution of Gaussian and Cauchy function. The equivalent analytical expressions for Warren-Averbach size-strain separation are then obtained. The Fourier coefficients $F(L)$ in terms of a distance, $L$, perpendicular to the diffracting planes is obtained by Fourier transform of the Voigt function [26] and can be written as

$$F(L) = \left(-2L\beta_C - \pi L^2 \beta_G^2\right) \tag{8}$$

where, $\beta_C$ and $\beta_G$ are respectively the Cauchy and Gauss components of total integral breadth $\beta$.

$\beta_C$ and $\beta_G$ can be written as:

$$\beta_C = \beta_{SC} + \beta_{DC} \tag{9}$$

$$\beta_G^2 = \beta_{SG}^2 + \beta_{DG}^2 \tag{10}$$

where, $\beta_{SC}$ and $\beta_{DC}$ are the Cauchy components of size and strain integral breadth respectively and $\beta_{SG}$ and $\beta_{DG}$ are the corresponding Gaussian components.

The size and distortion coefficients are obtained considering at least two reflections from the same family of crystallographic planes. The surface weighted average domain size $D_S$ and strain $\left\langle \varepsilon_L^2 \right\rangle^{\frac{1}{2}}$ are given by the equations:

$$D_S = 1/2\beta_{SC} \tag{11}$$

$$\left\langle \varepsilon_L^2 \right\rangle = \left[ \beta_{DG}^2 / (2\pi) + \beta_{DC} / (\pi^2 L) \right] / S^2 \quad \text{where} \quad S = \frac{2\sin\theta}{\lambda} \tag{12}$$

The volume weighted domain size [16] is given by:

$$D_V = \frac{1}{\beta_S} \quad \text{where} \quad \beta_S = \frac{\beta\cos\theta}{\lambda}, \text{ integral breadth in the units of } S, (\text{Å})^{-1}.$$

The volume weighted column-length distribution functions are given by:

$$P_v(L) \propto L \frac{d^2 A_S(L)}{dL^2} \tag{13}$$

For a size-broadened profile, the size coefficient is given as:

$$A_S(L) = \exp(-2L\beta_{SC} - \pi L^2 \beta_{SG}^2) \tag{14}$$

From equation (8), we get,

$$\frac{d^2 A_S(L)}{dL^2} = [(2\pi L \beta_{SG}^2 + 2\beta_{SC})^2 - 2\pi\beta_{SG}^2] A_S(L) \tag{15}$$

Selivanov and Smislov [26] showed that equation (15) is a satisfactory approximation of size distribution functions.

## 3.5. MarqX Method

MarqX [27] is a computer program for the modeling of diffraction patterns; it mostly addresses problems of materials science which typically include lattice-parameter determination and line profile analysis.

Since the analytical profiles are pseudo-Voigt functions, the integral breadth ($\beta$) can be obtained as [27]

$$\beta_{pV} = HWHM[(1-\eta)(\pi/\ln 2)^{1/2} + \eta\pi]$$ (16)

The knowledge of HWHM, $\eta$, $\beta$ permits the calculation of Voigt parameter $\phi$ [28]. The corresponding Lorentz and Gauss components of the integral breadth ($\beta_{gL}$, $\beta_{fL}$, $\beta_{gG}$, $\beta_{fG}$ respectively) can be calculated from [29]

$$\beta_{hL} = \beta_{gL} + \beta_{fL}$$ (17)

$$\beta_{hG} = (\beta_{gG}^2 + \beta_{fG}^2)^{1/2}$$ (18)

where $\beta_{hL}$, $\beta_{hG}$ are the Lorentz and Gauss component of the total $\beta$. $\beta_{gL}$, $\beta_{fL}$ are the Lorentz components of instrument and sample contribution respectively and $\beta_{gG}$, $\beta_{fG}$ are the corresponding Gauss components.

Using this program, the Williamson-Hall plot (WH plot) i.e. $\beta^* = \dfrac{\beta\cos\theta}{\lambda}$ vs $d^* = \dfrac{2d\sin\theta}{\lambda}$, where $d$ is the interplanar spacing, $\lambda$ is the wavelength of the X-ray, and the SS plot i.e. $\left(\dfrac{\beta^*}{d^*}\right)^2$ vs $\dfrac{\beta^*}{(d^*)^2}$ [28] can be obtained. Moreover, profile parameters are used to calculate the Fourier coefficient ($A_L$) of each modeled peak corrected for instrumental broadening. $A_L$ is the product of a size coefficient ($A_s(L)$) and distortion coefficient ($A_D(L)$) can be represented as [16,30]

$$A_L = A_s(L) \times A_D(L)$$ (19)

The WA plot i.e. $\ln(A_L)$ vs. $(d^*)^2$ is plotted to separate size and strain component. WA analysis can be performed on all the reflections assuming isotropy or by selecting specific (hkl) reflection.

## 3.6. Evaluation of Dislocation Density in a Material

A relationship between the average domain size $D_{av}$ and the microstrain quantity $\langle\varepsilon_L^2\rangle^{\frac{1}{2}}$ obtained from the line profile analysis can now be established to evaluate one more important parameter namely, the dislocation density $(\rho)$. Considering the dislocation arrangements to be embedded in a mosaic structure, Williamson and Smallman (31) have derived the following expression for $\rho$.

Assuming absence of extensive polygonisation or dislocation pile-ups, the following expression for dislocation density $\rho$ can be written as:

$$\rho = \left(\rho_D \rho_S\right)^{\frac{1}{2}}$$

where

$$\rho_D \text{ (due to domain size broadening)} = \frac{3}{D_{av}^{2}} \tag{20}$$

$$\text{and } \rho_S \text{ (due to strain broadening)} = \frac{K\left\langle \varepsilon_L^2 \right\rangle}{\vec{b}^{2}} \tag{21}$$

Here, $\vec{b}$ is the Burgers vector for the slip plane of the system and $K$ is a constant given by

$$K = \frac{6\pi EA}{\mu} \ln\left(\frac{r}{r_0}\right) \tag{22}$$

where $\mu$ is the shear modulus, $E$ is the Young's modulus, $A = \frac{\pi}{2}$ for Gaussian and $A \approx 2$ for Cauchy strain distributions respectively. The function $\ln\dfrac{r}{r_0}$ is a function of r with a slow variation in nature, r being radius of the crystal containing dislocation and $r_0$ an integration limit giving dislocation core radius.

Thus, an estimation of $D_{av}$ and $\left\langle \varepsilon_L^2 \right\rangle^{\frac{1}{2}}$ from the line profile analysis enables us to determine $\rho$ from the above equations.

## 4. RESULTS AND DISCUSSION

### 4.1. Analysis of Deformed Powdered Sample

Typical diffraction profiles for zircaloy-2 in cold-worked and annealed state are shown in Figure 1 which demonstrates only hexagonal $\alpha$ phase [32]. The $\beta$-phase (in Zr-2.5% Nb and Zr-1%Nb-1%Sn-0.1%Fe) could not be traced because of its small concentration undetectable by X-ray diffraction. The lattice parameters a and c of these alloys using Cohen's least – squares method [33,34] have been listed in Table 1 [35]. The fault-unaffected reflections considered for these alloys for the determination of average domain size $D_{av}$ and the microstrain $\left\langle \varepsilon_L^2 \right\rangle^{\frac{1}{2}}$ at $L$= 50Å are shown in Table-1.

The effective domain size $D_e$ (Å) obtained from fault-affected reflections for these alloys have been calculated and are shown in Table-2 [35] along with the compound fault parameters $(3\alpha+\beta)$ when $L_0$ is odd and $(3\alpha+3\beta)$ when $L_0$ is even. The deformation fault $\alpha$ and the growth fault $\beta$ calculated from the compound fault values for various reflections for all the alloys are shown in Table-3 [35]. Considering the fact that growth fault is absent in hexagonal alloy systems [16], as there is no peak shift or peak asymmetry due to growth fault in hcp structures, the deformation fault $\alpha$ was recalculated taking $\beta=0$, and has been included in Table-3 [35].

The difference in observations that has been obtained in these alloy systems compared to pure zirconium was the low average domain size $D_{av}$ (Å) from fault-unaffected reflections. Pure zirconium [36] showed average domain size $D_{av}$ as 280Å, which was nearly six times larger than the present values (Table-1) [35]. This may be due to the refinement of the domains within the matrix due to the formation of various intermetallics and/or second phase (β-phase) precipitates. Previous investigations [37] have shown that Zr-2.5%Nb contains two phases α-Zr and β-Zr whereas Zr-1%Nb contains predominantly α-Zr with little β-Zr [38]. The increase in Nb from 1 to 2.5 % clearly revealed that the grain growth was restricted and average domain size ($D_{av}$) became finer (Table-1) due to the possible presence of higher volume fraction of β phase in the matrix of Zr-2.5%Nb than Zr-1%Nb as Nb is the β stabilizer [39].

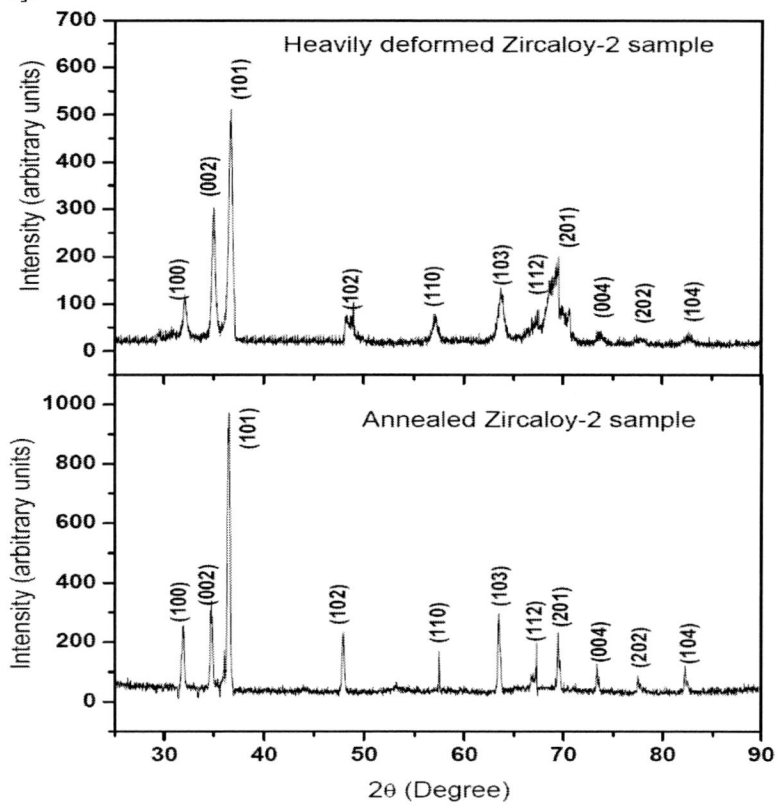

Figure1. Diffraction profiles for Zircaloy-2 in cold-worked and annealed state [32].

**Table 1. Values of lattice parameters (a and c), average domain size ($D_{av}$) from fault-unaffected reflections and microstrain $\langle \varepsilon_L^2 \rangle^{\frac{1}{2}}$ for zirconium base alloys.[35]**

| Alloy composition | Lattice Parameter a (Å) | Lattice Parameter c (Å) | Fault-unaffected Reflections $H\text{-}K$=3N | Average Domain size ($D_{av}$) (Å) | microstrain $\langle \varepsilon_L^2 \rangle^{\frac{1}{2}} \times 10^3$ at $L$=50Å |
|---|---|---|---|---|---|
| Zircaloy-2 | 3.235 | 5.157 | 100<br>002<br>110<br>112<br>004 | 46 | 7.4 |
| Zirlo | 3.220 | 5.137 | 100<br>002<br>110<br>112<br>004 | 68 | 9.0 |
| Zr-2.5%Nb | 3.233 | 5.151 | 100<br>002<br>110<br>112<br>004 | 36 | 5.0 |

**Table 2. Effective domain size ($D_e$) from fault affected reflections and compound fault parameter values [35]**

| Alloy | Fault-affected reflections $H\text{-}K$=3N±1 | Effective Domain size ($D_e$) (Å) | Compound fault parameters $3\alpha + \beta\,3\alpha + \beta$ $L_0$ = odd $L_0$ = even ($10^{-3}$) |
|---|---|---|---|
| Zircaloy-2 | 101<br>103<br>201<br>102<br>202 | 110<br>51<br>59<br>19<br>58 | -136.85<br>-12.89<br>-31.40<br>216.67<br>-34.06 |
| Zirlo | 101<br>103<br>203<br>102<br>104<br>202 | 67<br>104<br>36<br>48<br>84<br>92 | 2.38<br>-30.74<br>106.25<br>42.88<br>41.28<br>-15.83 |
| Zr-2.5%Nb | 101<br>103<br>201<br>102<br>202 | 65<br>45<br>64<br>44<br>56 | -133.84<br>-36.49<br>-79.58<br>-35.36<br>-107.08 |

**Table 3. Fault Parameter Values and Dislocation Densities
of the three Zr- based alloys [35]**

| Alloy | Fault densities $(10^{-3})$ | | Deformation fault density $\alpha(10^{-3})$    taking $\beta = 0$ | Dislocation density $\rho\,(m^{-2})$ $(10^{16})$ |
|---|---|---|---|---|
| | $\alpha$ | $\beta$ | | |
| Zircaloy-2 | -90.89 | -151.66 | 0.10 | 4.65 |
| Zirlo | 0.37 | -1.21 | 3.54 | 3.84 |
| Zr-2.5%Nb | -29.78 | 6.03 | -26.16 | 3.99 |

## 4.2. Analysis of Cold Rolled Sample

The microstructural parameters like the average domain size, domain size along the different crystallographic directions, microstrain within the domain, dislocation density and the stacking fault probabilities of the cold rolled Zircaloy-2 and Zr–2.5%Nb alloys at different deformations have been characterized by the Simplified Breadth Method using the Voigt function modeling and the Modified Rietveld Method using the whole powder pattern fitting technique. All these techniques are based on the analysis of the shapes of the broadened diffraction profiles. In the Simplified Breadth Method we have fitted the individual peak by a Voigt function and a Lorentzian function respectively to find out the volume weighted average domain size ( $D_v$ ) and the microstrain ( $\varepsilon$ ). In the Modified Rietveld Method, all the diffraction peaks were fitted simultaneously by a pseudo-Voigt (pV) function. The surface weighted average domain size ( $Ds$ ), the effective domain size ( $De$ ) along the different crystallographic directions and the average microstrain values $\langle \varepsilon_L^2 \rangle^{\frac{1}{2}}$ within the domain were then evaluated.

We have carried out the single peak analysis using Simplified Breadth Method for both Zircaloy-2 and Zr-2.5% Nb alloys at each deformation stage i.e. at 30%, 50%, 60% for Zr-2.5% Nb and 30%, 50% and 70% for Zircaloy-2 for all the reflections. The variations of $D_v$ and $\varepsilon$ with deformation for the individual peak have been shown in Figure 2 and Figure 3 respectively [40].

If the domains were isotropic, $D_v$ would have been identical along the different crystallographic directions. For Zircaloy-2 samples with 30% deformation, $D_v$ varies within a range of 123 Å to 206 Å in different directions, indicating that the domains are not spherically isotropic. The peaks of the reflections (002), (110), (101) could not be fitted as the shape factor $\phi$ of these peaks had fallen beyond the permitted values [23]. At 50% deformation, the range of $D_v$ lies between 135Å and 266Å. The domain size did not change much with deformation along certain crystallographic directions like <103>, <102>, <112>, <104> and <004>. At 70% deformation, the domain size, which remained almost invariant along <004>, <112> and <103> showed significant variations along <100>, <102> and <110>. This clearly indicates the elongation of the domains along <100> and <102> and contraction along <110> directions.

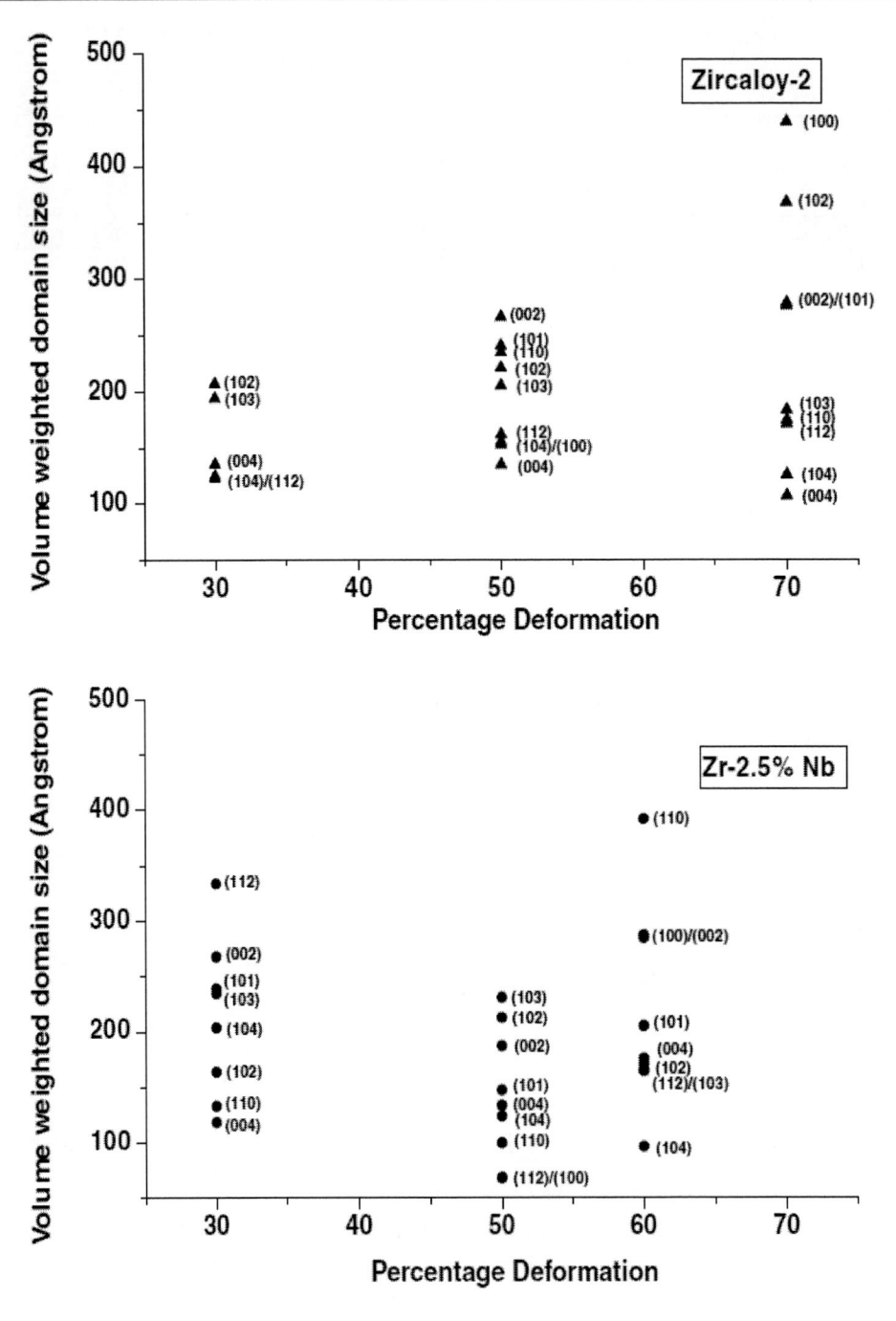

Figure2. Variation of volume weighted average domain size as a function of percentage deformation[40].

For Zr-2.5%Nb, the domains were found to be anisotropic at 30% deformation. At 50% deformation, $D_v$ is strongly reduced along <112>, but the size did not vary significantly in the other directions. At 70% deformation, $D_v$ is enhanced along <110> and <100> directions, indicating that the domains are preferentially elongated along certain crystallographic directions, similar to Zircaloy-2.

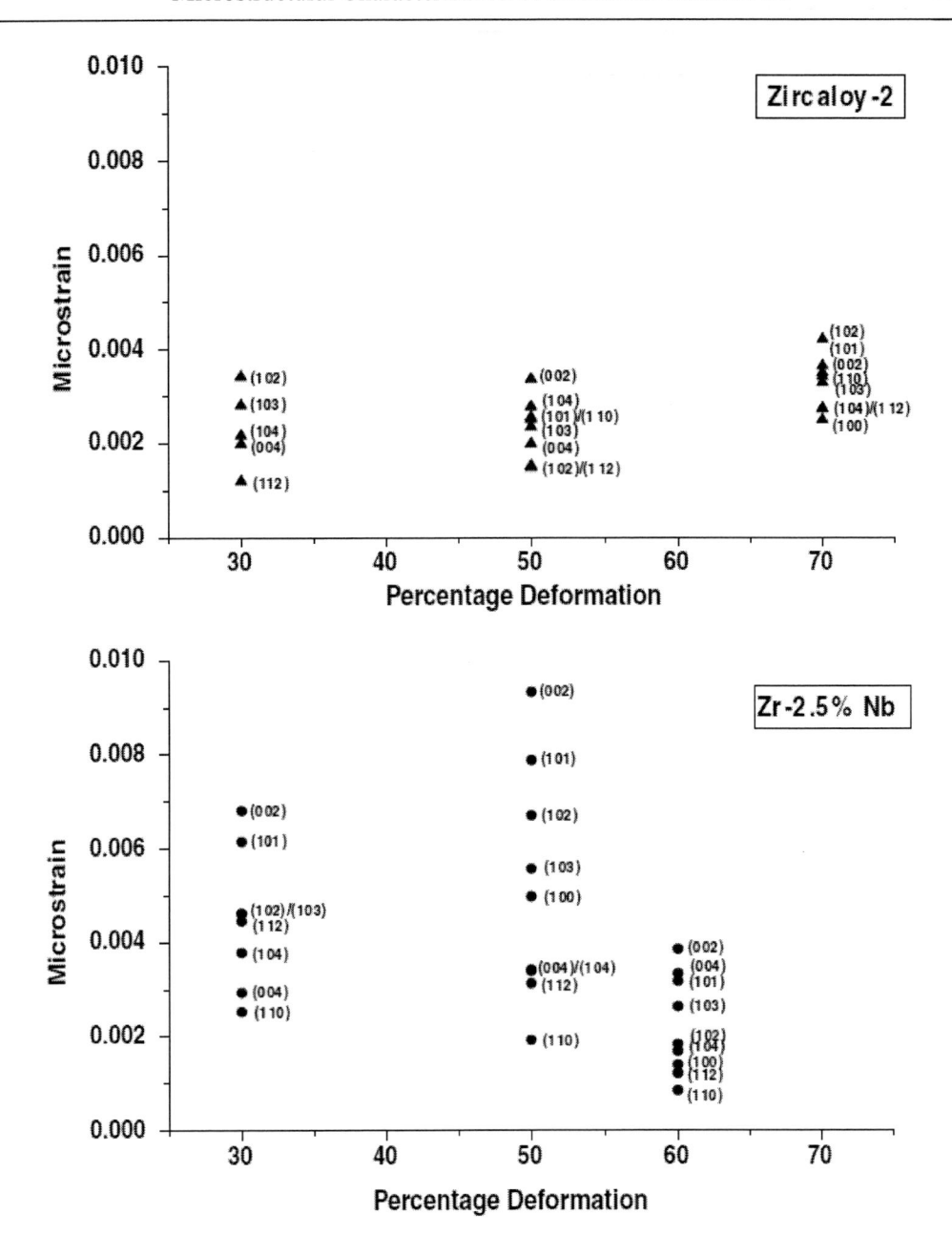

Figure3. Variation of microstrain as a function of percentage deformation [40].

We have carried out Modified Rietveld Method for analyzing microstructural parameters using program LS1 on XRD patterns of Zircaloy-2 and Zr-2.5%Nb. The surface weighted average domain size, $D_s$ was found to remain almost unchanged for Zircaloy-2, but it decreased with increasing deformation for Zr-2.5% Nb (Table 4) [40]. The refinement in each case was found to be satisfactory based on the values of the weighed pattern index $R_{wp}$, $R_{exp}$ and Goodness Of Fit (GOF). The values of $R_{wp}(\%)$, $R_{exp}(\%)$ and GOF have been incorporated in Table 4.

**Table 4. Surface weighted average domain size $\left(D_s\right)$, average microstrain $\left\langle\varepsilon_L^2\right\rangle^{\frac{1}{2}}$ and average dislocation density ($\rho$) with deformation [40]**

| Sample | Percentage cold-deformation | Surface weighted average Domain Size $\left(D_s\right)$ (Å) | Average Microstrain $(10^{-3})$ at $L$=50Å $\left\langle\varepsilon_L^2\right\rangle^{\frac{1}{2}}$ | Average Dislocation Density $\rho$ ($m^{-2}$) | $R_{wp}$ (%) | $R_{exp}$ (%) | GOF |
|---|---|---|---|---|---|---|---|
| Zircaloy-2 | 30 | 204 | 3.22 | 3.73 x $10^{15}$ | 3.27 | 2.48 | 1.31 |
|  | 50 | 196 | 3.03 | 3.65 x $10^{15}$ | 3.44 | 2.60 | 1.32 |
|  | 70 | 185 | 3.38 | 4.32 x $10^{15}$ | 2.90 | 1.89 | 1.53 |
| Zr-2.5% Nb | 30 | 182 | 2.94 | 3.82 x $10^{15}$ | 4.09 | 2.92 | 1.39 |
|  | 50 | 120 | 4.39 | 8.65 x $10^{15}$ | 2.73 | 1.98 | 1.38 |
|  | 60 | 98 | 5.58 | 13.4 x $10^{15}$ | 3.11 | 2.18 | 1.42 |

## 4.3. Analysis of Irradiated Samples

Figure 4 shows the damage profiles of 11 MeV proton in Zircaloy-2, 15 MeV proton and 116 MeV $O^{5+}$ beam in Zr-1Sn-1Nb-0.1Fe obtained using the program SRIM 2000 [41,42]. The range of 11 MeV and 15 MeV proton in Zircaloy-2 and Zr-1Sn-1Nb-0.1Fe was found to be approximately 440 μm and 742 μm respectively. In both the cases the range of the projectile is more than the thickness (400 μm) of the samples irradiated. Hence, the proton completely penetrated the sample and only a fraction of the elastic energy of the proton beam was deposited on it. Thus, the damage produced in the samples was a bulk phenomenon due to complete penetration of the proton beam.

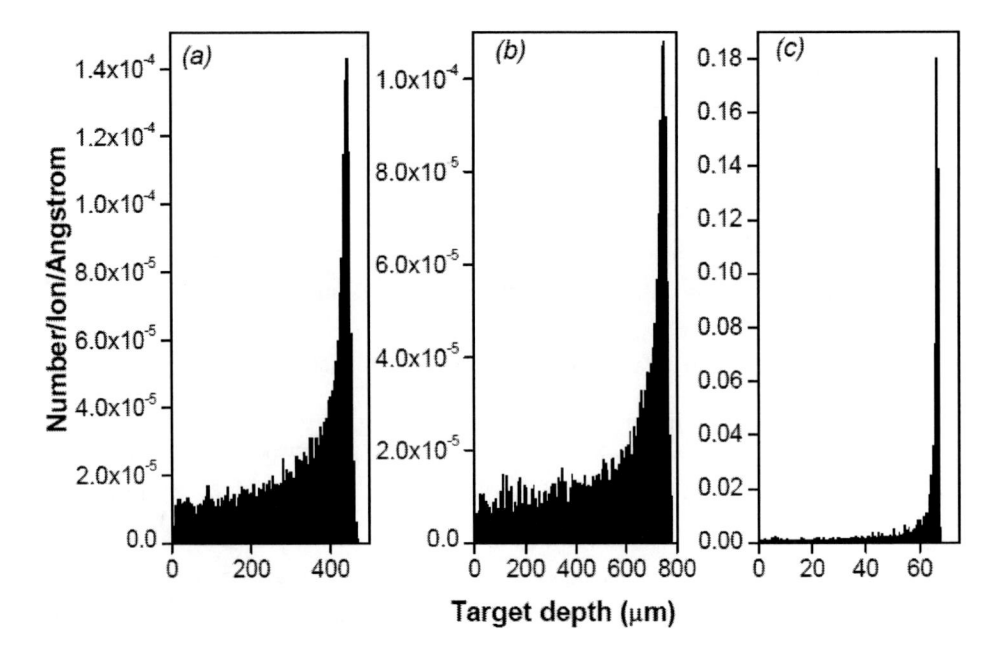

Figure 4. Damage profile of (a) 11 MeV proton in Zircaloy-2, (b) 15 MeV proton in Zr-1Nb-1Sn-0.1Fe, and (c) 116 MeV $O^{5+}$ in Zr-1Nb-1Sn-Fe. [42].

From the damage profile of 11 MeV and 15 MeV proton in zirconium based alloys, it is seen that the Bragg peak of the target displacements falls beyond the thickness of the samples, which means that only a fraction of the total elastic energy is being utilized for causing displacement of atoms and as a result isolated point defects are created in the irradiated samples of zirconium based alloys. These isolated point defects do not cause broadening of the peaks or change in the shape of the profile but contribute to the background values close to the diffraction peaks due to diffuse scattering (Huang scattering). As a result the domain size, microstrain and dislocation density (characterized by Warren Averbach analysis and MarqX technique) as observed in the unirradiated sample remain almost unaltered even with increasing dose of irradiation in case of proton irradiated samples.

Table 5a and 5b shows the variation of domain size and microstrain as a function of dose in Zr-1%Nb-1%Sn.1Fe. [42]

**Table 5a. Results obtained from Size-Strain (SS) plot of the XRD profiles of Zr–1Nb–1Sn–0.1Fe irradiated by 15 MeV proton [42]**

| Dose (Protons/m$^2$) | $D_v$ (Å) | Microstrain ($10^{-4}$) |
|---|---|---|
| Unirradiated | $1547 \pm 63$ | $7.2 \pm 0.5$ |
| $1 \times 10^{20}$ | $1449 \pm 86$ | $7.6 \pm 0.4$ |
| $5 \times 10^{20}$ | $1532 \pm 76$ | $7.0 \pm 0.4$ |
| $5 \times 10^{21}$ | $1517 \pm 79$ | $7.9 \pm 0.3$ |

**Table 5b. Results of Warren-Averbach (WA) analysis of the XRD profiles of Zr–1Nb–1Sn–0.1Fe irradiated by 15 MeV proton [42]**

| Dose (Protons/m$^2$) | $D_s$ (Å) | Microstrain ($10^{-4}$) |
|---|---|---|
| Unirradiated | $971 \pm 52$ | $8.6 \pm 0.3$ |
| $1 \times 10^{20}$ | $923 \pm 49$ | $6.9 \pm 0.3$ |
| $5 \times 10^{20}$ | $986 \pm 56$ | $7.3 \pm 0.4$ |
| $5 \times 10^{21}$ | $1012 \pm 53$ | $7.7 \pm 0.3$ |

On the contrary, the range of 116 MeV $O^{5+}$ beam in Zr-1Sn-1Nb-0.1Fe is around 67 μm as calculated by SRIM 2000 [44]. Oxygen being a heavy ion impart so much energy to the primary knock on atoms that a displacement cascade is produced consisting of highly localized production of interstitials and vacancies associated with a single initiated event. In contrast, the reaction pathways of proton beam through the sample of thickness 400 μm produces fairly uniform radiation damage. Moreover, the damage energy deposition with distance i.e. dE/dx is found to be almost two orders less than that of oxygen beam as shown in Figure 5 [42,43]. Thus, during the travel of oxygen ion the total elastic energy deposited within the material of 67 μm is much larger than that of proton. Besides, as the primary recoil proceeds through the sample, loosing energy in successive collisions, the displacement cross-section increases [45]. Thus the distance between successive displacements decreases and at the end of its track, the recoil collides with practically every atom in its path, creating a very high local concentration of vacancies and interstitials. The microstructural parameters of the oxygen-irradiated samples at doses $1 \times 10^{17}$, $5 \times 10^{17}$, $1 \times 10^{18}$ and $5 \times 10^{18}$ $O^{5+}$ions/m$^2$ have been characterized by XRDLPA using modified Rietveld technique. The best fit values of the domain size and micro-strain for unirradiated and oxygen-irradiated samples have been plotted against the dose and are shown in Table 6 [43,44]

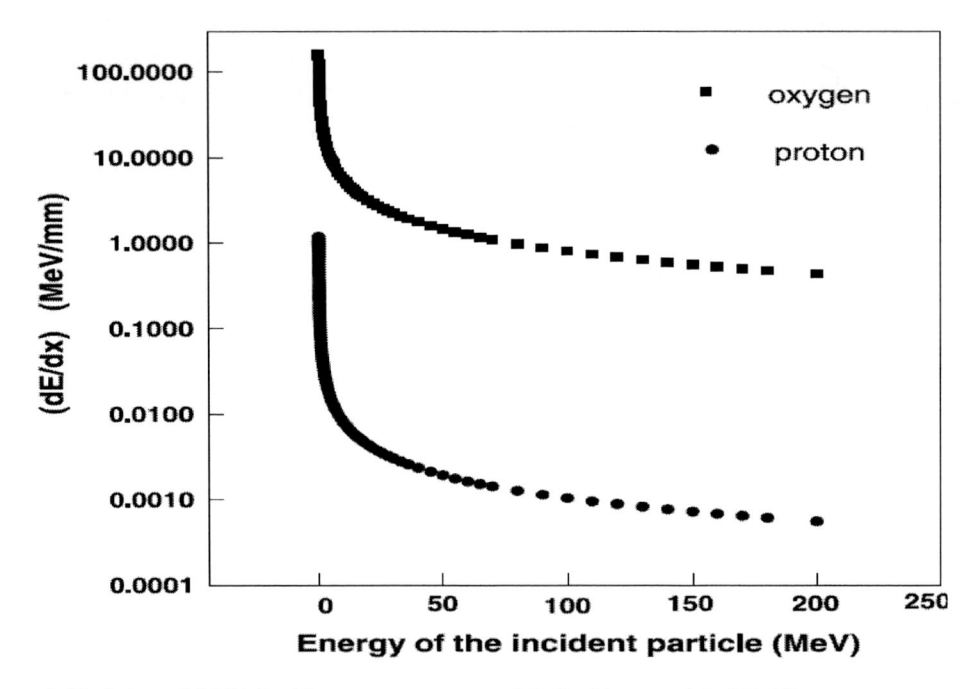

Figure 5. Variation of (dE/dx) with respect to energy of the incident particle [42,43].

**Table 6. Variation of average domain size ($D_{av}$), microstrain ($\langle\varepsilon_L^2\rangle^{\frac{1}{2}}$) and dislocation density ($\rho$) as a function of dose for oxygen-irradiated Zr-1%Nb-1%Sn-0.1Fe [43,44]**

| Dose ($O^{5+}/m^2$) | Average Domain Size ($D_{av}$) (Å) | Microstrain $\langle\varepsilon_L^2\rangle^{\frac{1}{2}}$ ($10^{-3}$) | Dislocation Density ($\rho$) ($m^{-2}$) |
|---|---|---|---|
| 0 (unirradiated) | 1070 | 0.10 | $1.5\times10^{14}$ |
| $1\times10^{17}$ | 590 | 1.23 | $4.9\times10^{14}$ |
| $5\times10^{17}$ | 372 | 1.33 | $9.1\times10^{14}$ |
| $1\times10^{18}$ | 340 | 1.70 | $1.18\times10^{15}$ |
| $5\times10^{18}$ | 310 | 2.34 | $1.8\times10^{15}$ |

It is seen from Table 6 that there was a drastic decrease in domain size from unirradiated samples to the samples at a dose of $5\times10^{17}$ $O^{5+}/m^2$ but these values saturated with increasing dose of irradiation. The values of microstrain were found to increase with dose. The dislocation density increased almost by an order of magnitude for the samples irradiated with $1\times10^{18}$ $O^{5+}/m^2$ and $5\times10^{18}$ $O^{5+}/m^2$ as compared to the unirradiated samples. These values were also found to saturate with dose (Figure 6) [43]. The dislocation density was found to be of the order of $\sim10^{15}$ ($m^{-2}$) for the samples irradiated with $1\times10^{18}$ and $5\times10^{18}$ $O^{5+}$ ions/$m^2$.

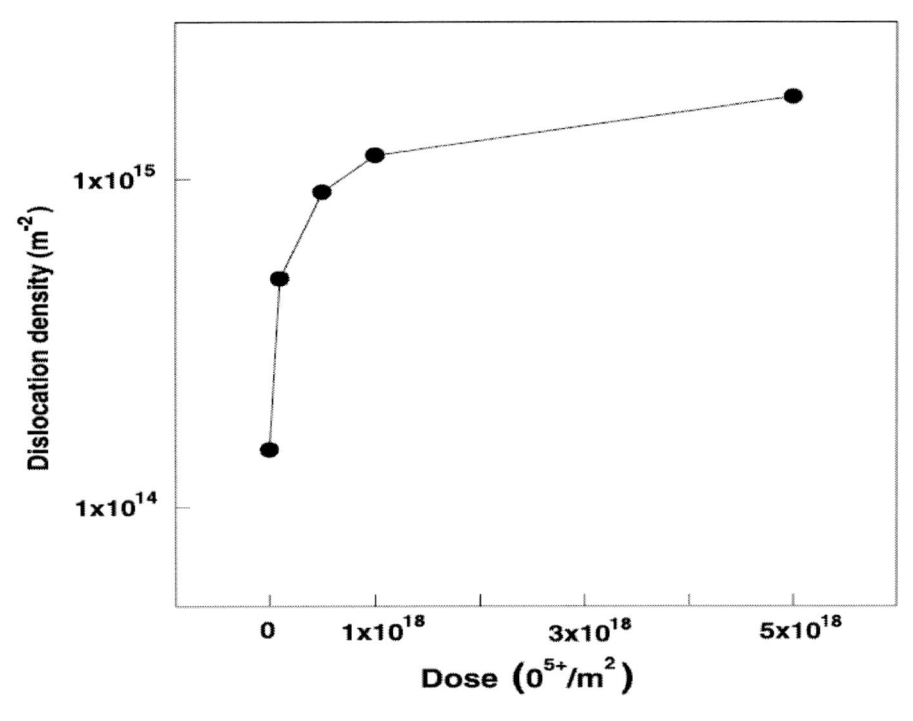

Figure 6. Variation of dislocation density with irradiated dose of $O^{5+}$ particles [43].

For $O^{5+}$ ion irradiation in zirconium based alloys, the damage is maximum within a distance of 2-3 μm at the end of the reaction path. A concentration gradient of defects in the sample was thus created within a small reaction path of 67 μm, which helped in the migration of defects. Moreover, the diffusion coefficient $D_a$ of a particular lattice atom is enhanced due to irradiation [13] and is given by the following equation:

$$D_a = f_v D_v Cv + f_{2v} D_{2v} C_{2v} + f_i D_i C_i + \ldots..$$

where v, 2v and i stand for vacancy, di-vacancy and interstitial respectively. D and C values are the corresponding diffusion coefficients and concentrations, f values are the correlation factors. The migration of defects by radiation-enhanced diffusion resulted agglomeration into defect clusters which collapsed in the form of dislocation loops. As a result, the order of dislocation density increases almost by an order with increase in dose. But, at higher doses, though more vacancies are created, the annihilation rate of vacancies also increases with increasing sink density. Hence, saturation was observed in the density of dislocations with the increase in the dose of irradiation (as seen in Figure6) with decrease in domain size (Figure 7) [43] in zirconium based alloys.

In the $Ne^{6+}$ irradiated samples of Zr-1%Sn-1%Nb-0.1Fe, the effective domain size $D_e$, along different crystallographic directions were also found to decrease with dose as compared to the unirradiated material but the shape of the domains were almost isotropic. We have plotted the projections of $D_e$ (along different directions) on the plane containing the directions <002> and <100>. Only the projections in the first quadrant are shown in (Figure 8) for unirradiated and irradiated Zr–1.0%Nb–1.0%Sn–0.1%Fe [46].

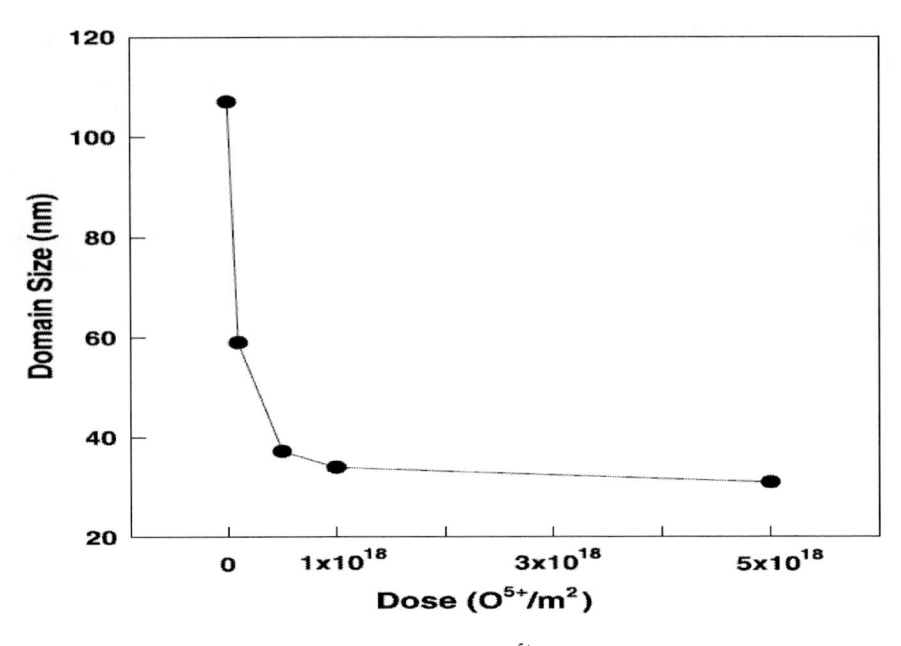

Figure 7. Variation of domain size with irradiated dose of $O^{5+}$ particles [43].

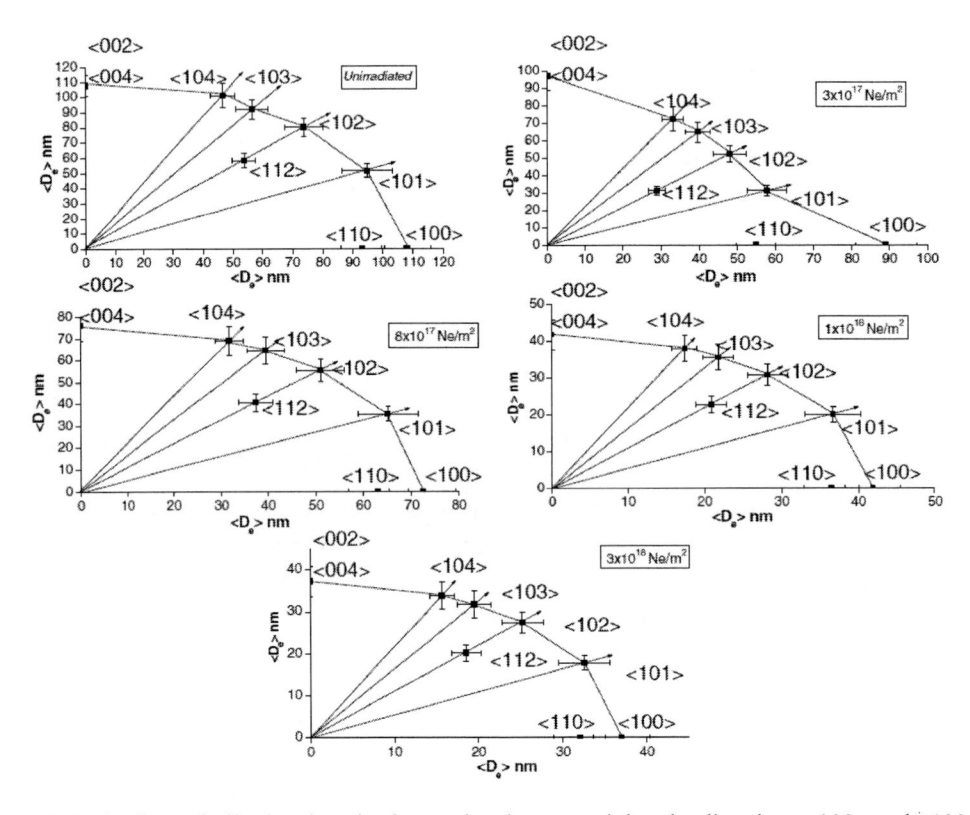

Figure 8. Projections of effective domain size on the plane containing the directions <002> and <100> (First quadrant) for unirradiated and $Ne^{6+}$ irradiated Zr–1.0%Nb–1.0%Sn–0.1%Fe at different doses. [46].

It was clearly observed that $D_e$ was almost isotropic (spherical) with values $<D_e>_{002} \sim$ 1077 Å and $<D_e>_{100} \sim 1070$ Å for the unirradiated sample, $<D_e>_{002} \sim 969$Å and $<D_e>_{100} \sim 891$ Å at a dose of $3 \times 10^{17} Ne^{6+}/m^2$, $<D_e>_{002} \sim 760$ Å and $<D_e>_{100} \sim 724$ Å at a dose of $8 \times 10^{17}$ $Ne^{6+}/m^2$, $<D_e>_{002} \sim 417$ Å and $< D_e >_{100} \sim 419$ Å at a dose of $1 \times 10^{18} Ne^{6+}/m^2$ and $< D_e >_{002} \sim$ 372 Å and $< D_e >_{100} \sim 370$ Å at a dose of $3 \times 10^{18} Ne^{6+}/m^2$. These values clearly signify that the shape of the domains did not change with dose though the variations in the size of the domains were significant with the increasing dose as compared to the unirradiated sample.

In order to understand the mechanism of evolution of defects in the transient regime, Zr-1Nb samples were irradiated at room temperature with 116 MeV $O^{5+}$ at low doses and at low dose rate. The microstructural parameters of the irradiated samples have been characterized by X-Ray Diffraction Line Profile Analysis (XRDLPA). Zr-1Nb samples were irradiated with 116 MeV $O^{5+}$ ions at different doses ranging from $5 \times 10^{17}$ to $8 \times 10^{18} O^{5+}/m^2$. X-ray Diffraction Line Profile Analysis was performed to characterize the microstructural parameters of these samples. Average domain size, microstrain and dislocation density were estimated as a function of dose and are shown in Table 7. Anomaly was observed in the value of these parameters at a specific dose of $2 \times 10^{18} O^{5+}/m^2$ which proves that the irradiation induced microstructure in this alloy is very much dose dependent in the low dose regime. This anomaly could be accounted for the suppression of vacancy agglomeration by forming solute atom-vacancy complex stages of annealing, which gets dissociated at higher annealing temperature.

**Table 7. Average domain size, microstrain and dislocation density for 116MeV $O^{5+}$ irradiated Zr-1Nb sample as a function of dose**

| Dose ($O^{5+}/m^2$) | Average Domain size ($D_{av}$) (Å) | Average Microstrain ($10^{-3}$) | Average Dislocation density $\rho$ ($m^{-2}$) |
|---|---|---|---|
| 0 (unirradiated) | 1070 | 0.6 | $1 \times 10^{14}$ |
| $5 \times 10^{17}$ | 267 | 1.85 | $4.90 \times 10^{15}$ |
| $1 \times 10^{18}$ | 250 | 1.91 | $5.44 \times 10^{15}$ |
| $2 \times 10^{18}$ | 454 | 1.57 | $2.44 \times 10^{15}$ |
| $8 \times 10^{18}$ | 282 | 1.90 | $4.77 \times 10^{15}$ |

# 5. CONCLUSION

Different model based approaches of X-ray Diffraction Line Profile Analysis have been used to characterize the microstructure of the deformed and irradiated zirconium based alloys. These techniques are suitable as the analysis is much easier, reliable and quick and can be easily adopted in the shop floor practice where characterization of the average microstructural parameters is mandatory. These techniques help to achieve the information of the microstructural parameters like cell size or domain size, dislocation density, microstrain within the domains, averaged over a volume of $10^9 \mu m^3$. In characterizing the microstructure of the material, these methods are complimentary to each other. Though the parameters obtained by different techniques are differently defined and thus not necessarily comparable, the values of domain size and microstrain show similar trend obtained from different techniques.

# REFERENCES

[1] N.R.McDonald, *J.Aust. Inst. Metals,* (1971), Volume 16, 179.

[2] G.J.C.Varpenter, *Can. Metall. Quart.,* (1985), Volume 24, 251

[3] R.A. Holt, *J. Nucl. Mater.* (1976), Volume 59, 234.

[4] G.J.C. Carpenter, D.O. Northwood, *J Nucl. Mater.,* (1975), Volume 56, 260.

[5] G.J.C. Carpenters, J.F. Watters, *Acta Met.,* (1973), Volume 21,1207.

[6] W.K.Alexander, V.Fidleris, R.A.Holt, *ASTM STP* (1977), Volume 633, 344

[7] R.A.Holt, *J. Nucl. Mater.,* (1980), Volume 90, 193.

[8] S.K. Chatterjee, S.K. Halder, S.P. Sengupta, *J. Appl. Phys.* (1976),Volume 47,411.

[9] S.K. Chatterjee, S.P. Sengupta, *J. Mater. Sci.* (1974), Volume 9, 953.

[10] R. Sen, S.K. Chattopadhyay, S.K. Chatterjee, *Jpn. J. Appl. Phys.* (1997) Volume 36, 364.

[11] J. I. Langford in Defects and microstructure analysis by diffraction ( Ed. R. L. Snyder, J. Fiala and H. J. Bunge) Oxford University press, (1999) 59.

[12] B. E. Warren. in X-ray diffraction. Reading (MA): Addision-Wesley; 1969, p. 264.

[13] G.K. Williamson, W.H. Hall, *Acta Metall.* (1953) Volume 1, 22 .

[14] L. Lutterotti, P. Scardi, *J. Appl. Crystallogr.* (1990) Volume 23, 246

[15] D. Balzar, H. Ledbetter, *J. Appl. Cryst.* (1993) Volume 26, 97

[16] B.E. Warren, B.L. Averbach, *J. Appl. Phys.* (1950), Volume 21, 595.

[17] P. Scherrer, *Nachr Gott* (1918) Volume 2, 98.

[18] V.N. Selivanov, E.F. Smislov, *Zavod Lab.* (1991), Volume 57, 28.

[19] G.K. Williamson, W.H. Hall. *Acta Metall.* (1953), Volume 1, 22.

[20] March, Z. Kristallogr (1932), Volume 81, 285.

[21] W.A. Dollase, *J. Appl. Crystallogr.* (1986) Volume 19, 267.

[22] G. Will, M. Bellotto, W. Parrish, M. Hatr, *J. Appl. Crystallogr.* (1988), Volume 21, 182.

[23] Th H de Keijser, J.I, Langford, E. J. Mittemeijer, A.B.P.Vogels, *J. Appl. Cryst.* (1982), Volume 15, 308.

[24] J. I. Langford, *J. Appl. Cryst.,* (1978) Volume 11, 10.

[25] D. Balzar and H. Ledbetter, *J. Appl. Crystallogr.* (1993), Volume 26, 97

[26] V.N. Selivanov and E.F. Smislov, *Zavod Lab* (1991), Volume 57, 28.

[27] Y. H. Dong and P. Scardi: *J. Appl. Cryst.,* (2000), Volume 33, 184.

[28] J. I. Langford: Accuracy in Powder Diffraction II. NSIT special publication 846, edited by E. Prince and J. K. Stalick, pp. 110-126. Wasihington: USG Printing Office, 1992.

[29] M. Ahtee, L. Unonius, M. Nurmela and P. Suortti: *J. Appl. Cryst.,* (1984), Volume 17, 352.

[30] B. E. Warren, B. L. Averbach: *J. Appl. Phys.,* (1952), Volume 23, 497.

[31] G. K. Williamson, R. E. Smallman, (1956), *Phil. Mag.,* Volume 1, 34.

[32] P.Mukherjee, Study of Microstructural Imperfections Using X-Ray Diffraction Line Profile Analysis on Deformed and Irradiated Zirconium Base Alloys, Thesis, Jadavpur University (2003).

[33] R. Sen , Chattopadhyay S. K., Chatterjee S. K., *Jpn. J. Appl. Phys.,* (1997), Volume 36, 364.

[34] R. Sen, Chattopadhyay S. K., Chatterjee S. K., *Met. Trans. A.,* (1998), Volume 29, 2639.

[35] P. Mukherjee, S.K. Chattopadhyay, S.K. Chatterjee, A.K. Meikap, P. Barat, S.K. Bandyopadhyay, Pintu Sen, and M.K. Mitra, *Met. Trans. A.,* (2000), Volume 31A, 2405.

[36] S.K.Chatterjee, S.P. Sen Gupta, *J. Mater. Sc.,* (1974),Volume 9, 953.

[37] S,A. Aldridge, B.A.Cheadle, *J. Nucl. Mater.,* (1972), Volume 42, 32.

[38] W.R.Thorpe, I.O.Smith, *J. Nucl. Mater.,* (1978), Volume 75, 209.

[39] X.Meng, D.O., Northwood, *ASTM STP,* (1989), Volume 1023, 478.

[40] P. Mukherjee, A. Sarkar, P. Barat, S.K. Bandyopadhyay, Pintu Sen, S.K. Chattopadhyay, P. Chatterjee, S.K. Chatterjee, M.K. Mitra, *Acta Materialia* (2004), Volume 52, 5687.

[41] J.P. Biersack, L.G. Haggmark. Nucl. Instrum Methods (1980) Volume174, 257 [The Stopping and Range of Ions in Matter (SRIM 2000) software developed by J.Ziegler and J.P.Biersack is available on the Website http://www.srim.org/].

[42] A. Sarkar, P. Mukherjee and P. Barat, *Met. Trans. A.,* (2008), Volume 39, Number 7, 1602.

[43] P. Mukherjee, P. Barat, S.K. Bandyopadhyay, P. Sen, S.K. Chattopadhyay, S.K. Chatterjee, M.K. Mitra, *J. Nucl. Mater.* 305 (2002) 169–174.

[44] P. Mukherjee , A. Sarkar, P. Barat, *Materials Characterization*, (2005) Volume 55, 412.

[45] R.W.Weeks et al. *J. Nucl. Mater.* (1970), Volume 36, 223.

[46] A. Sarkar, P. Mukherjee, P. Barat, *Journal of Nuclear Materials* 372 (2008) 285–292.

In: Nuclear Materials
Editor: Michael P. Hemsworth, pp. 157-176

ISBN 978-1-61324-010-6
© 2011 Nova Science Publishers, Inc.

*Chapter 6*

# CURRENT TRENDS IN MATHEMATICAL MODELING AND SIMULATION OF FISSION PRODUCT TRANSPORT FROM FUEL TO PRIMARY COOLANT OF PWRs

*Nasir M. Mirza and Sikander M. Mirza*
Department of Physics and Applied Mathematics,
Pakistan Institute of Engineering and Applied Sciences,
Nilore, Islamabad, Pakistan

## ABSTRACT

The primary coolant activity has been of great concern for many decades mainly due to the persisting high post-shutdown radiation levels resulting in prolongation of reactor down time entailing substantial economic repercussions. Along with the activation products, the fission products released from defective fuel pins contribute towards these radiation fields. Both static and kinetic models have been prepared for the estimation of fission product activity in the primary coolant in steady-state as well as for power and flow rate transients. These models are based on two-stage processes: fuel-to-gap, and gap-to-coolant releases. Here, we present a review of multi-step models based on mass transport equations and models with details on the fission product activity transport to coolant using the FPCART code. In these simulations, the primary circuit has been treated using one-dimensional nodal scheme. The resulting set of governing kinetics equations have been implemented in the FPCART code which uses the LEOPARD and the ODMUG codes as subroutines.

For steady-state operation, the predictions of the FPCART code have been found in good agreement with the results obtained using the ODIGEN-2.0 code with maximum difference in the corresponding values below 0.36% for over 39 dominant fission products studied. Also, the FPCART predictions agree within 2.4% with the corresponding data from the ANS-18.1 standard as well as with the experimentally measured power plant data.

According to FPCART simulations, the fission product activity in primary coolant has strong dependence on flow rate. In the case of slow flow rate coast-downs ($\tau = 2000$ h), up to 8.6% increase in the fission product total specific activity has been observed. The power transients result in substantial changes in the fission product activity. The

FPCART predicted values of [131]I activity have been found in good agreement with the corresponding measured data from the Beznau and the Surry PWRs.

Due to inherent randomness of the fuel failure process, the stochastic simulation technique has also been found suitable for predicting the fission product activity in primary coolants of PWRs. For this purpose, three-step based model has been developed and implemented in the FPCART-ST code which simulates the burst release of fission products from random fuel failure. The predicted values of specific activities using stochastic simulations have been found in good agreement with the corresponding measured data of the ANGRA-I nuclear power plant as well as that of EDITHMOX-1 experiments.

## GLOSSARY

| Symbol | Stands for |
|---|---|
| $\phi$ | Neutron flux ($\#/cm^2.s$) |
| $N_{X,i}$ | Number of i[th] radionuclide atoms in the region: Fuel (X=F), Gap (X=G) and Coolant (X=C) |
| $\lambda_i$ | Decay constant of the i[th] radionuclide ($s^{-1}$) |
| $F$ | Average fission rate (fissions/W.s) |
| $P$ | Thermal power of reactor (W) |
| $D$ | Failed fuel fraction (#) |
| $\tau$ | Core-to-circuit primary coolant resident time ratio (#) |
| $f_{ij}$ | Branching ratio ($i \rightarrow j$) in the i[th] chain |
| $\beta$ | Bleed-out fraction of the primary coolant for boron chemical control |
| $L$ | The coolant leakage rate (g/s) |
| $W$ | The total primary coolant mass (g) |
| $\eta_i$ | Resin purification efficiency for the i[th] radionuclide |
| $Q$ | Let-down flow rate (g/s) |
| $\nu_i$ | Escape rate coefficient of the i[th] radionuclide |
| $\sigma_i$ | Microscopic absorption cross section for the i[th] radionuclide |
| $Y_i$ | Fission yield of the i[th] radionuclide |
| $D_F$ | Number of failed fuel rods |
| $\xi$ | Characteristic decay constant for the escape rate ($s^{-1}$) |

## 1. INTRODUCTION

According to recent statistics by the IAEA [1] there are 437 nuclear power plant currently generating a total of 370705 MW(e) worldwide. Out of these, 265 are pressurized water reactors (PWRs); while the boiling water reactors (BWRs), pressurized heavy water cooled and moderated reactors (PHWRs), gas cooled graphite moderated reactors (GCRs), light water cooled graphite moderated reactors (LWGRs) and fast breeder reactors (FBRs) have current operational units 92, 45, 18, 15, 2 respectively. The PWRs contribute about 66% of the total power generated. The number of under-construction PWRs, BWRs, PHWRs, LWGRs and FBRs is 46, 3, 3, 1, 2 respectively. The operational lifespan of nuclear power plant ranges up to 42 years with almost half of them are over 25 years old. The plant aging implies more frequent failures and consequently longer down times along with smaller values of plant availability factors.

The pressurized water reactors constitute dominant majority of both the operational as well under-construction units. They have one- to two-orders of magnitude higher steam generator dose rates in comparison with their competitors GCRs and FBRs. This results in prolongation of their maintenance schedule entailing loss of revenue amounting to several million dollars annually [2]. The severity of this problem is increased by plant aging. In case of PWRs, the high levels of dose rates stem from primary coolant activity which is essentially constituted by the corrosion, fission and the activation products [3-5]. Current trend of extending fuel cycle length along with development of high burn-up cores for existing as well as under-construction reactors has further aggravated this problem.

In past, both experimental as well theoretical efforts have been made for the determination of fission product activity (FPA) in primary coolant circuits. The value of FPA, being indicative measure of fuel failure has been extensively used for estimation of number of failed fuel (FF) pins. While gross estimates of the failed fuel fraction are possible by measurement of the dominant FPA, Zänker et al [6] have developed a complementary cesium coolant concentration ratio which provides information about the failed fuel fraction as well as the possible location of failed fuel pins. This technique is based on utilizing $^{134}Cs/^{136}Cs$ ratio in addition to the usual $^{134}Cs/^{137}Cs$ ratio. Radiochemical analysis of PWR primary coolant challenges traditional approaches especially when analysis of radioisotopes emitting low energy radiations is required in the presence of high background or matrix effects [7]. Koo et al [8] have developed first-order processes based analytical method for release of fission products from fuel to gap and to the coolant in the primary circuit. Their model incorporates the relevant physical processes including pallet oxidation and consequent enhancement of diffusivity, formation of gas bubbles at grain boundaries and estimation of their inter-linkage to the open surface stochastically, the fission product migration from fuel to gap and finally their leakage into the primary coolant stream. Their model shows good agreement with the corresponding experimental measurement in the intermediate linear heating regime only and in the remaining range, under-prediction has been reported. Avanov [9] proposed a model explaining the peculiarities of fission gas release from reactor fuel. A good comparison of the predicted values and the corresponding experimental measurements were reported.

A historical review of the research effort towards detection of failed fuel in LWRs has been presented by Kneider and Schneider [10]. They point-out that automated non-destructive methods for failed fuel detection are preferred industry-wide. Commonly gamma-ray radio-quantification of $^{133}Xe$ is carried out for this purpose, while blocking the $^{13}N$ interference using anti-coincidence techniques. Lately, ultrasound pulse echo technique employing matched filters has been developed which has acceptably small value of false alarm [11]. Clink and Freeburn have developed computer program IODYNE [12] which carries out estimation of failed fuel fraction using on-line coolant activity measurements. Such codes typically use combined failure models for low power and empirical failure models for high power failures. They are designed for steady-state, constant power regimes which tends to limit their usefulness in transient cases. Reportedly, under-prediction of failed fuel fraction has been observed even for steady-state operation [13]. Lewis [14] has developed an analytic model which distinguishes the release of fission products from failed fuel pins, from the ones from the fissions taking place from tramp uranium acting as surface contamination. Later, he developed this model further by using experimental data from placing purposely defective fuel elements in the reactor in order to gather data regarding fundamental physical processes involved [15]. The improved model extended the range of applicability to low temperature releases.

Tucker and White [16] developed a theoretical model for estimating release of unstable gaseous and volatile fission products from irradiated $UO_2$ fuel. This model assesses the probability of leakage of an unstable radioactive atom from interior of palettes out to the exterior before decaying. This probability was found strongly dependent on the extent of interconnectivity of grain-edge tunnel pores. Fission product release predictions based on this model were found compatible with the corresponding experimental measurements.

A detailed multi-step microstructure dependent model has been developed for estimating the release of fission products from $UO_2$ fuel [17]. This model includes diffusion, bubble and grain boundary movement, coalescing of bubbles etc., and the computed values of gap activities of $^{133}I$ and $^{135}I$ have been found in excellent agreement with the corresponding experimental measurements. The finite volume modeling of fission gas release from $UO_2$ fuel has been carried out by considering the gas diffusion inside a spherical grain under time-varying conditions [18]. The results show good agreement with the corresponding data computed using semi-analytical approximation.

Development of a new model for the chemical and volume control system (CVCS) and its implementation along with an existing primary coolant activity model has resulted in improved estimates of $^{129}I$ and $^{137}Cs$ activities. The removal efficiencies of CVCS resins for $^{129}I$ and $^{137}Cs$ have been found to influence the $^{129}I$ / $^{137}Cs$ ratio in a linear manner [19]. The coolant activity analysis program has been used for estimating the number of failed fuel rods, their degree of failure and their location in reactor core. This technique has been reported to outperform the conventional ultrasound based non-destructive technique.

## 2. EXPERIMENTAL EFFORTS

During past four decades, extensive experimental efforts have been made for the determination of fission product releases in normal as well as during severe accidents. For this purpose, both in-pile and out-of-pile tests were conducted. At the start, various fission product release studies were carried out using hot cells. During these tests, defective fuel rods were studied for fission product releases as models of leakages during normal operation of reactors. Later, specifically designed facilities were used for fission product release studies.

### 2.1. In-Pile Tests

In 1980, the Risø National Labs in Denmark started a series of experimental measurements of in-pile fission product released from irradiated fuel under power transients. Their investigation indicated the key role played by the bubbles at the grain boundaries for fission product releases from the fuel matrix. The other important contributors included mechanical restraint pressure, fuel geometry, initial pore structure of the ceramic fuel and the amount of fuel burnup [20].

The collaborative effort of the French utility (EDF) with the European Commission (EC), named as Phebus project, started in late 1980s with the aim to study the physical phenomena of fission product releases from degraded fuel. This involved transport of fission products from fuel matrix into the fuel-clad gap; their retention, buildup and leakage into the primary coolant circuit; deposition and role of water chemistry for their buildup; and finally, their leakage into the containment building [21].

Experimental tests were also performed in order to study the release of fission products from degraded fuel under severe accidents. The severe fuel damage (SFD) tests, at the Idaho National Engineering Laboratory [22], the ACRR reactor tests at the Sandia National Laboratory [23], the source term experiments project (STEP) at the Argonne national laboratory [24] and the loss-of-flow transients (LOFT-FP) [25] were essentially designed for the experimental determination of FP releases from the irradiated fuel. The experimental condition simulated severe accidents where reactor fuel remained at very high temperatures.

## 2.2. Out-of-Pile Tests

With the objective to study the release of fission products from overheated fuel rods and from the molten core in the case of severe accidents, a series of experimental tests were conducted at the SASCHA facility, Karlsruhe Nuclear Research Center, Germany during late 1970s and early 1980s [26]. Similar, out-of-core tests were conducted by the French CEA at Grenoble under the HEVA program between 1983—89 [27]. This was later extended as the VERCORS program which continued up to 1996 [28]. While the first part was more focused on release of fission products from degraded by mostly intact fuel, the later series extended it to study the releases from essentially molten fuel. The Japan Atomic Energy Research Institute (JAERI) initiated out-of-pile test program as Verification Experiment of Gas / Aerosol release (VEGA) in 1999 [29]. In these tests, the release studies were performs over much broader temperature range.

## 3. REVIEW OF FISSION PRODUCT ACTIVITY SIMULATION CODES

In past, a variety of computer programs have been developed for the estimation of fission product activity(FPA) in the PWRs. These include point-depletion codes which ignore spatial effects and treat the system as a homogeneous entity. These calculations are generally restricted to the core and no attempt is made on modeling or simulation of transport of fission products from the fuel matrix into the fuel-clad gap and subsequently to the coolant. The ORIGEN2 and the WIMS computer programs fall into this category. The ORIGEN2 [30] program performs point depletion and isotopic buildup of 950 nuclides with up to 120 actinides but lack spatial details. Lately, this deficiency has been partly taken care of in the MONTEBURNS [31] which links the ORIGEN2 code with the MCNP code and the spatial details are incorporated by the MCNP [32] Monte Carlo code.

The Winfrith Improved Multigroup Scheme (WIMS) [33] computer code has extensively been used for macroscopic group constant generation especially in the worldwide reduced enrichment for research and test reactors (RERTR) [34] effort. This program performs depletion and isotope buildup computations using one-dimensional transport theory employing DSN or Stochastic scheme. The active set of fission products is limited to 35 only while the remaining ones are lumped into a single hypothetical material. This code does not handle transport of fission products from fuel to gap, and out to the coolant regions. The CASMO4 [35] and the DARWIN [36] codes belong to the same category as WIMS. Both of these are pin-cell codes and carryout depletion/buildup of radionuclides and do not perform isotope transport calculations for various region.

The specifically designed computer codes for conducting radionuclide transport from fuel, to gap and to coolant can be divided into three basic categories: (a) empirical, (b) semi-empirical and (c) mechanistic. The *empirical* codes use experimental data to get fitted relationships that work as models for transport of radionuclides under various operating conditions. While the codes based on such models show excellent agreement with the corresponding experimental measurement, they do lack the basic physics behind these processes and as a result, these codes may fail completely when the conditions deviate only slightly from the original range of values. The MELCOR [37] and CORSOR [38] essentially belong to the empirical category of such codes. These are extensively used for safety and risk analysis and are based on grouping of similar elements. Lumped models are typically used for inter-granular transport of fission products and it tends towards solving the problem by using algebraic expressions. These codes do not have explicit models for transport of fission products into the coolant.

The *semi-empirical* codes tend to use empirical models for a part of the problem while the remaining one is solved by using detailed modeling. The FIPREM [39] belongs to this category and solves the inter-granular problem using Booth equivalent sphere model while the transport of fission products into the gap is solved based on finite difference approximation of the diffusion model.

The *mechanistic* codes offer much broader range of applicability. The VICTORIA [40] and ECART [41] codes are examples of such effort. These are designed for analysis of post-severe-accident state. They offer handling of wide range of radioisotopes and incorporate their transport, dispersion and deposition under various physical conditions. However, since they were not designed for normal steady state, and for transient analysis, therefore, they do not predict the transport of fission product out of fuel, into the fuel-clad gap and eventually into the coolant regions.

To bridge this gap, efforts have been made for the development of computer codes designed specifically for modeling the transport of fission products from fuel matrix into the fuel-clad gap and subsequently into the primary coolant region. In this regard, development of the FPCART code and its improved version FPCART-ST are important developments. Both deterministic as well as stochastic approaches have been used in these programs. Details of this research effort are provided in the following sections.

## 4. KINETIC MODELING

A schematic diagram of the reactor core and the primary circuit of a typical PWR is given in Figure 1. We consider a 300 MW(e) PWR with the design specification given in Table 1 [42].

**Table 1. Design specification of typical pressurized water reactor**

| Parameter | Value* |
|---|---|
| Specific power (MWth/Kg. U) | 33 |
| Power density (MWth/m$^3$) | 102 |
| Core height (m) | 4.17 |
| Core diameter (m) | 3.37 |
| Number of fuel assemblies | 194 |
| Fuel pins (rods) per assembly | 264 |
| Fuel material | UO$_2$ |
| Clad material | Zilcoloy |
| Lattice pitch (mm) | 12.6 |
| Fuel pin outer diameter (mm) | 9.5 |
| Average enrichment (%) | 3.0 |
| Coolant flow rate (Mg/s) | 18.3 |
| Linear heat rate (kW/m$^3$) | 17.5 |
| Coolant pressure (MPa) | 15.5 |
| In-let coolant temperature ($^{\circ}$C) | 293 |
| Out-let coolant temperature ($^{\circ}$C) | 329 |

* Reference [42].

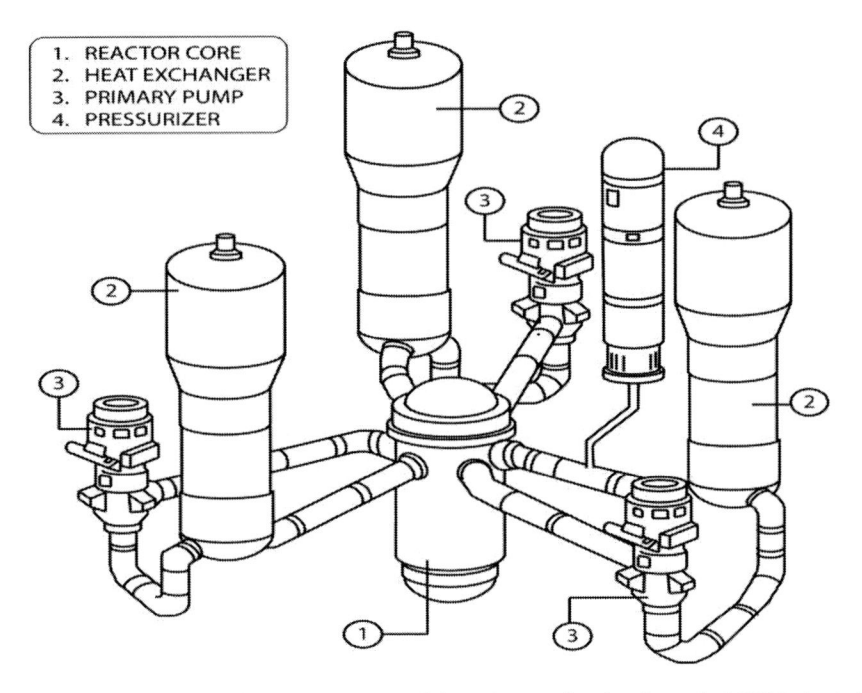

1. REACTOR CORE
2. HEAT EXCHANGER
3. PRIMARY PUMP
4. PRESSURIZER

Figure 1. Schematic diagram of the reactor core and the primary circuit of a typical PWR depicting the primary pumps, heat exchanges and pressurizer.

The reactor is assumed to be in cold clean state at the start. The fission product activity in the fuel (F), gap (G) and in the primary coolant (C) is governed by the extending the standard set of governing equations [43]:

$$\frac{dN_{F,i}}{dt} = F\,Y_i\,P + \sum_{j=1}^{i-1} f_{ij}\,\lambda_j\,N_{F,j} - [\lambda_i + \nu_i + \sigma_i\,\phi]\,N_{F,i} \tag{1}$$

for the gap region:

$$\frac{dN_{G,i}}{dt} = \nu_i N_{F,i} + \sum_{j=1}^{i-1} f_{ij}\,\lambda_j\,N_{G,j} - [\lambda_i + D\,\epsilon_i + \sigma_i\,\phi]N_{G,i} \tag{2}$$

and, for the coolant region:

$$\frac{dN_{C,i}}{dt} = D\,\epsilon_i\,N_{G,i} + \sum_{j=1}^{i-1} f_{ij}\,\lambda_j\,N_{C,j} + \left[\lambda_i + Q\frac{\eta_i}{W} + \beta + \tau\,\sigma_i\,\phi + \frac{L}{W}\right]N_{C,i}. \tag{3}$$

where, the subscript 'i' refers to the index of isotope in the decay chain composed of four isotopes: $i = 1, 2, \cdots, 4$. The corresponding values of various parameters used in the simulations along with the values of basic physics parameters for various fission products and their decay products are given in Tables 2 and 3 respectively [44].

**Table 2. Standard values of operational parameters used in simulations**

| Parameter | Value* |
| --- | --- |
| $D$ | $2.5 \times 10^{-3}$ |
| $L\ (g/s)$ | 2.3 |
| $Q\ (g/s)$ | 470 |
| $\beta$ | 0.001 |
| $W\ (g)$ | $1.072 \times 10^9$ |
| $V\ (cm^3)$ | $1.485 \times 10^9$ |
| $\tau$ | 0.056 |
| $P_o\ (MW_{th})$ | 998 |
| $F\ (Fissions/W.s)$ | $3.03 \times 10^{10}$ |

* Reference [45].

## 4.1. Computational Scheme

The coupled set of ODEs, given as Eqs. (1), (2) and (3), has been solved numerically using fourth order Runge-Kutta method. The corresponding computer program FPCART [45] uses the LEOPARD [46] and the ODMUG [47] as subroutines. The LEOPARD code performs the cell averaged group constants for the PWR core region and the ODMUG solved the corresponding one-dimensional multigroup diffusion equation and finds the group fluxes. Using this data, the FPCART program performs numerical computation of the time dependent fission product concentration in the fuel, gap and in the primary coolant of PWRs. These calculations are performed for both the steady-state as well as for power transients. A block diagram of various computational steps involved in the FPCART program is shown in Figure 3.

**Table 3 Values\* of the basic physics parameters for various fission products and the associated decay products**

| Isotope (Location in Chain) | $T_{1/2}$ | η | Y | ν | σ (b) | Branching Ratios, $f_{ij}$ | | | | |
|---|---|---|---|---|---|---|---|---|---|---|
| | | | | | | 1→2 | 1→3 | 2→3 | 2→4 | 3→4 |
| Se-85(1) | 31.1 s | 0.9 | 5.56E-3 | 6.5E-8 | 0.0 | 1.0 | 0.0 | | | |
| Br-85(2) | 2.9 m | 0.9 | 1.28E-2 | 6.5E-8 | 0.0 | | | 0.9984 | 0.0016 | |
| Kr-85M(3) | 4.48 h | 0.4538 | 1.3E-2 | 6.5E-8 | 0.0 | | | | | 1.0 |
| Kr-85(4) | 10.752y | 3.958E-5 | 1.3E-2 | 6.5E-8 | 1.84E-1 | | | | | |
| SE-87(1) | 5.29s | 0.9 | 7.60E-3 | 6.5E-8 | 0.0 | 1.0 | 0.0 | | | |
| Br-87(2) | 55.7s | 0.9 | 2.03E-2 | 6.5E-8 | 0.0 | | | 1.0 | 0.0 | |
| Kr-87(3) | 1.272 h | 0.7454 | 2.56E-2 | 6.5E-8 | 5.518E+1 | | | | | 1.0 |
| Rb-87(4) | LL | 0.9 | 2.56E-2 | 6.5E-8 | 7.690E-2 | | | | | |
| Kr-88 (1) | 2.84 h | 0.5674 | 3.55E-2 | 6.5E-8 | 0.0 | 1.0 | 0.0 | | | |
| Kr-89 (1) | 3.15 m | 0.986 | 4.63E-2 | 6.5E-8 | 0.0 | 1.0 | 0.0 | | | |
| Rb-89 (2) | 15.4 m | 0.9 | 2.05E-3 | 1.3E-8 | 0.0 | | | 1.0 | 0.0 | |
| Sr-89 (3) | 50.7 d | 0.986 | 1.75E-4 | 1.0E-11 | 5.264E-2 | | | | | 1.0 |
| Sr-90 (1) | 28.79 y | 0.9 | 7.58E-2 | 1.0E-11 | 8.739E-2 | 1.0 | 0.0 | | | |
| Y-90 (2) | 2.671 d | 0.9 | 8.97E-8 | 1.6E-12 | 5.93E-1 | | | 1.0 | 0.0 | |
| Sb-129(1) | 4.36 h | 0.9 | 5.43E-3 | 1.0E-11 | 0.0 | 0.166 | 0.834 | | | |
| Te-129M(2) | 33.6 d | 0.9 | 1.4E-7 | 1.0E-9 | 2.90E-1 | | | 0.63 | | |
| Te-129(3) | 1.16 h | 0.9 | 5.11E-3 | 1.0E-9 | 0.0 | | | | 0.37 | |
| I-129(4) | 1.61E7 y | 0.99 | 5.0E-6 | 1.3E-8 | 5.225 | | | | | 1.0 |
| Te-131M(1) | 1.25 d | 0.9 | 1.1E-2 | 1.0E-9 | 0.0 | 0.21 | 0.79 | | | |
| Te-131(2) | 25.0 m | 0.9 | 1.71E-2 | 1.0E-9 | 0.0 | | | 1.0 | 0.0 | |
| I-131 (3) | 8.023 d | 0.99 | 3.29E-5 | 1.3E-8 | 3.229E-1 | | | | | 0.0109 |
| Xe-131M(4) | 11.93 d | 8.923E-3 | 4.05E-4 | 6.5E-8 | 0.0 | | | | | |
| Sb-132M(1) | 4.1 m | 0.9 | 1.07E-2 | 1.0E-11 | 0.0 | 0.0 | 1.0 | | | |
| Sb-132 (2) | 2.8 m | 0.9 | 1.67E-2 | 1.0E-11 | 0.0 | | | 1.0 | 0.0 | |
| Te-132 (3) | 3.204 d | 0.9 | 1.54E-2 | 1.0E-9 | 4.89E-4 | | | | | 1.0 |
| I-132 (4) | 2.28 h | 0.99 | 2.06E-4 | 1.3E-8 | 0.0 | | | | | |
| Te-133M (1) | 55.4 m | 0.9 | 3.06E-2 | 1.0E-9 | 0.0 | 1.0 | 0.0 | | | |
| I-133 (2) | 20.8 h | 0.99 | 4.9E-2 | 1.3E-8 | 0.0 | | | 0.0285 | 0.9715 | |
| Xe-133M (3) | 2.19 d | 4.663E-2 | 2.67E-5 | 6.5E-8 | 0.0 | | | | | 1.0 |
| Xe-133 (4) | 5.244 d | 2.001E-2 | 4.9E-02 | 6.5E-8 | 2.44E+1 | | | | | |
| Te-134 (1) | 41.8 m | 0.9 | 6.97E-2 | 1.0E-9 | | 1.0 | 0.0 | | | |
| I-134 (2) | 52.8 m | 0.99 | 7.83E-2 | 1.3E-8 | | | | 1.0 | 0.0 | |
| Te-135 (1) | 19.0 s | 0.9 | 3.11E-2 | 1.0E-9 | 0.0 | 1.0 | 0.0 | | | |
| I-135(2) | 6.57 h | 0.9 | 2.98E-2 | 1.3E-8 | 2.119E-3 | | | 0.1651 | 0.8349 | |
| XE-135M(3) | 15.29m | 0.9098 | 1.10E-2 | 6.5E-8 | 0.0 | | | | | 0.994 |
| XE-135(4) | 9.14 h | 0.2205 | 6.54E-2 | 6.5E-8 | 2.445E+5 | | | | | |

\* References [27, 44].

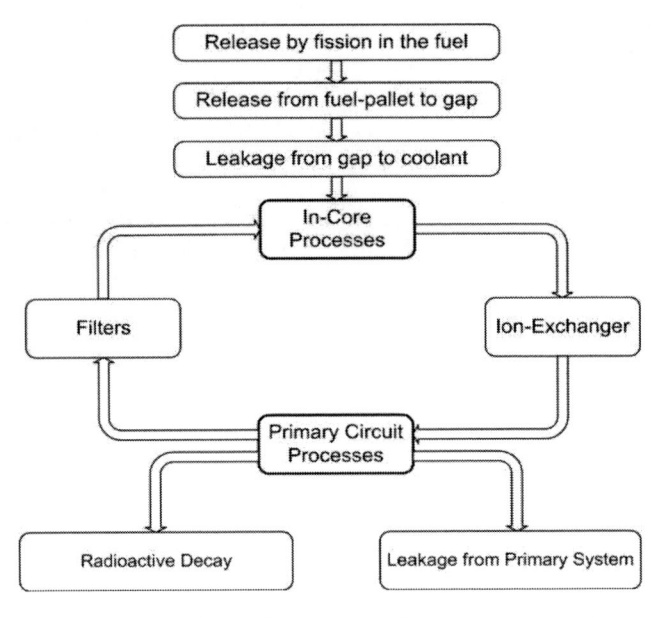

Figure 2. Block diagram of the production and loss mechanism of various fission products in the primary circuit of a typical PWR.

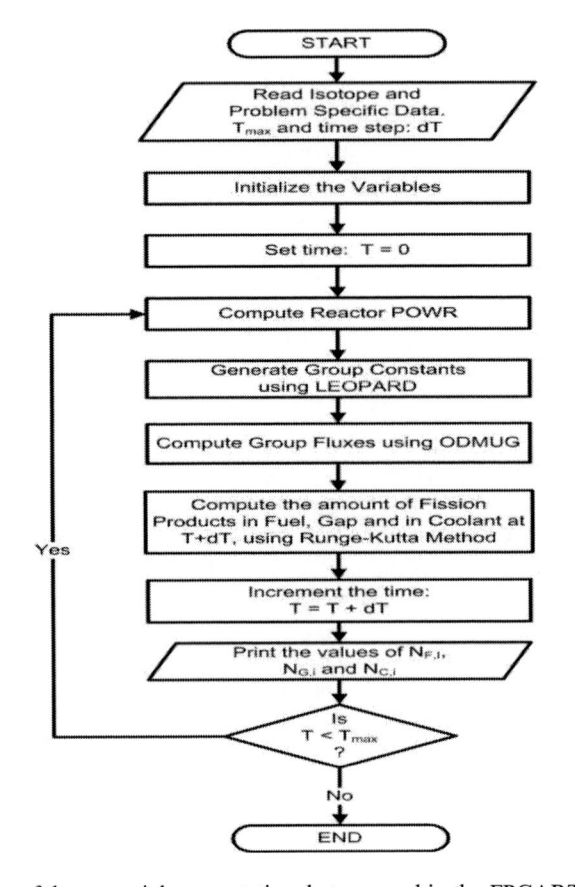

Figure 3. Block diagram of the essential computational steps used in the FPCART computer program for kinetic simulation of fission product in steady-state as well as power transients [45].

## 4.2. Steady State Analysis

The FPCART computer program has been used for the simulation of activity buildup in a typical 1000 MWth PWR taken as clean system initially. Compared with the ORIGEN2 code calculated values of the saturation activities of various dominant radionuclides in the core region, the FPCART predictions have excellent agreement as shown in Figure 4. Differences, of the order of a few percent, are found in the two corresponding values in the cases of $^{134}$Te, $^{137}$Xe, $^{88}$Rb and $^{88}$Kr, which may be attributed primarily to the difference in the fission yields.

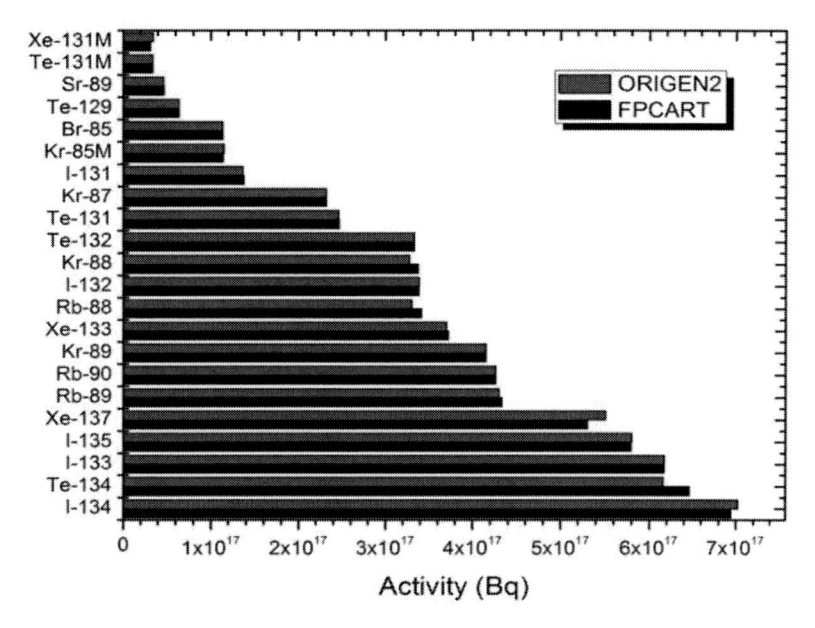

Figure 4. Comparison of the FPCART predicted values of activities of various isotopes in PWR fuel with the corresponding computed data using the ORIGEN2 code.

The saturation values of various radioisotopes in the fuel, gap and coolant regions can be estimated using direct analytical expression:

$$N_{C,i}^{sat} = \left[ D\, \epsilon_i N_{G,i}^{sat} + \sum_{j=1}^{i-1} f_{ij} \lambda_j N_{C,j}^{sat} \right] / \left[ \lambda_i + \frac{Q}{m} \eta_i + \beta_i + \tau \sigma_i \phi + \frac{L}{m} \right], \quad (4)$$

with,

$$N_{G,i}^{sat} = \left[ \nu_i N_{F,j}^{sat} + \sum_{j=1}^{i-1} f_{ij} \lambda_j N_{G,j}^{sat} \right] / [\lambda_i + D\, \epsilon_i + \sigma_i\, \phi], \quad (5)$$

and

$$N_{F,i}^{sat} = \left[ F\, Y_i\, P + \sum_{j=1}^{i-1} f_{ij}\, \lambda_j\, N_{F,j}^{sat} \right] / [\lambda_i + \nu_i + \sigma_i\, \phi]. \quad (6)$$

The saturation values of the specific activities computed by using the FPCART program have been found in excellent agreement with the corresponding results found by using direct analytical expressions given as Eq. (4) through (6).

## 4.3. Power Perturbations

Normally PWR power levels do change depending on the grid load requirements. As a result, the activity concentrations of various radioisotopes also change with perturbations in power. In the FPCART code, the time dependent power is given in terms of power factor $f(t)$ such that:

$$P(t) = P_0 \, f(t) \tag{7}$$

where,

$$f(t) = \begin{cases} 1, & t \leq t_{in} \\ 1 - \alpha \, [t_m - t], & t_{in} \leq t \leq t_m \\ w_2/w_o & t > t_m \end{cases} \tag{8}$$

with, $\alpha$ = slope of the linear increase; $t_{in}$ = time when the power perturbation starts; $t_m$ = time when the power changes by factor $w_2/w_o$. The power transients may result in dramatic changes in radioisotope concentrations and some of them do have strong influence on the reactor operation. The $^{135}$Xe is an important radioisotope from this perspective. The FPCART calculated time variation of the specific activity of $^{135}$Xe for step change in reactor power is shown in Figure 5. Just after the start of the over-power transient, the $^{135}$Xe specific activity first drops off and then starts to rise to a higher saturation level. This is essentially due to higher removal rate of $^{135}$Xe initially, followed by higher production rate from its parent. When the reactor is scrammed, the $^{135}$Xe specific activity soars as its removal by neutron absorption is absent at this stage.

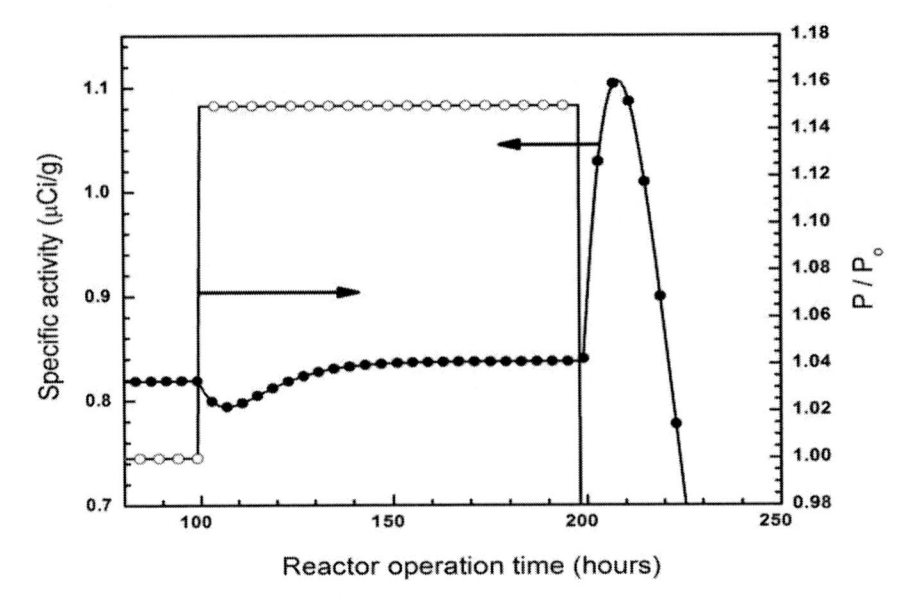

Figure 5. The FPCART calculated variation of $^{135}$Xe specific activity with time for step power transients.

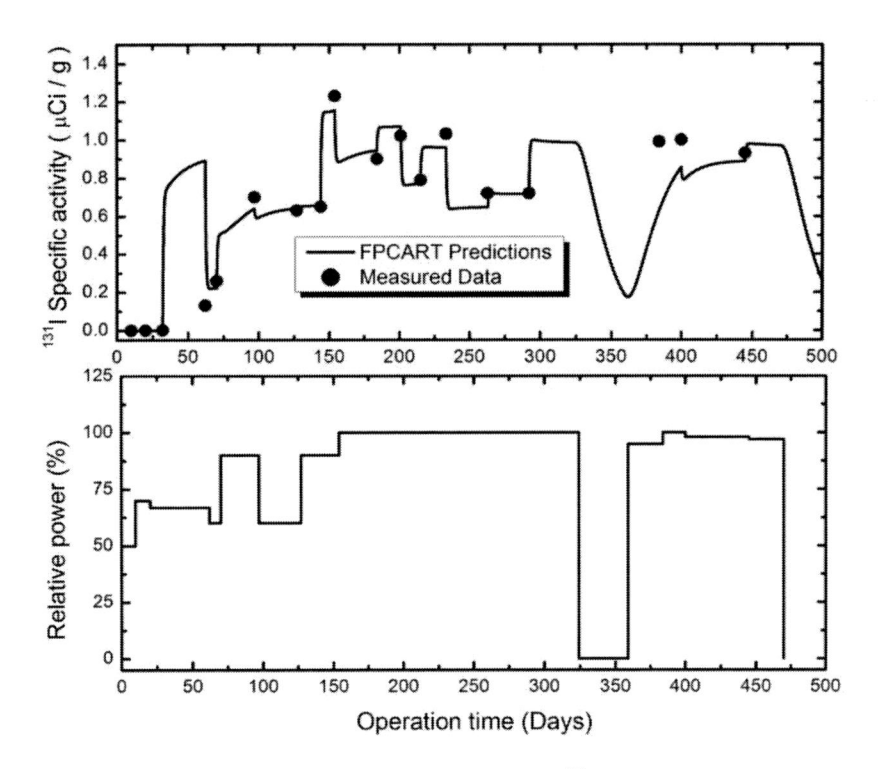

Figure 6. Comparison of the FPCART predicted values of $^{131}$I specific activity variations with the corresponding experimental data for the Beznau (unit-1) power plant undergoing power transients.

The quantitative predictions of the FPCART code have also been found reliable. The FPCART computed specific activities of various dominant radioisotopes have been compared with the corresponding experimental measurements carried out at various nuclear reactors including Beznau (unit-1), Surry (unit-1) and the Zorita power plant. The results for the specific activity of $^{131}$I in the Beznau (unit-1) are shown in Figure 6. The FPCART predicted values are in close agreement with the corresponding experimental measurements throughout the power transient. Similar agreement has been observed in the other cases as well.

## 4.4. Flow-Rate Transients

The primary pump coast down has been modeled in the FPCART code by accounting for the momentum and the frictional retardation [48]:

$$l \, \rho \, \frac{dv}{dt} = -\frac{1}{2} C_f \, \rho \, v^2 \tag{9}$$

where, $l$ = total length of the loop; $\rho$ = the fluid density; $C_f$ = total pressure loss coefficient; and $v$ = the fluid speed. In terms of flow rate ($w(t)$), the corresponding solution is:

$$w(t) = w_o / [1 - t/t_p], \tag{10}$$

where, $w_o$ = the steady state value of flow rate; $t_p = 2\,l/(C_f\,v_o)$ which has value typically above 2000 h for transients without boiling crisis.

The transient response of primary coolant activity has been studied for $t_p = 2000\,h$ pump coast down in the case of 1000 MWth PWR [49]. The FP activity has been found to increase due to pump coast-down. For this scenario, the FPCART computed variation of the primary coolant total specific activity is shown in Figure 7. From the on-set of pump coast-down, the value of coolant total specific activity starts rising in a ramp style with a characteristic value of the slope of about 1.0 pCi/g. s.

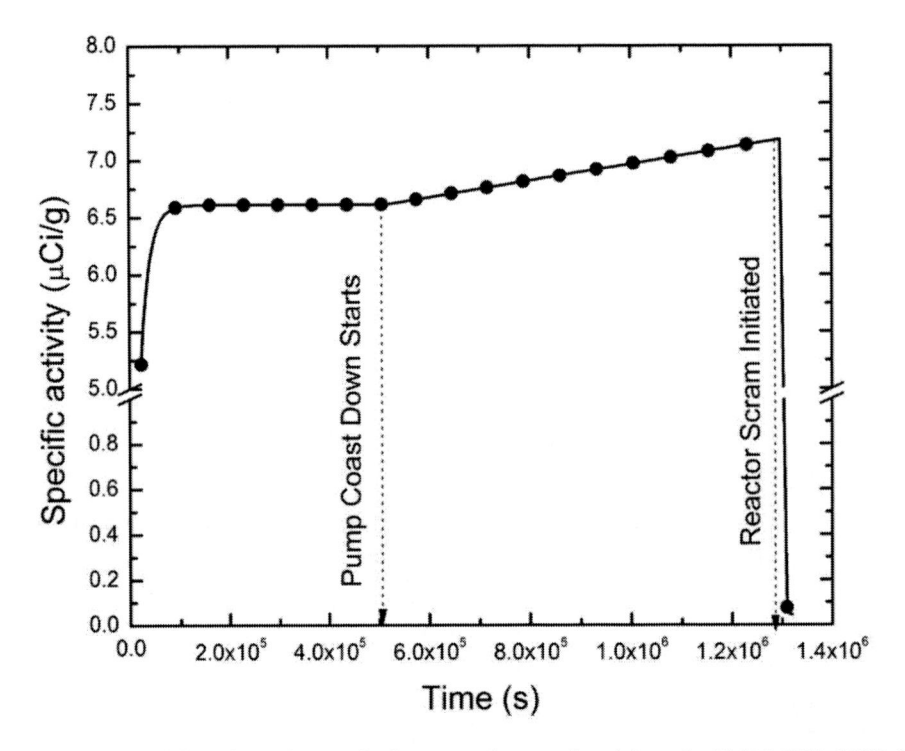

Figure 7. Simulation of time dependence of primary coolant total activity of a 1000 MWth PWR for a $t_p = 2000\,h$ pump coast-down flow rate transient.

The dependence of the value of maximum total specific activity on the pump coast-down characteristic time shows a rising followed by approach to saturation trend as given in Figure 8. The corresponding data has been found to follow Sigmoidal Logistic function (type 1):

$$A_{tot}(t_p) = a\,\{1 + exp[-k\,(t_p - t_{pc})]\}^{-1} \qquad (10)$$

where, the fitted parameters have the following values:

$$a = 7.18918 \pm 3.50645 \times 10^{-4}\ \ \mu Ci/g \qquad (11)$$

$$t_{pc} = -2532.03896 \pm 109.80616\ \ hours \qquad (12)$$

$$k = 0.00171 \pm 6.07739 \times 10^{-5} \; hours^{-1} \tag{13}$$

The fitted function along with the FPCART computed values of the maximum total specific activity are shown in Figure 8.

## 5. STOCHASTIC MODELING

The three-stage deterministic model of the FPCART code has been extended by incorporating Monte Carlo based random fuel failure and radioisotope release rates. This methodology has been implemented in the computer code FPCART-ST. The fuel rod failure time sequence is generated by random sampling based on a probability distribution function $g(t)$. The corresponding intensity function is:

$$\psi(t) = g(t)/G(t), \tag{14}$$

where, $G(t)$ represents the corresponding cumulative probability distribution:

$$G(t) = exp\left(-\int_0^t \psi(s) \; ds\right) \tag{15}$$

The probability of accepting a fuel failure at $t_k$ after $t_j$ is found by using the rejection technique [50]. First a random number '$\eta$' is generated and it is compared with the ratio '$q$':

$$q = \frac{g(t_k)}{g(t_j)}, \tag{16}$$

and, if $\eta > q$, then $t_k$ is accepted as failure event time, otherwise, the procedure is repeated. In this model, the fuel matrix to gap escape rate coefficient is taken as:

$$\epsilon = D_o \; \epsilon_o \exp[-\xi \; (t - t_o)] + D_F \; \epsilon_o \tag{17}$$

where, $\epsilon_o$ is the initial burst release rate from a punctured fuel rod; $\xi$ is the decay constant for the escape rate ; $t_o$ is time of initiation of the fuel rod fuel rod failure; $D_F$ represents the number of failed fuel rods so far while $D_o = 1$ represents the failure of the current fuel rod. Typical values of these parameters are: $D = 60$ ; $\epsilon_o = 10^{-8} \; s^{-1}$ ; $\xi = 7.2 \times 10^{-5} \; s^{-1}$.

The Angra-1 PWR power plant (657 MWe), of Westinghouse design, was shut-down before the completion of its 4th cycle. The main reason, for this premature shutdown, was abnormally high primary coolant activity. Later investigation revealed that one sixth of its fuel assemblies had failed and the fission products had leaked into the primary circuit. This has been simulated with the FPCART-ST code and the variation of $^{131}$I activity in primary coolant is shown as Figure 8. The predicted values of the specific activity of $^{131}I$ in primary coolant shows good agreement with the Angra-1 plant measurements. Generally, the theory and experiment agree with in 15% of each other and the observed spikes in the experimental values can be attributed to the coupled power-flow rate transients while in simulations, only the power transients were incorporated.

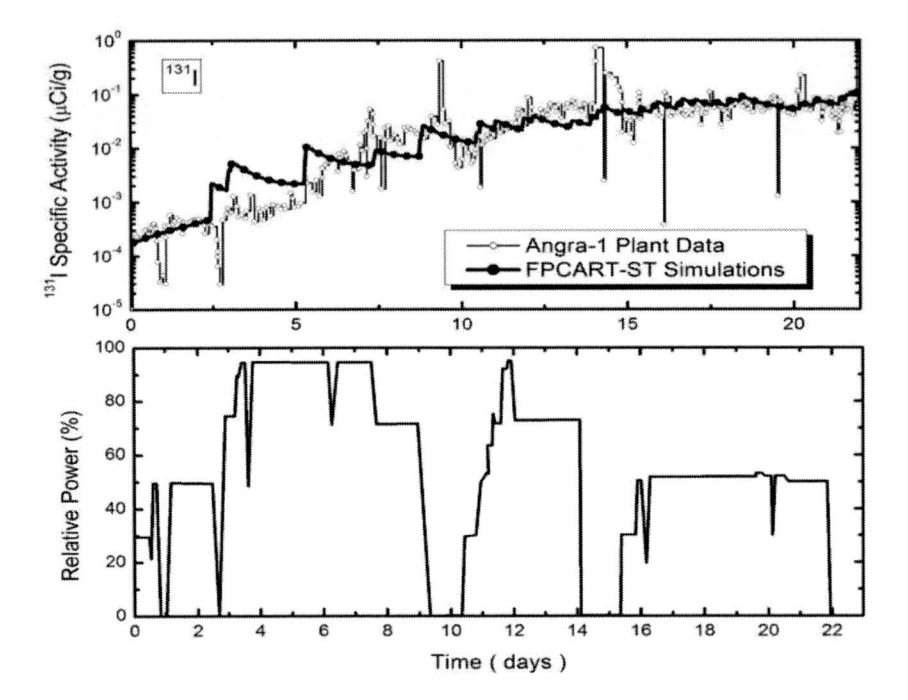

Figure 8.

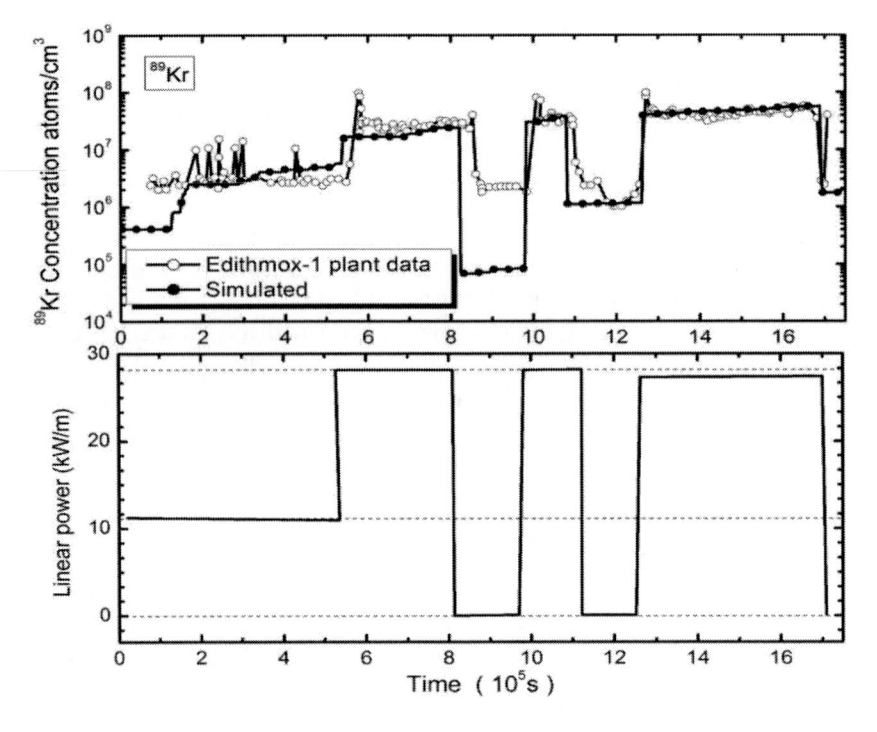

Figure 9.

The stochastic simulation of FP releases in PWRs, have also been carried out for $^{89}Kr$ specific activity for the Edithmox-1 primary loop. A comparison of the simulation results with the corresponding experimental measurements is shown in Figure 9. It is clear from this graph that stochastic simulations are in good agreement with the measurements throughout the power transient [51]. Again, the spikes in activity concentrations at the on-set of each power transient may possibly be due to the corresponding flow-rate perturbations.

## 6. CONCLUSIONS

Fission product activity continues to play a dominant role during normal reactor operation as well during transients. A review of research efforts towards modeling and simulation of the fission product release from fuel, to gap and subsequently to the primary coolant circuit has been presented. Typically, multi-stage models have proved adequate for the simulation of fission product activity in the primary circuits of PWRs. In contrast with the empirical and the semi-empirical models, the mechanistic codes offer much wider scope of applications. Kinetic modeling has been found effective for simulating fission product activity during power as well as in flow-rate transients. Also, due to inherent probabilistic nature of the fuel failure phenomenon, stochastic modeling has been found suitable for burst release studies from randomly failing fuel rods.

In view of the current trend towards high burn-up cores and extending core life-times, coupled with the fact that the current fleet of nuclear power system is aging, the importance, of issues related to fuel failure, and annual cumulative dose received by workers, has increased by many folds. For PWRs to remain economically feasible against their competitors, continued experimental and theoretical efforts are needed for bringing the primary circuit radiation dose levels to much lower values.

## REFERENCES

[1]    IAEA, Nuclear power reactors in the world, Reference Data Series No. 2, International Atomic Energy Commission, *IAEA-RDS-2/30*, Vienna, 2010.

[2]    T. Moore, The challenge of nuclear fuel reliability, *EPRI, J.* (Fall) (2005).

[3]    C. J. Wood, Recent developments in LWR radiation field control, *Progress in Nuclear Energy,* 19(3), (1987), 241-266.

[4]    G. C. Comley, The significance of corrosion products in water reactor coolant circuits, *Progress in Nuclear Energy*, 16(1), (1985) 41—72.

[5]    D. H. Lister, Corrosion-product release in Light Water Reactors, *EPRI Report NP-3460*, March (1984).

[6]    H. Zänker, H. Müller, R. Berndt, A complementary cesium coolant concentration ratio for localizing defective PWR fuel rods during reactor operation, *J. Radioanalytical and Nuclear Chemistry,* 152(1), (1991) 47-54.

[7]    L. Lazare, C. Crestey, C. Bleistein, Measurement of 90Sr in primary coolant of pressurized water reactor, *J. Radioanalytical and Nuclear Chemistry*, 279(2), (2009) 633—638.

[8]   Y. H. Koo, D. S. Sohn, Y. K. Yoon, Release of unstable fission products from defective fuel rods to the coolant of a PWR, *J. Nuclear Materials*, 209(3), (1994) 248—258.

[9]   A. S. Avanov, The model of the fission gas release out of porous fuel, *Annals of Nuclear Energy*, 25(15), (1998) 1275-1280.

[10]  S. D. Kreider, A. Schneider, The detection of failed fuel in LWRs: A historical review, *Trans. Am. Nucl. Soc.* 61 (1990) 46-47.

[11]  J. M. Seixas, F. P. Freeland, W. Soarce-Filho, Matched filters for identifying failed fuel rods in nuclear reactors, Proc. *IEEE Intl. Conf. Electronics, Circuits and Systems* - Vol. 2, Sant Julians, Malta, (2—5 September 2001) 643-646.

[12]  L. Clink, H. Freeburn, Estimation of PWR fuel rod failure throughout a cycle, *Trans. Am. Nucl. Soc.*, 54(1), (1987), 13.

[13]  J. D. B. Lambert, K. C. Gross, E. V. Depiante, Adaptation of gas tagging for failed fuel identification in light water reactors, *Proc. ASME 4th Intl. Conf. Nuclear Engineering.* CONF-960306-25, (1996).

[14]  B. J. Lewis, Fission product release from nuclear fuel by recoil and knockout, *J. Nuclear Materials,* 148(1), (1987) 28- 42.

[15]  B. J. Lewis, Fundamental aspects of defective nuclear fuel behavior and fission product release, *J. Nuclear Materials*, 160(2-3), (1988), 201—217.

[16]  M. O. Tucker, R. J. White, The release of fission products from UO2 during irradiation, *J. Nuclear Materials*, 87(1), (1979) 1-10.

[17]  M. J. F. Notley, I. J. Hastings, A microstructure –dependent model for fission gas release and swelling in UO2 fuel, *Nuclear Engineering and Design*, 56(1) (1980) 163-175.

[18]  L. C. Bernard, E. Bonnaud, Finite volume method for fission gas release modeling, *J. Nuclear Materials,* 244(1), (1997) 75—84.

[19]  K. H. Hwang, K. J. Lee, Modeling the activity of 129I and 137Cs in the primary coolant and CVCS resin of an operating PWR, *J. Nuclear Materials*, 350(2) (2005) 153—162.

[20]  P. Kundsen, et al., Fission products behavior in high burn up reactor fuel subjected to slow power increases, *Nucl. Tech.,* 72 (1986) 258—267.

[21]  P. Hardt, A. Tattegrain, The FEBUS fission product project, *J. Nucl. Mater.*, 188 (1992) 115—130.

[22]  Z.R. Martinson, M. Gasparini, R.R. Hobbins, D. A. Petti, C. M. Allison, J.K. Hohorst, D.L. Hagrman, K. Vinjamuri, PBF (Power Burst Facility) severe fuel damage test 1--3 test results report, *NUREG/CR-5354; EGG-2565*, (October 1989).

[23]  M. D. Allen, H. W. Stockman, K. O. Reil, J. W. Fisk, Fission product release and fuel behavior of irradiated light water reactor fuel under severe accident conditions: The ACRR ST-Experiment, U.S. Nuclear Regulatory Commission *NUREG/CR-5345*, (1991).

[24]  L. Baker, Source term experiments project (STEP): A summary, *EPRI NP-5753M,* (1988).

[25]  D. L. Reeder, LOFT system and test description, *USNRC NUREG/CR-0247*, (1978).

[26]  H. Albrecht, Release of fission and activation products at LWR core-melt: Final report of the SASCHA program, Kernforschungszentrum Karlsruhe, Federal Republic of Germany *KFK-4264*, (1987).

[27] J. P. Leveque, B. André, G. Ducros, G. L. Marois, G. Lhiaubet, The HEVA experimental program, *Nuclear Technology*, 108 (1994) 33.

[28] Y. Pontillon, P. P. Malgouyres, G. Ducros, G. Nicaise, R. Dubourg, M. Kissane, M. Baichi, Lessons learnt from VERCORS tests: Study of the active role played by UO2-ZrO2-FP interactions on irradiated fuel collapse temperature, *J. Nucl. Mater.*, 344 (2005) 265—273.

[29] T. Nakamura, A. Hidaka, K. Hashimoto, Y. Harada, Y. Nishino, H. Kanazawa, H. Uetsuka, J. Sugimoto, Research program (VEGA) on the fission product release from irradiated fuel, Department of Reactor Safety Research, Nuclear Safety Research Center, Tokai Research Establishment, Japan Atomic Energy Research Institute, *JAERI-Tech.* 99-036, (1999).

[30] A. G. Groff, ORIGEN2: A versatile computer code for calculating the nuclide compositions and characteristics of nuclear materials, *Nucl. Technol.*, 62 (1983) 335—352.

[31] W. S. Charlton, R. T. Perry, B. L. Fearey, T. A. Parish, Calculated actinide and fission product concentration ratios for gaseous effluent monitoring using MONTEBURNS 3.01, *Nucl. Technol.*, 131 (2000) 210—227.

[32] J. F. Briesmeister, (ed.), MCNP- A general Monte Carlo N-particle transport code, User's manual: Los Alamos National Laboratory, *LA–13709–M*, (2000).

[33] M. J. Halsall, A summary of WIMSD4 input options, Atomic Energy Establishment, Winfrith, Dorchester AEEW-M-1327, (1980).

[34] A. Tavelli, The U.S. RERTR program status and progress, Proc. Intl. meeting on Reduced Enrichment for Research and Test Reactors, *ANL/TD/CP-95339*, Jackson Hole, Wyoming, October 5-10, (1997).

[35] D. Knott, B. H. Forssén, M. Edenius, CASMO-4, A fuel assemblyburn-up program methodology, *STUDSVK/SOA-95/2*, (1995).

[36] A. Tsilanizara, DARWIM: An evolution code system for a large range of applications, *J. Nucl. Sci. Technol*, 37, (2000) 845—849.

[37] R.M. Summers, et al., MELCOR 1.8.0: A Computer Code for Nuclear Reactor Severe Accident Source Term and Risk Assessment Analyses, *NUREG/CR-5531, SAND90-0363* (1991).

[38] M. R. Kuhlmann, D. J. Lehmicke, R. O. Meyer, CORSOR user's manual, Battelle Columbus Labs. *USNRC Report NUREG/CR-4173*, (1985).

[39] C. Ronchi and C. T. Walker, Determination of Xenon Concentration in Nuclear Fuels by Electron Microbe Analysis, J. *Phys. D: Applied Physics*, 13, (1980) 2175.

[40] T. J. Haemes, D. A. Williams, N. E. Bixler, A. J. Grimley, C. J. Wheatley, N. A. Johns, N. M. Chown, VICTORIA: A mechanistic model of radionuclide behavior in the reactor coolant system under severe accident conditions, Sandia National Laboratory, Albuqurque, *NM 87185, NUREG/CR-5545*, (1990).

[41] F. Parozzi, "Computer Models on Fission Product and Aerosol Behavior in the LWR Primary System, Part II," Commission of the European Communities, *EUR Report #* 14676 EN, (1992).

[42] S. Glasstone, A. Sessonske, Nuclear Reactor Engineering, 4th ed. Von Nostradam, New York (1994).

[43] M. H. Chun, N. L. Tak, S. K. Lee, Development of a computer code to estimate the fuel rod failure using primary coolant activities of operating PWRs, *Annals of Nuclear Energy*, 25(10) (1998) 753-763.

[44] JEFF 3.1, Joint evaluated nuclear data library for fission and fusion applications, *NEA Databank, AEN, NEA*, (2005).

[45] M. J. Iqbal, N. M. Mirza, S. M. Mirza, Kinetic simulation of fission product activity in primary coolant of typical PWRs under power perturbations, *Nuclear Engineering and Design* 237(2), (2006) 199-205.

[46] R. F. Barry, LEOPARD: A Spectrum Dependent Non-Spatial Depletion Code for IBM-7094, *WCAP-3269-26*, Westinghouse Electric Corporation (1963).

[47] Thomas, J.R., Edlund, H.C., Reactor Statics Module - Multigroup Criticality Calculations, *Proc. Conf. ICTP*, Trieste (1980).

[48] E. E. Lewis, *Nuclear power reactor safety*, John Wiley and Sons, (1977).

[49] S. M. Mirza, M. J. Iqbal, N. M. Mirza, Effect of flow rate transients on fission product activity in primary coolant of PWRs, *Progress in Nuclear Energy*, 49 (2007) 120-129.

[50] M. J. Kalos, P. A. Whitlock, Monte Carlo Methods, vol. 1., Wiley and Sons, (1986).

[51] M. J. Iqbal, N. M. Mirza, S. M. Mirza, Stochastic simulation of fission product activity in primary coolant due to fuel rod failures in typical PWRs under power transients, *J. Nuclear Materials*, 372 (2008) 132-140.

In: Nuclear Materials
Editor: Michael P. Hemsworth, pp. 177-216

ISBN 978-1-61324-010-6
© 2011 Nova Science Publishers, Inc.

*Chapter 7*

# RECENT ADVANCES IN MOLECULAR DYNAMICS MODELLING OF RADIATION EFFECTS IN α-ZR

## *Roman Voskoboinikov*[*]

Australian Nuclear Science and Technology Organisation,
New Illawarra Road, Lucas Heights, NSW 2234, Australia

## ABSTRACT

Environmental sustainability and higher burn-up of nuclear fuel rely on new cladding materials capable of withstanding harsh environmental influence without exhibiting significant degradation of service properties. In order to optimize these materials for safe operating conditions it is important to quantify possible detrimental effects. The experimental work necessary to reach this objective is often time-consuming and costly. At the same time, due to their linear and/or temporal scale not all phenomena are amenable to experimental investigation. It is therefore essential to complement experiments by using mathematical modelling and scientific computing.

In our research we have focused on systematic study of radiation effects in zirconium that is widely used as a cladding material in modern commercial light water nuclear reactors. Large-scale molecular dynamics (MD) modelling is conducted for an investigation of primary damage creation, self-interstitial and vacancy clusters formation, and their stability in high energy displacement cascades. The simulations are carried out for a wide range of temperatures ($100 \text{ K} \leq T \leq 600 \text{ K}$) and primary knock-on atom (PKA) energies $5 \text{ keV} \leq E_{pka} \leq 25 \text{ keV}$. This study of over 300 cascades is the largest yet reported for this metal. At least 25 cascades for each ($E_{pka}$, $T$) pair are simulated in order to ensure statistical reliability of the results. The high number of simulations for each condition of temperature and energy has revealed the wide variety of defect clusters that can be created in cascades. Mobile or sessile, two-dimensional or three-dimensional clusters of both vacancy and interstitial type can be formed. The number of Frenkel pairs, population statistics of clusters of each type, the fraction of vacancies and self-interstitial atoms (SIA) in point defect clusters, cluster per cascade yield etc. are obtained and their dependence on the temperature and PKA energy is investigated. Strong spatial and size correlations of SIA and vacancy clusters formed in displacement cascades are observed.

---

[*] roman.voskoboynikov@gmail.com

Both vacancy and SIA clusters can be mobile. However, depending on their type self-interstitial clusters exhibit one-dimensional, planar or three dimensional motions, whereas vacancy clusters of only one type can glide and in one dimension only. Separate MD simulations of some SIA and vacancy clusters are performed to study their thermal stability and possible transformations.

Typical clusters of point defects found in displacement cascades are extracted for investigation of radiation hardening of zirconium. Two vacancy loops (in the basal plane and prism plane) and three SIA clusters (a SIA dislocation loop in prism plane, a small triangular extrinsic fault in the basal plane and a disordered three dimensional SIA cluster) are considered. Atomic-scale details of the interaction of typical vacancy and SIA clusters placed in the gliding planes of two edge dislocations, $1/3[11\bar{2}0](0001)$ and $1/3[11\bar{2}0]\{1\bar{1}00\}$, are studied by large-scale MD modelling. Interaction mechanisms ranging from cluster dragging to partial or complete absorption of point defect clusters accompanied by dislocation climbing up or down are revealed. Stress–strain curves and the critical stress for a dislocation breakaway from a cluster are obtained for all the configurations studied.

# 1. INTRODUCTION

Zirconium and Zr-based alloys belong to the family of the basic structural materials of modern commercial nuclear power plants. Fuel cladding of light water reactors is made of zirconium alloy and being the primary barrier between fuel pellets and the coolant, it separates nuclear fuel and radioactive products from the environment. Heat and radiation resistance of the material and the integrity of fuel cladding determine the lifetime of fuel assemblies and safety during operation, transportation and storage. The influence of long-term exposure to irradiation with neutrons and fission fragments can be evaluated experimentally but such experimental work is very expensive and time consuming. Optimization of the material for safe usage and prediction of evolution of its service properties under irradiation can be assisted by multiscale materials modelling, an approach which brings together different techniques suitable at different time and length scales of the phenomenon under consideration. Large strides in simulating damage production and defect properties have been made over the past decade as a result of the increasing power of computational facilities. The most important properties of defects, which directly affect the microstructure evolution and therefore changes in service properties, are mechanisms of point defects and defect cluster formation in high energy collision cascades, clusters stability and mobility.

Displacement cascades are formed by the recoil of primary knock-on atoms (PKAs) with a kinetic energy of more than ~ 1keV. The collision cascade process is characterised by the very short temporal and length scales of the order of picoseconds and nanometers, respectively, and therefore cannot be studied directly by existing experimental techniques. Only atomic-scale computer simulation by the method of molecular dynamics (MD) modelling is suitable for studying primary damage creation, see reviews [1-5] and references therein for further details.

Studying stability, mobility and interaction between point defect clusters is of primary importance since they are formed directly in the cascade process and their behaviour has important consequences for microstructure evolution and service property changes under cascade damage conditions. For example, an imbalance in the proportion of clustered defects of one species versus another, e.g. self-interstitial atoms (SIAs) versus vacancies, created a complementary imbalance in point defect fluxes to the bulk of material which are responsible for solute transport, swelling by void formation, creep etc. Furthermore, SIA and vacancy clusters formed in close proximity create a weaker stress field than either would individually, and they also can facilitate mutual recombination of radiation defects as a result of cascade overlap in high-dose irradiation conditions or due to interaction with gliding dislocations.

A significant fraction of MD studies of displacement cascades has treated only a few (~5) cascades at a given PKA energy and matrix temperature, and so firm statistics on cluster type and size distribution, particularly as needed to computational models of damage evolution, is lacking. Also, and possibly for the same reason, there appears to be insufficiency of cluster types observed in earlier simulations, so it is not easy to compare results of modelling with available experimental data. Thus, an in-depth study of collision cascades in Zr is undertaken. More than 300 displacement cascades for a wider range of temperature, T=100-600 K, and PKA energy Epka=10-25 keV are simulated and more rigorous structural and statistical analysis is conducted. Special attention is paid to reliable and comprehensive statistical treatment of the results. Data on defect production in cascades are presented here, and typical point defect clusters are shown and their properties, stability and mechanisms of motion are investigated.

The second part of the review considers hardening of Zr exposed to irradiation. Modelling of the interaction of $1/3[11\overline{2}0](0001)$ and $1/3[11\overline{2}0](1\overline{1}00)$ edge dislocations with typical residual radiation defects found in displacement cascades is carried out. A wide variety of interaction mechanisms of the gliding dislocations with typical vacancy and self-interstitial atom (SIA) clusters placed in the glide planes are observed ranging from sessile segment formation, cluster drag without absorption and/or cluster shear to partial or complete absorption of point defect clusters with corresponding dislocation climb up or down. Stress–strain curves and the critical resolved shear stress for a dislocation breakaway from a residual defect cluster are obtained for all considered dislocations and clusters.

The study is presented in the following manner. The computational methods used for displacement cascade modelling and analysis are summarised in section 2. The results for the number of defects created under the different irradiation conditions are presented in Section 3. A significant proportion of both vacancies and SIAs form clusters in the collision process and statistics for vacancies and interstitials that are not single point defects are presented in Section 4. Typical cluster configurations that arise in cascade regions are described in Section 5. Fundamentals of atomic-scale modelling of dislocations in α-Zr are presented in Section 6. Interaction peculiarities of $1/3[11\overline{2}0](0001)$ and $1/3[11\overline{2}0](1\overline{1}00)$ edge dislocations with typical point defect clusters found in displacement cascades are provided is sections 7 and 8, respectively. And the results are discussed and conclusions drawn in Section 9.

# 2. SIMULATION TECHNIQUE

## 2.1. MD Method

The simulations are performed for a model of pure $\alpha$-zirconium based on the equilibrium short-range many-body potential of Ackland et al. [6] modified by fitting to the screened Coulomb repulsion [7] at short distances. However, due to an error the published potential does not provide necessary smoothness at the knot point $x_2$ defined in [6]. We have therefore used the correct version from [8] which, for a crystal at temperature, T, equal to 0 K gives a basal lattice parameter, a=0.32490 nm; a lattice parameter ratio c/a=1.5952; a cohesive energy of 6.25015 eV; a vacancy formation energy of 1.786 eV, a self-interstitial atom formation energy of 3.76 eV (the latter is for the basal dumbbell, which is essentially degenerate with the basal crowdion), the melting temperature T=1778K and the energy of the intrinsic stacking fault in basal plane $I_2$=80 mJ/m$^2$. The intrinsic stacking fault in the prism plane is unstable. The threshold displacement energy along the <11$\bar{2}$0> crystallographic direction is equal to 27.5±0.5 eV at $T$=0K. This interatomic potential is the same as that used in the cascade simulations of bulk Zr in [9,10] and of Zr with a surface [11], and for modelling of point defect diffusion and SIA cluster mobility in Zr [12,13].

Three ambient crystal temperatures, namely 100, 300 and 600K are considered. A low temperature (such as 100 K) is used for most previous computer simulation studies, whereas the higher temperature range is more consistent with conditions met by metals in practical situations.

The MD simulation box is close to cubic in shape with (0001), (11$\bar{2}$0) and (1$\bar{1}$00) faces. Periodic boundary conditions at constant volume are applied along all three principal axes. The box size is chosen in accordance with primary knock-on atom energy. For Epka equal to 10 keV and 15 keV it contains 1 016 736 atoms and for Epka equal to 20 keV and 25 keV it contains 2 038 400 atoms. These numbers are approximately four times those used in the earlier studies reported in [9,10] for a similar Epka and T condition, thereby leading to a smaller increase in lattice temperature once the PKA energy has dissipated. No energy/temperature damping is employed. The temperature increase does not exceed 50K in any of the simulations.

To ensure zero total pressure, the lattice parameter is fitted to the ambient temperature, see Table 1. The same c/a ratio of 1.5952 is taken for the whole temperature range. Prior to initiation of the PKA the model crystals are equilibrated for about 40-70 ps depending on the crystal temperature.

**Table 1. Lattice parameter as a function of crystal temperature**

| T, K | 0 | 100 | 300 | 600 | 900 |
|---|---|---|---|---|---|
| a , Å | 3.2490 | 3.2505 | 3.2540 | 3.2623 | 3.2730 |

From 25 to 30 cascades are simulated for each PKA energy and temperature in order to ensure statistical reliability of the results and obtain as many point defect clusters of various configurations, shapes and sizes as possible. In the same set, PKAs are introduced along high index crystallographic directions like <4$\bar{3}$1$\bar{3}$>, <5$\bar{3}$2$\bar{4}$>, <7$\bar{4}$3$\bar{5}$> and <2$\bar{1}$$\bar{1}$1> in order to avoid recoil atom channeling events which are quite possible especially at low temperatures and high PKA energies. To study channelling by MD it would be necessary to use boxes with unreasonably large numbers of atoms. However, a channelled atom experiences a low rate of energy loss until it is defocused and so channelling would not have a significant effect on the data reported here. This would not be the case at higher energy where sub-cascade formation becomes dominant. Within the same series, PKAs are created in different places of MD cell after different equilibration times so that (quasi)random spatial and temporal PKA distributions are simulated. The centre of the MD box is shifted to the centre of gravity of kinetic energy of the constituent zirconium atoms to avoid boundary effects. Furthermore, the likelihood of a cascade extending beyond the box boundary and re-entering via periodic conditions is minimised. This is reflected in the small number of simulations abandoned in this research: only three events in more than 300 cascade simulations had to be stopped due to adverse channelling. It also lessens the selectivity effect in which there is a tendency for debris from the more compact cascades to be over-represented in the final statistics.

## 2.2. Modelling the Cascade Ballistic Stage

In the early stage of a displacement cascade, a small proportion of atoms recoil with high velocity while the rest satisfy the equilibrium velocity distribution for the ambient temperature. However, convergence of the MD algorithm for integrating the equations of motion of the atoms is governed by the time-step, $t_s$, of the fastest atom, and the time-step for a PKA with energy $E_{PKA} = 20$ keV is several orders of magnitude lower than that for atoms equilibrated at $T = 100$ K, for example. Use of small $t_s$ over all atoms during the initial stage of a cascade is computationally inefficient, and in order to accelerate calculations in this stage a variable time step algorithm [14, 15] is implemented, as follows.

Atoms in MD cell are divided into subsets 'hot' and 'cold'. The former consists of atoms with velocity $v$ in the range $pv_{max}$ to $v_{max}$, where $v_{max}$ is the velocity of the fastest atom at that time, and atoms in layers of thickness $d_{hot}$ surrounding these fast atoms. The latter comprises the rest. This designation allows separate integration of two micro-canonical ensembles that have essentially different parameters (temperature distribution and time-step). Time-step $t_s$ is determined by the condition that the fastest atom does not travel by more than $qa_0$ in one iteration. Integration of the equation of motion of atoms in the hot region(s) is applied with $t_s$ calculated repeatedly in this way for $n_{hot}$ steps while the cold atoms remain stationary. The total time, $t_{cold}$, of the $n_{hot}$ steps is calculated, following which the cold atoms are moved with time-step $t_{cold}$ and their new velocities obtained. The velocity distribution in the system is then analysed, new hot and cold volumes are determined and separate integration is repeated for the two subsets. The shape of the hot region(s) depends on the PKA energy, crystal temperature and the stage of cascade evolution. It is rather irregular at the beginning, especially for high PKA energy and low temperature when processes such as channelling and sub-cascade initiation occur. As the primary and higher-order knock-on atoms exchange kinetic energy with surrounding cold matrix atoms, the hot volume grows and eventually the distinction between hot and cold disappears. When the whole simulated volume is assigned as hot, the equations of motion of all atoms are integrated with the same time-step. The velocity-verlet algorithm, see e.g. [16], is applied for integration throughout.

The parameter set for this technique is chosen from trial simulations to test for total energy conservation. The settings are as follows $p = 1/30$, $q = 0.007$, $n_{hot} = 10$ and $d_{hot}$ is chosen to be equal to the effective range of the interatomic potential. This choice errs on the conservative side and defines the two ensembles of hot and cold atoms such that they stay in thermodynamic equilibrium with each other. Depending on the energy of the fastest atom the time step varied from 0.0015 fs just after introduction of the PKA to 4.2 fs by the end of simulation. The kinetic energy of the PKA absorbed by the atoms in the box is not extracted or dumped: the corresponding temperature increase depends on PKA energy and box size, but does not exceed 50K, and does not affect the simulation results [17, 18].

Both ensembles of hot and cold atoms are micro-canonical and total energy conservation within the usual fluctuations is therefore automatically fulfilled. There cannot be direct coincidence of damage caused by displacement cascades evaluated using accelerated and conventional techniques because no two cascades are exactly the same, even with the same simulation method. However, the conditions for energy conservation with the same interatomic potential ensure that the spectra of total number and clusters of defects are equivalent (within the statistical variations) for the two techniques.

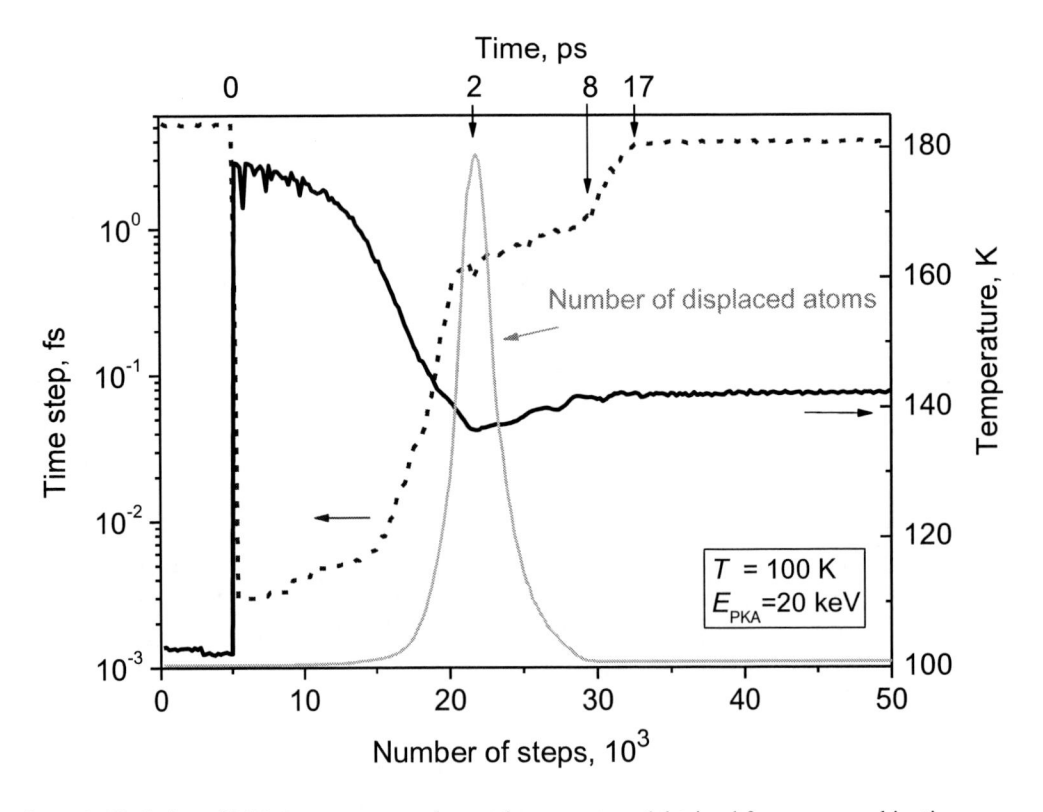

Figure 1. Variation of MD time-step, $t_s$, and crystal temperature (obtained from average kinetic energy of the atoms) with number of time steps for a 20keV cascade in a model initially at 100K. The number of displaced atoms obtained by the equivalent spheres analysis is superimposed.

Figure 1 provides an example of the evolution of the time-step (broken line), average temperature (black solid line) over the MD cell and number of atoms displaced from their lattice sites (thick grey line) on the number of integration steps for a typical 20 keV cascade in a crystal at $T = 100K$. The time-step is relatively large (~4fs) during initial equilibration of the crystal at 100 K, but then drops by more than three orders of magnitude (to ~0.001fs) when the PKA is introduced. It increased gradually in the ballistic phase as $E_{PKA}$ is distributed by collisions among more and more atoms, and then faster as more kinetic energy is converted to potential energy by displacement of these atoms. After the number of displaced atoms, and thus the potential energy, reached a maximum, most atoms return to lattice sites, resulting in release of the potential energy and corresponding increase in $T$. The time-step increases slowly in this thermal-spike phase until the whole crystal became 'hot' and $T$ stabilises at the final value of 142 K. At this point (~20-30ps) a uniform time-step of about 4 fs is applied. The MD routine is stopped when the number of point defects remains constant for 50 ps ($T = 100K$, 300K) or 30 ps ($T = 600K$). In some cases, short-term evolution of damage is studied by continuing the simulation to 150 ns or more.

## 2.3. Identification of Point Defects and Point Defect Clusters

Unambiguous identification of point defects and their clusters is challenging when a large number of cascades has to be treated. Three different methods are applied here. One is based on analysis of Wigner-Seitz (WS) cells and is used to identify vacant sites (= cells with no atoms) and SIAs (= cells with more than one atom). This method is adequate[1] for quantitative determination of the number of surviving vacancies and SIAs, but is not appropriate for identification of structures with correlated atomic displacements typical of stacking faults and interstitial dislocation loops with perfect Burgers vector. The second is the equivalent spheres (ES) analysis - sometimes called the displaced-atom analysis - based on the position of atoms with respect to lattice sites. For each lattice site the closest distance to an atom is evaluated, and if larger than a threshold value the site is treated as vacant. Contrariwise, for each atom the closest distance to a lattice site is determined, and if larger than the threshold value the atom is treated as 'displaced'. The threshold displacement used here is $0.32a_0$. Evaluation of the number of point defects is not straightforward in this approach. The third method, cluster configuration analysis, uses the output of ES analysis to identify a point defect cluster in the following manner. Since the highest value of the di-vacancy binding energy in zirconium occurs when the two vacancies are in first-nearest-neighbour coordination, and similarly for the two SIAs in a di-interstitial configuration, we assume that, irrespective of its type, a point defect belongs to a cluster provided there is another point defect belonging to the cluster in the first coordination sphere. Cluster type is defined by the imbalance between constituent vacant sites and displaced atoms assigned to the cluster. Provided the number of vacancies exceeds the number of displaced atoms, the cluster is treated as vacancy cluster and vice versa. (If the number of vacancies coincides with the number of displaced atoms, the object under consideration is a fragment of crystal displaced by a shock wave and is excluded from further consideration.) The overall number, $N_{FP}$, of surviving Frenkel pairs is the sum of vacancies (not vacant sites) over all single vacancies and vacancy clusters, $N_v$, which equals the sum of SIAs (not displaced atoms) over all single SIAs and SIA clusters, $N_i$. (Conservation of atom number is assured by checking that $N_v = N_i$.)

---

[1] In certain cases, particularly at elevated temperatures in strained regions when three atoms are in the same cell, Wigner-Seitz cell analysis can produce an incorrect number of Frenkel pairs.

In order to identify stacking faults, the local geometry analysis has been performed. For this the number and position of the first neighbours are checked for each atom. 12 neighbours in HCP coordination correspond to a regular atom of the perfect HCP structure; 11 or 10 first neighbours occur near point defects, at the edge of point defect clusters or dislocaitons. 9 HCP atoms of 12 first neighbours in HCP coordination arise for FCC atom. Lower coordination numbers can occur, e.g. in a dislocation core.

## 3. NUMBER OF FRENKEL PAIRS

The variation of $N_{FP}$, with $E_{PKA}$ four all three temperatures is shown in Fig. 2, where each point is the result from one cascade. The Figure emphasizes the dispersion of the $N_{FP}$ values for a given condition and points to the importance of our intention to obtain reliable and statistically comprehensive values for the population of point defects by conducting a large number of simulations for each $E_{PKA}$ and $T$ combination.

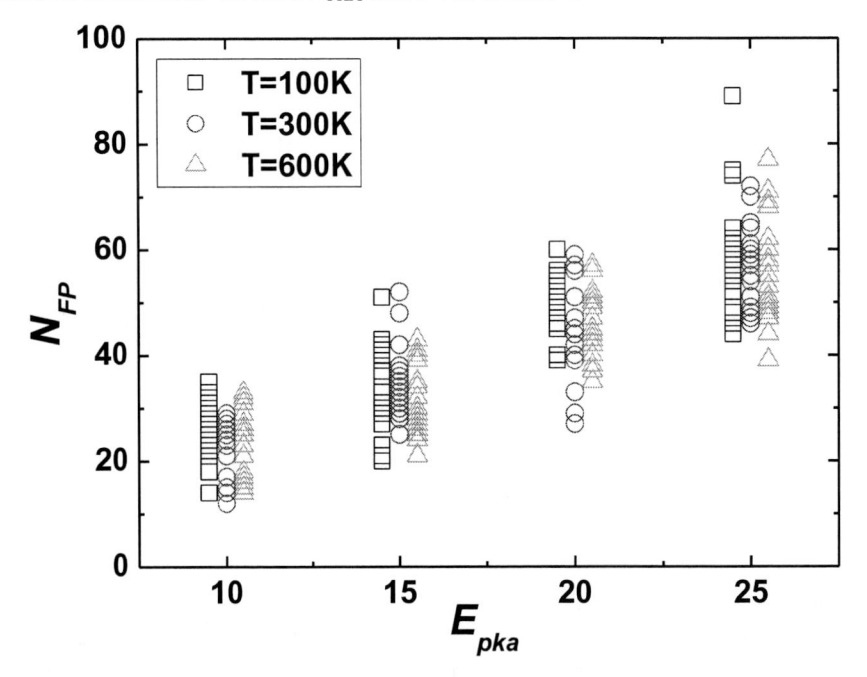

Figure 2. Number of Frenkel pairs vs PKA energy for the three irradiation temperatures. Each point is the result from one cascade and the data are displaced slightly from their actual $E_{PKA}$ values to assist clarity.

The mean value of $N_{FP}$ for all 12 irradiation conditions is shown in Fig. 3. Standard deviation is shown by bars. At the energy and temperature values where they overlap (10 to 20 keV, and 100 and 600 K), the $N_{fp}$ values of Fig.3 are close to those found in the work of Gao et al. [10], who carried out only eight simulations per $(E_{pka}, T)$ pair. Thus the two sets of results can be considered to offer firm data for $N_{fp}$ at 100 and 600 K over the wide PKA energy range from 0.5 to 25 keV.

In contrast to copper, see [19], the number of point defects created in Zr does not depend on the temperature. Recombination of point defects during the cooling phase of a cascade depends on several factors. High temperature facilitates migration of point defects and decreases the cooling rate of the cascade region. However dense cascades are more likely to occur at elevated temperatures due to defocusing of the recoil atoms and can lead to formation of clusters of point defects and therefore reduce recombination. At the same time, stability of point defect clusters depends on temperature. Therefore, we now consider features of point defect cluster formation in displacement cascades in α-zirconium.

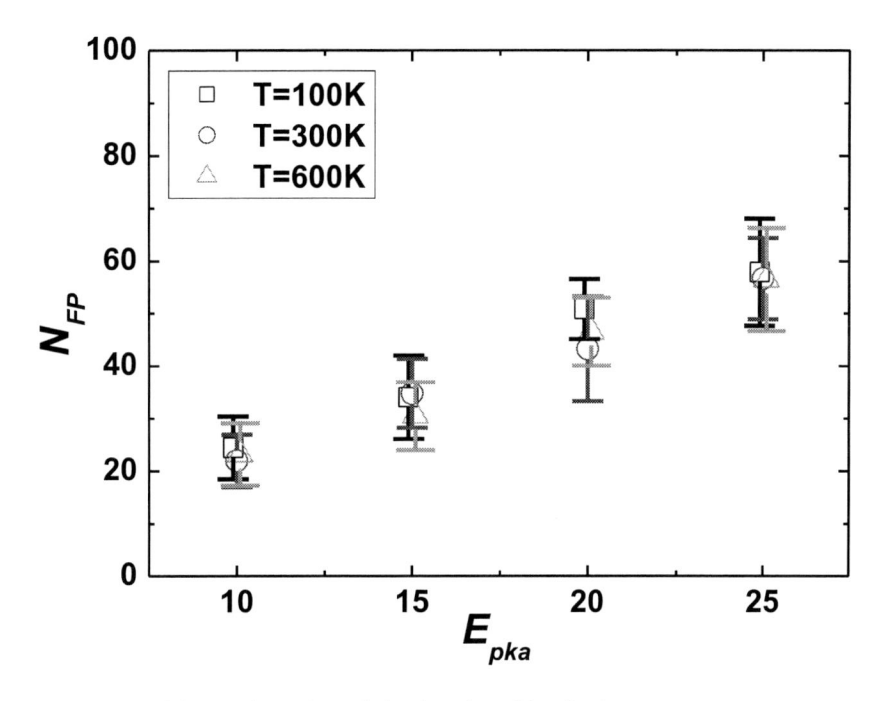

Figure 3. Mean value of the number of Frenkel pairs plotted in Fig. 2.

## 4. FRACTION OF POINT DEFECTS IN POINT DEFECT CLUSTERS

The fraction of self-interstitial atoms, $\varepsilon_i$, in SIA clusters of four and more and of vacancies, $\varepsilon_v$, in vacancy clusters of three and more are shown in Figs.4 and 5, respectively. We consider interstitial clusters of four or more SIAs because clusters of this size range, when in the form of a close-packed set of $<11\bar{2}0>$ crowdions, exhibit rapid one-dimensional glide without change of direction in MD simulations. The minimum size of three for vacancy clusters is made because the trivacancy cluster composed of three vacancies making a regular triangle in one basal plane and one vacancy created by the displacement of an equidistant zirconium atom from one of the nearest neighbour basal planes has high binding energy and produces the smallest possible stacking fault similar to stacking fault tetrahedron in FCC crystal structure.

Roman Voskoboinikov

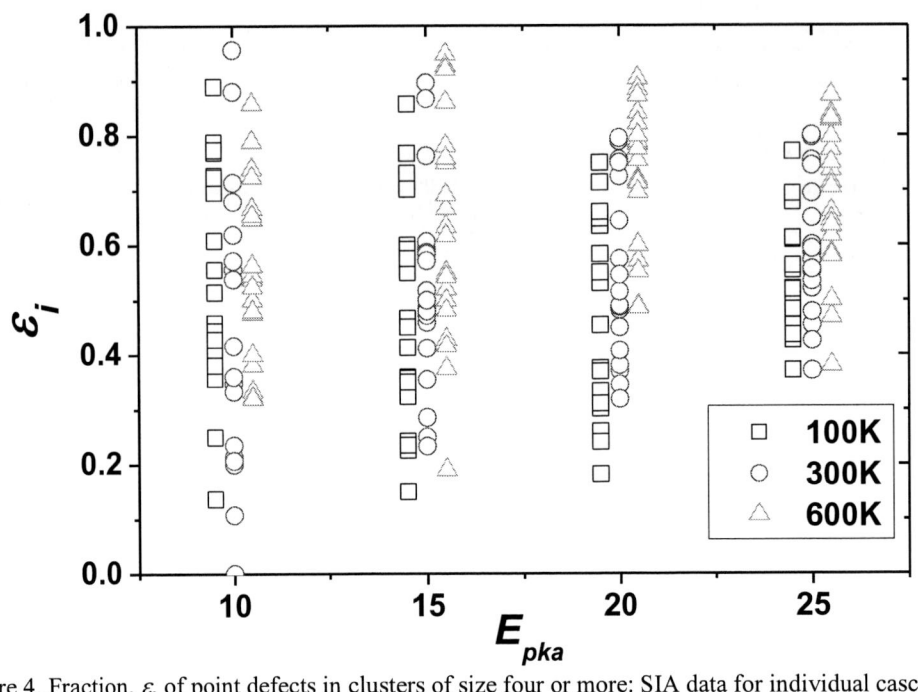

Figure 4. Fraction, $\varepsilon$, of point defects in clusters of size four or more: SIA data for individual cascades.

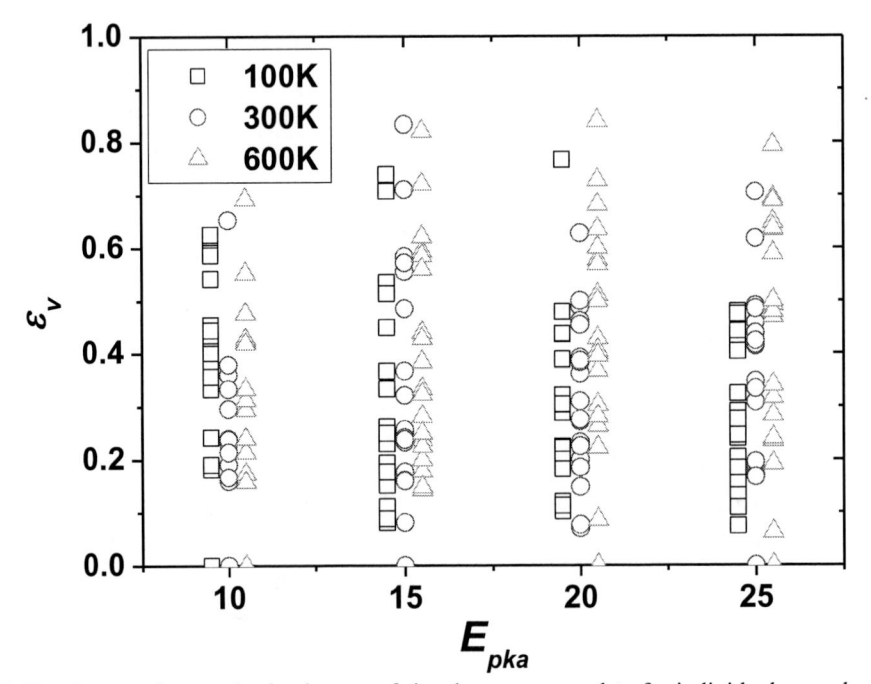

Figure 5. Fraction, $\varepsilon$, of vacancies in clusters of size three or more: data for individual cascades.

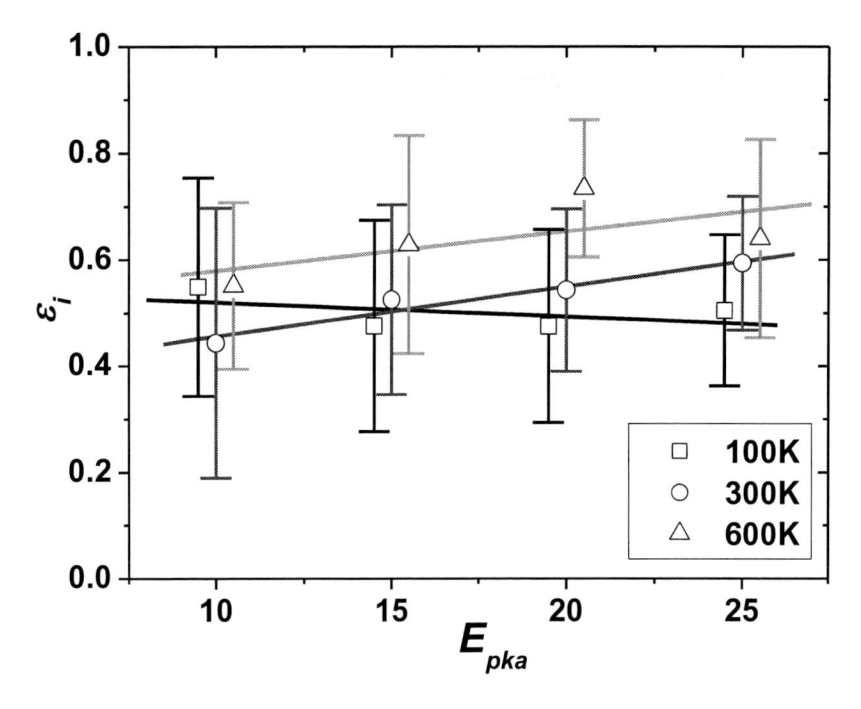

Figure 6. Mean value of $\varepsilon_i$ plotted in Fig. 4 with standard deviations.

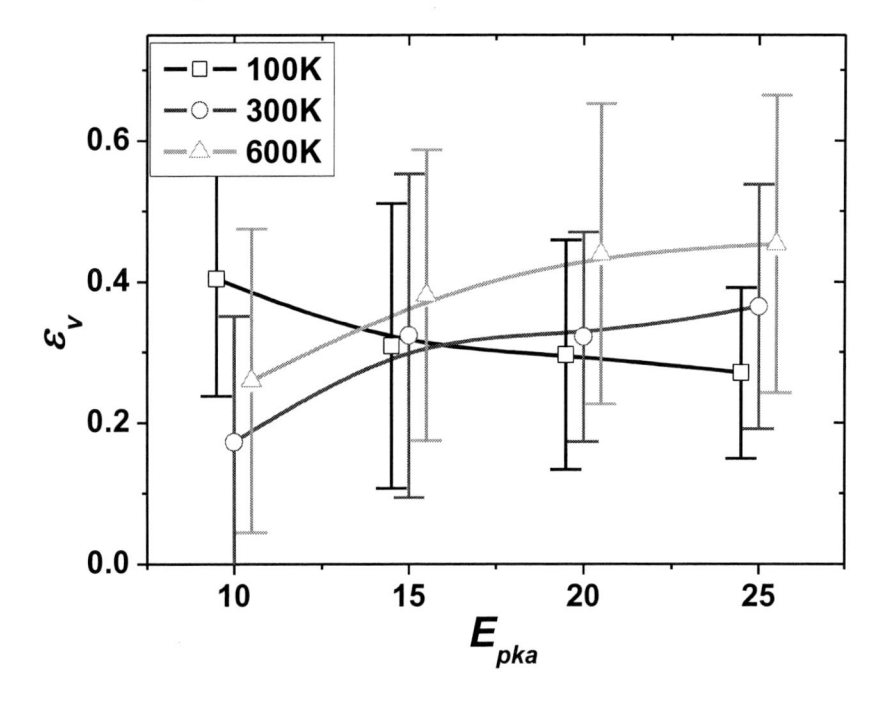

Figure 7. Mean value of $\varepsilon_v$ plotted in Fig. 5 with standard deviations.

The mean values of $\varepsilon_v$ and $\varepsilon_i$ are shown in Figs. 6 and 7. Standard deviations of the mean values are shown by bars. The dispersions of $\varepsilon_v$ and $\varepsilon_i$ are quite wide. Nevertheless, it is clear that both $\varepsilon_v$ and $\varepsilon_i$ manifest similar behaviour. Moreover, the clustered fraction of self-interstitials is only weakly dependent on $E_{pka}$ and $T$, that of vacancies more so. $\varepsilon_i$ is larger than

$\varepsilon_v$ for the same $E_{pka}$ and $T$ but both exhibit an increase with increasing $E_{pka}$ at 300 and 600 K and a decrease at 100 K. Several factors prevent clustering of point defects at low temperatures. First, displacement cascades are not localized. Collision cascades are extended along the projectile trajectory with a tendency to create sub-cascades. We registered formation of two sub-cascades in one simulation with PKA energy of as low as 15 keV and $T$=100 K. As a consequence point defects are created at relatively large distances from each other and do not produce clusters. Second, low temperature and fast cooling rate 'freezes' vacancies, and at the same time, self-interstitials are created in the form of crowdions along $<11\bar{2}0>$ and can move away from the cascade core region to the periphery.

It should be noted that SIAs have a higher affinity to cluster in comparison with vacancies for all temperature and PKA energy ranges. This is reflected by Fig.8, where each point corresponds to the $\varepsilon_v$ and $\varepsilon_i$ values for one cascade. It is clear that the great majority of cascades fall into the upper triangular region, i.e. the relation $\varepsilon_i / \varepsilon_v \geq 1$ is nearly always satisfied.

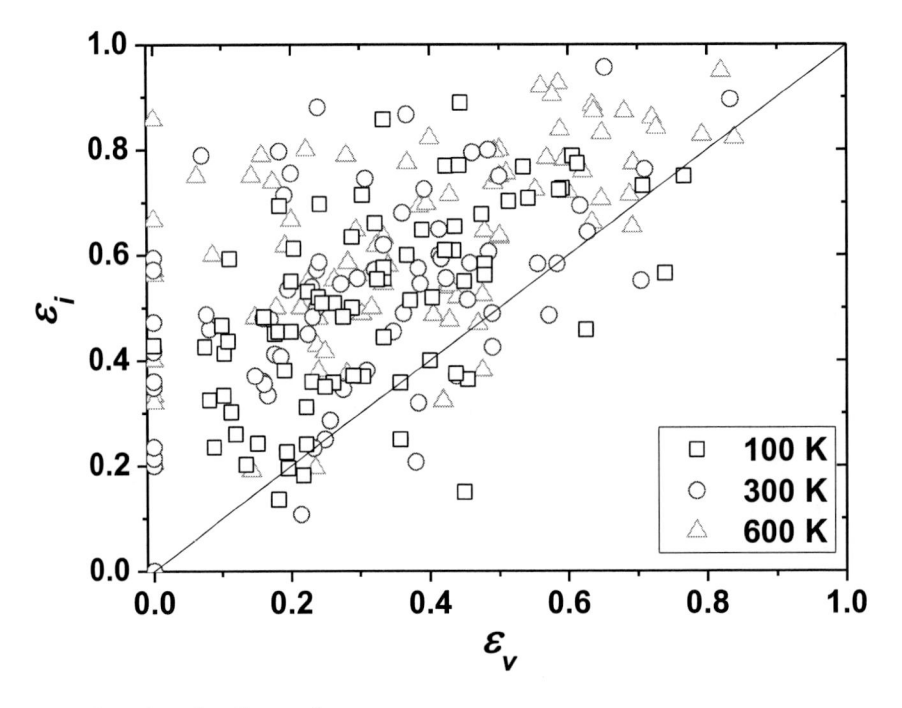

Figure 8. $\varepsilon_v$ and $\varepsilon_i$ values for all cascades.

More than 300 displacement cascades with different PKA energy and temperature have been simulated and they generated a large number of vacancy and self-interstitial clusters of various shapes and sizes. Despite the large number, the clusters can be categorized into only three typical configurations each for vacancy and SIA clusters. We consider SIA clusters first.

## 5. TYPICAL POINT DEFECT CLUSTERS FOUND IN DISPLACEMENT CASCADES

### 5.1. SIA Clusters

Of the three typical configurations of SIA clusters created in displacement cascades in pure α-Zr, the most common is the interstitial dislocation loop with Burgers vector $1/3<11\bar{2}0>$. Such loops contain from 55 to 75 % of all SIAs residing in clusters containing four or more self-interstitials. The fraction of SIAs in these dislocation loops is nearly constant over the whole temperature and PKA energy range. A perfect dislocation loop is formed by a set of closely-packed, parallel crowdions with their axes along a close-packed direction. Having crowdion nature such loops can glide one-dimensionally along the direction of their Burgers vector with very low migration energy [13]. They have high binding energy and are stable during MD annealing for 10 ns up to 1200 K. An example of three such dislocation loops is presented in Fig.9. This is the last frame (38 ps after PKA introduction) of a 25 keV cascade at 600 K. A vacancy cluster of 45 vacancies in the centre of cascade region is accompanied by three glissile dislocation loops of 13, 16 and 25 SIAs. The total number of SIAs in the loops is comparable with the number of vacancies in the vacancy cluster. Perfect interstitial loops are seen to oscillate in the vicinity of a vacancy cluster and do not glide away due to elastic interaction with the stress field of the vacancy cluster. A correlation in space and size of vacancy and SIA clusters has also been found to occur for another metal with a close-packed crystal structure, namely FCC copper, see [20].

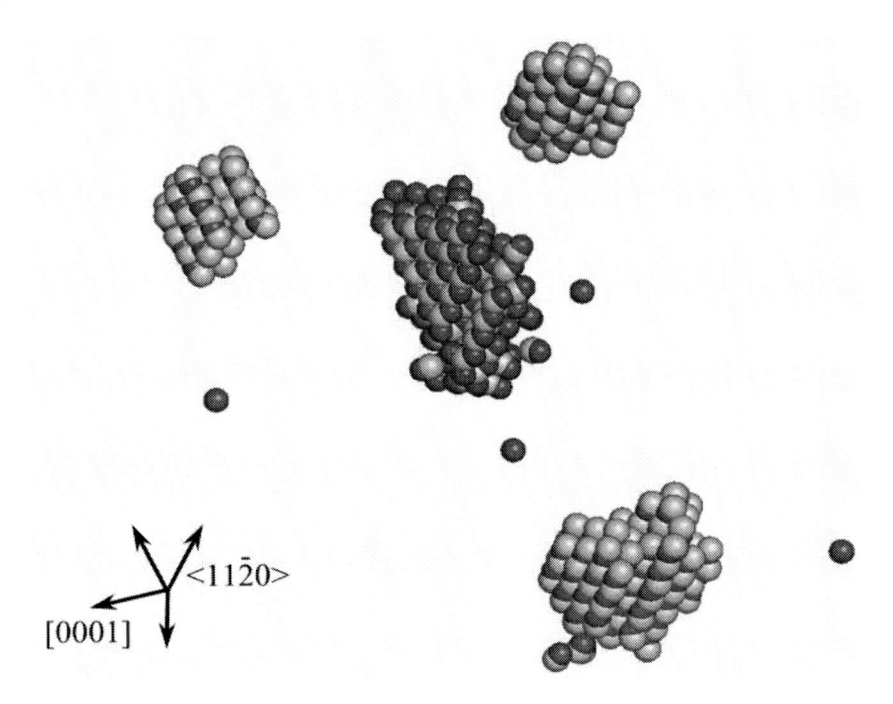

$<11\bar{2}0>$

[0001]

Figure 9. Three glissile SIA clusters of 13 SIAs (left), 16 SIAs (top) and 25 SIAs (bottom) with Burgers vector $1/3<11\bar{2}0>$ in the surrounding of a faulted prismatic vacancy loop formed in a 25 keV cascade at 100 K. Vacant sites are red (dark) spheres, and displaced atoms are cyan (light).

Another important type of cluster is a 3D arrangement of SIAs. A sample cluster of this type is shown in Fig.10. All such clusters are extended along the [0001] crystallographic direction. We did not detect any regular structure of displaced atoms in such clusters, and annealing at temperatures up to 900 K for 3 ns does not change the disordered internal structure. These clusters are sessile at 100 K and 300 K. However, at 600 K and above they start to diffuse in three dimensions. At 100 K approximately 20 % of the total number of SIAs existing in clusters is formed in these irregular 3D arrangements, whereas at 300 K and at 600 K this fraction is increased up to 40 %. There is no noticeable dependence of fraction of SIAs in irregular clusters on PKA energy.

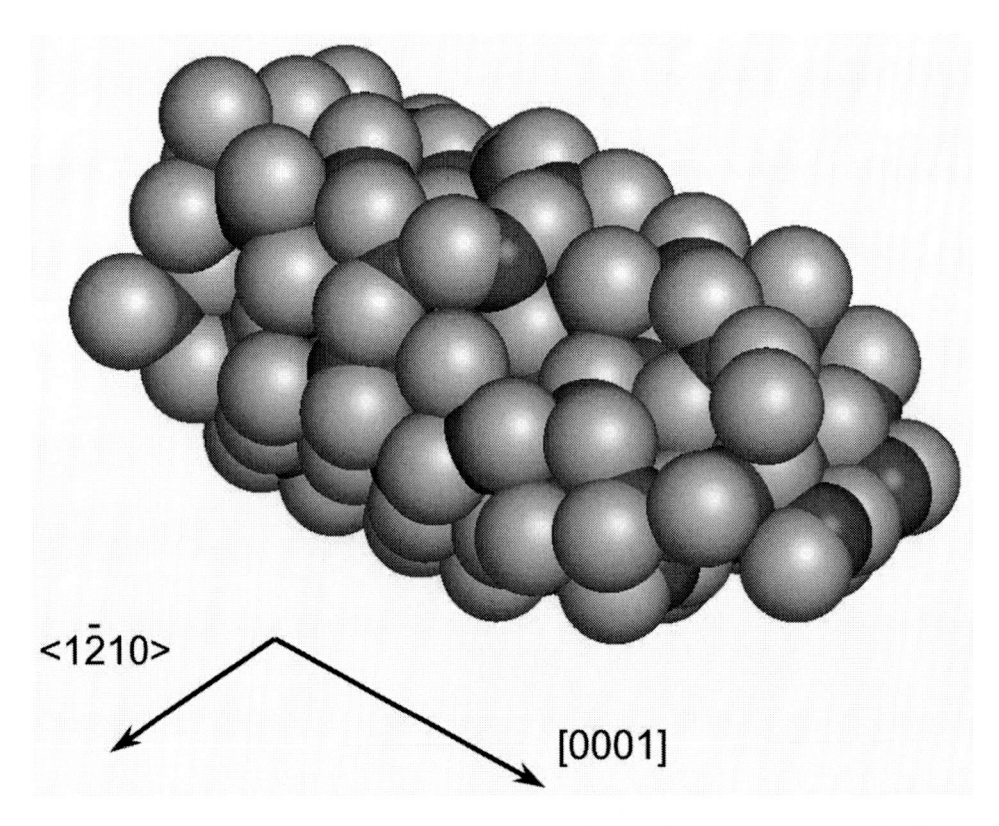

Figure 10. Irregular 3D cluster of 34 SIAs formed in a 25 keV cascade at 600 K. 122 displaced atoms share 88 lattice sites.

The third type of interstitial cluster is a triangular arrangement of SIAs within one basal plane. We found triangular clusters comprising four, five and six self-interstitials. An extended defect of this type was first seen with static simulations using a pair potential [21] and was a tri-interstitial consisting of six atoms symmetrically arranged around three lattice sites. Similar tetra- and penta-interstitials are found in low-energy displacement cascades in zirconium and titanium [9]. The fraction of SIAs in triangular clusters is shown in Fig.11. The maximum number is created in displacement cascades at 100 K and they arise at all PKA energies. The number of these clusters decreases at 300 K, but appear at all simulated PKA energies. At 600 K only a few triangular clusters are formed at PKA energies of 15 keV and 20 keV, and none arose at 10 and 25 keV.

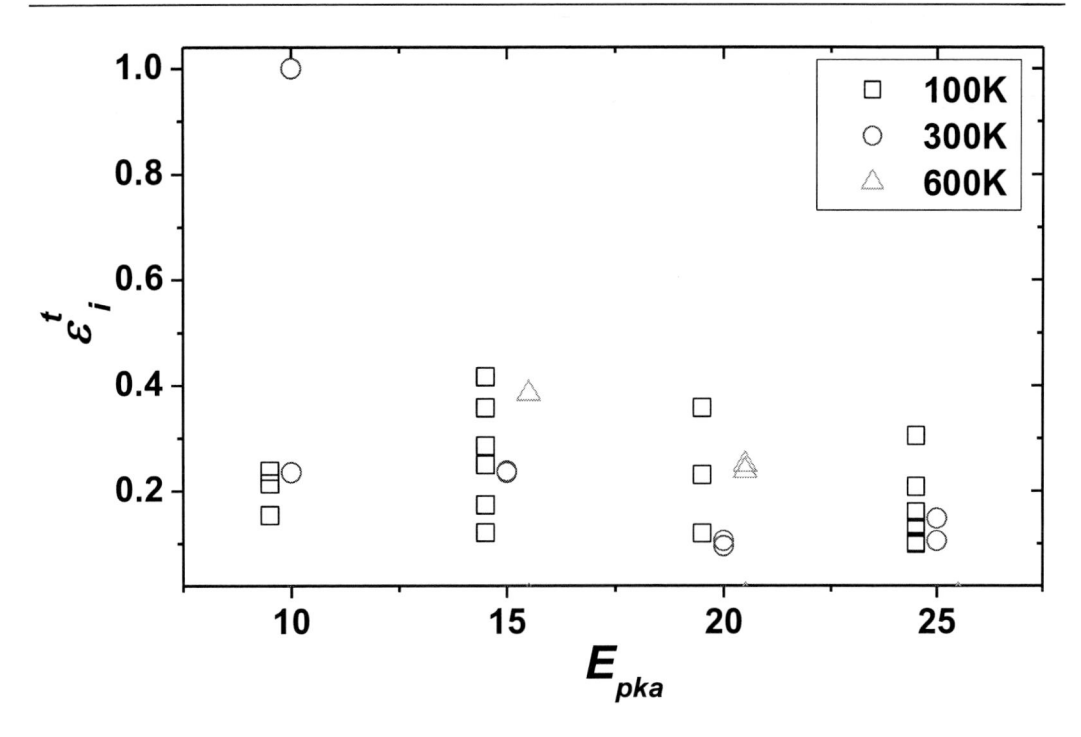

Figure 11. Fraction of SIAs in triangular SIA clusters vs PKA energy and crystal temperature.

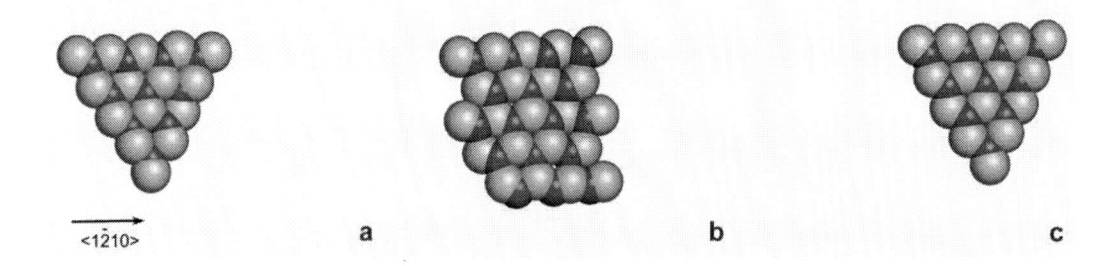

Figure 12. Planar diffusion of a triangular SIA cluster in the basal plane. (a) initial configuration, $\tau = 0$ ps; (b) intermediate glissile configuration, $\tau = 83$ ps; (c) reconstructed initial configuration, $\tau = 84$ ps. Ambient temperature $T = 600$ K.

An example of 5-SIA triangular cluster is shown in Figure 12a. The cluster consists of 15 displaced atoms sharing 10 vacant sites in one basal plane. We annealed this cluster at different temperatures for 7 to 10 ns in order to check its thermal stability. At temperatures from 100 K to 500 K the cluster remained immobile but at 600 K it starts planar motion. The motion occurs through an intermediate glissile configuration, see Fig.12. Due to thermal fluctuations the triangular cluster turns into a set of parallel crowdions lying in the same basal plane, see Fig.12b. Being in a glissile configuration, the cluster could glide along the $<11\bar{2}0>$ axis direction of its crowdions, although the free path is not large (typically only 3-5 lattice parameters). At the end of this 'jump' the cluster recovers its triangular shape, see Fig.12c. Although each particular move of such a cluster is one-dimensional, the macroscopic motion is planar because of the existence of three equivalent $<11\bar{2}0>$ directions in the HCP structure and stochastic choice of direction of jumps.

Jumps of the cluster are rare at 600 K, whereas at 900 K the jump frequency is noticeably higher and the triangular cluster exists in its intermediate glissile configuration for approximately half of the overall time. Above 900 K the cluster loses its triangular shape but still resides in one basal plane whilst undergoing stochastic jumps along one of the $<11\bar{2}0>$ directions. The highest annealing temperature was 1200 K. The cluster manifests high thermal stability and is not destroyed during an annealing time of 7 ns.

## 5.2. Vacancy Clusters

Large-scale MD modelling of collision cascades in α-Zr shows that vacancy clusters can be formed in a wide variety of shapes and sizes. However, on being annealed all vacancy clusters are found to transform into one of three main configurations. The most frequent vacancy cluster created in simulation of displacement cascades in zirconium has a shape close to a triangular prism with three $\{1\bar{1}00\}$ faces and a (0001) base, see Fig.13, or two triangular prisms rotated by 30° with respect to each other, see Fig.14. Such clusters arise at all temperatures and PKA energies, and contain from 50 to 70 % of all clustered vacancies. The dispersion of size and clustered fraction is extremely wide and we do not establish any clear dependence of the fraction of vacancies in prismatic clusters on the temperature and PKA energy.

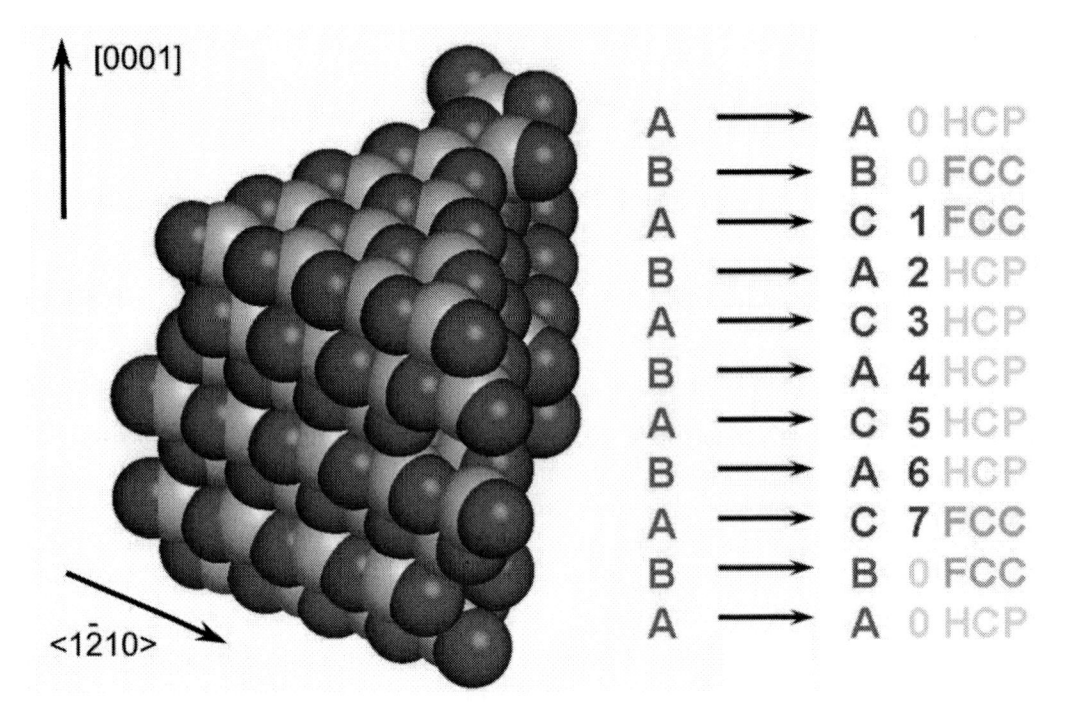

Figure 13. Prismatic vacancy cluster consisting of 42 vacancies (dark spheres). The diagram on the right illustrates the relaxation pattern of the cluster with *odd* number of the constituent close-packed planes. There are two FCC layers ($I_2$ intrinsic stacking fault) on the top and bottom of the cluster.

A distinctive feature of vacancy clusters of this type is that all displaced atoms (light spheres in Figs.13 and 14) lie in basal planes and are not displaced into interplanar positions. The particular shape of prismatic vacancy cluster depends on the number of close-packed planes it occupies. The cluster shown in Fig.13 contains an odd number of vacancy layers whereas that shown in Fig.14 consists of an even number. Relaxation of a prismatic cluster with an even number of layers, as in Fig.13, is prevented by the formation of a high energy fault where two identical close-packed layers arise in neighbouring positions.

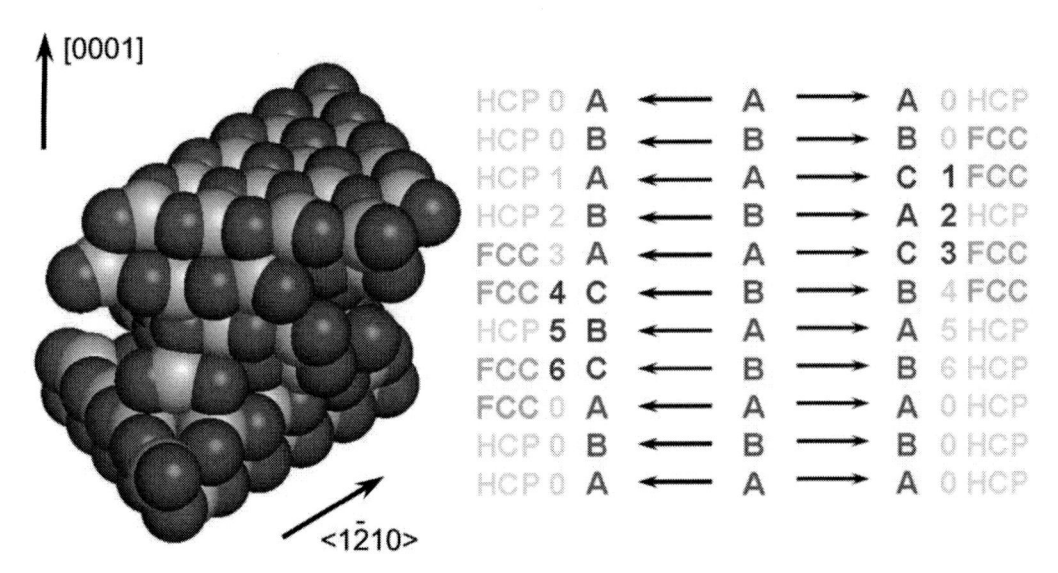

Figure 14. Prismatic vacancy cluster consisting of 36 vacancies (dark spheres). The diagram on the right shows the relaxation pattern of the cluster with *even* number of the constituent close-packed planes. There are two FCC layers ($I_2$ intrinsic stacking fault) on the top and bottom of each prism.

In order to check the thermal stability of prismatic vacancy clusters, annealing at different temperatures is carried out. A cluster similar to that shown in Fig.13 is created in an MD box, equilibrated at low temperature (10 K) with damping of kinetic energy and slowly heated up to a chosen annealing temperature. Annealing time is varied from 5 ns (at 1200 K) up to 12 ns (at 100 K). It is found that this vacancy cluster retains its form and immobility at temperatures below 600 K. At 600 K, however, it transforms into the configuration shown in Fig. 15b and can start gliding along the $<11\bar{2}0>$ direction indicated in a similar way to a SIA dislocation loop discussed in the previous section. In other words, the cluster is converted into a vacancy loop in a prism plane with perfect Burgers vector $1/3<11\bar{2}0>$. Like SIA dislocation loops of large size, such vacancy loops are not observed to change their glide direction, i.e. Burgers vector, to one of the other $<11\bar{2}0>$ directions during any of the undertaken MD simulations.

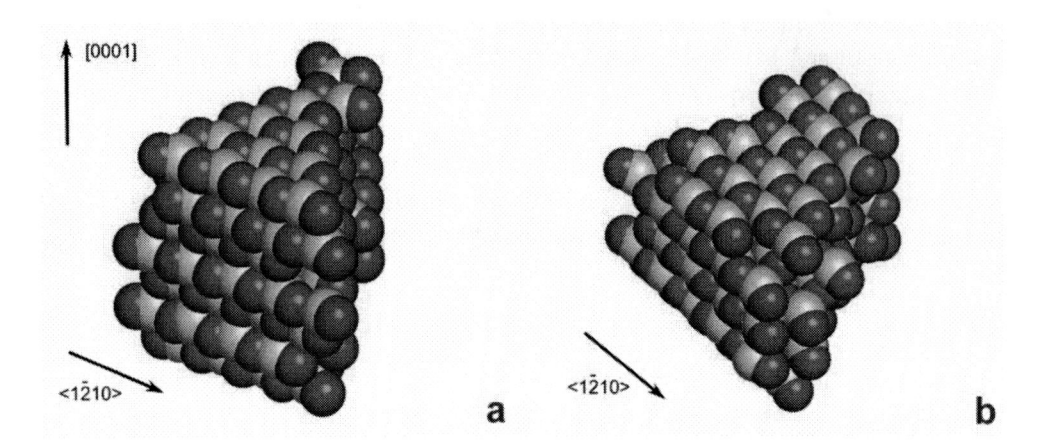

Figure 15. Prismatic vacancy cluster consisting of 42 vacancies. (Vacant sites: dark spheres; displaced atoms: light spheres.) General view (a) and configuration for glide along <$1\bar{1}20$> (b)

However, in contrast to SIA clusters, an increase in temperature does not facilitate glide motion. At 900 K the cluster is immobile and keeps its prism shape and at 1200 K the shape becomes less regular and the internal structure becomes more disordered. This may reflect weaker binding of vacancy clusters compared with interstitial ones, but the vacancy cluster does not lose its integrity entirely at any of the annealing temperatures considered.

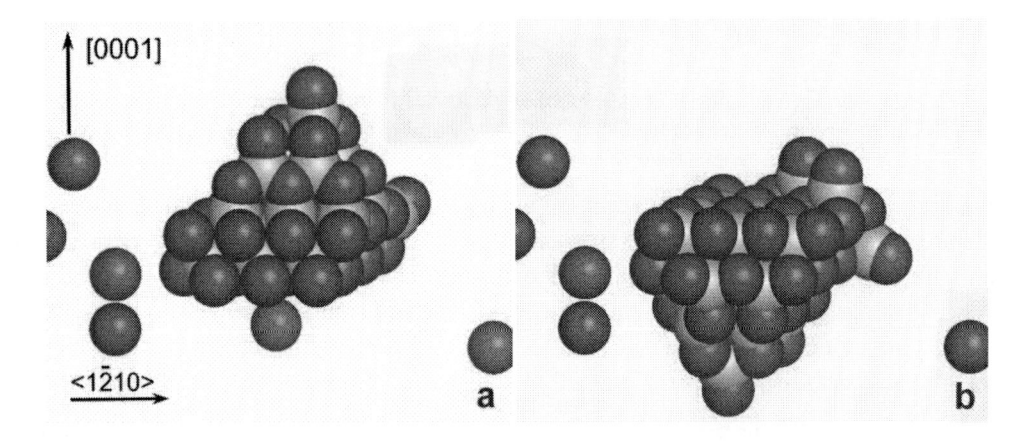

Figure 16. Pyramid-like vacancy cluster consisting of 21 vacancies created by 25 keV PKA at T=100 K. (Vacant sites: dark spheres; displaced atoms: light spheres.) Its form oscillates along [0001] with vertex of the pyramid directed upwards in the left figure (time $\tau$ = 20 ps since introduction of PKA) and, later, pointing downwards, $\tau$ = 25 ps (right figure).Vacancies in the surroundings provide reference points.

The second typical vacancy cluster is shown in Fig.16. It is formed at 100K and has a pyramid shape with the four sides of its (0001) base along two of the <$11\bar{2}0$> directions. The four inclined faces of such cluster tend to be on {$1\bar{1}01$} pyramidal planes. At the first sight it seems that such clusters are similar to the stacking fault tetrahedron in FCC metals with low stacking fault energy like copper. Indeed each displaced atom inside this cluster is situated below a vacant site. It looks like as with the FCC defect, the pyramid is formed by the collapse of crystalline planes above (Fig.16a) or below (Fig.16b) an agglomeration of vacancies in a basal plane. However, in contrast to the FCC case the high energy of stacking fault that would be formed by removal of one plane of atoms in the HCP structure is avoided by shift of one close-packed layer with formation of an extrinsic stacking fault. The transformation is not over during the course of a simulation, and it is observed that such a defect can oscillate between the 'up' and 'down' configurations (see Fig.16) over time.

Details of the relaxed structure arising from vacancy agglomeration in a basal plane depend on the number of vacancies, shape of agglomeration and actual temperature. Another pyramid-like vacancy cluster is shown in Fig.17. It is created in a displacement cascade at 600 K and then relaxed. Local geometry analysis of the cluster, Fig.18, unambiguously demonstrates that it is indeed created by the collapse of crystalline planes above an agglomeration of vacancies in the basal plane and the extrinsic stacking fault is created in the basal plane. Pyramid-like vacancy clusters, shown in Figs 16-18 are immobile.

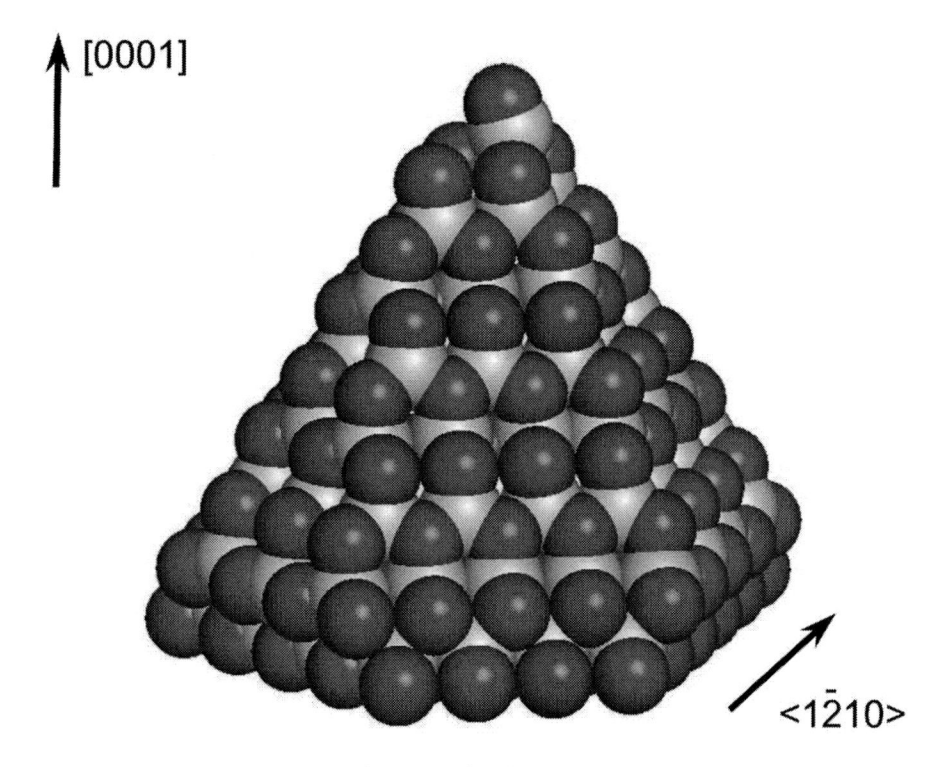

Figure 17. Large pyramid-like vacancy cluster. Dark spheres correspond to vacant sites; light spheres correspond to displaced atoms.

We have checked the stability of the cluster shown in Fig.17 against temperature increase and established that at 900 K it is turned into a prism-like vacancy cluster described earlier, see Fig 19 and compare it with Figs 13-14. The transformed cluster exhibits one-dimensional motion and internal atom arrangements typical for prismatic vacancy clusters. Thus, we can suggest the following chain of transformations of pyramid-like vacancy clusters. The pyramidal cluster with extrinsic fault is stable at low temperature and intermediate temperatures that depend on cluster size and shape; and at high temperatures pyramid-like vacancy clusters are transformed into prismatic vacancy loops.

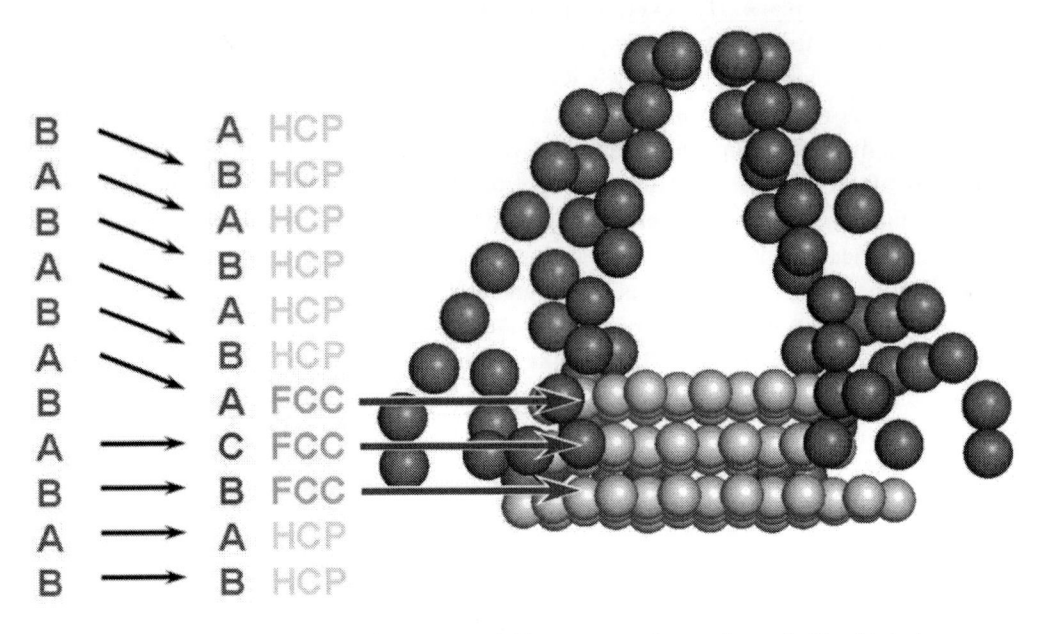

Figure 18. Local geometry analysis of the pyramid-like vacancy cluster shown in Fig.17. Only atoms at the edges of the clusters (dark blue spheres) and atoms in FCC coorditation (light grey spheres) are shown. Three FCC layers form extrinsic stacking fault. The corresponding sequence of close-packed planes is given on the left.

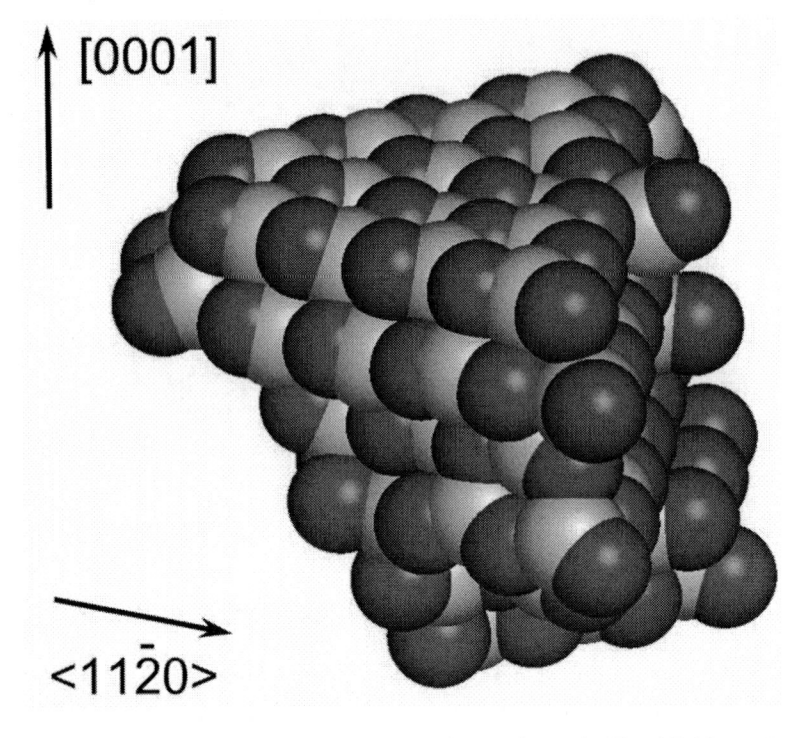

Figure 19. Transformation of the pyramid-like vacancy cluster shown in Figs 17-18 to prismatic vacancy cluster during annealing at 900 K.

In the framework of a statistical treatment of collision cascades it is established that the fraction of vacancies in pyramid-like vacancy clusters is approximately 20 % at 100 K and half this value at 300 and 600 K. We have not detected any statistically significant dependence of the fraction of vacancies in pyramid clusters on PKA energy.

Vacancy clusters of the third type are best described as two parts joint together. One part has the form of a prism-like cluster and the other part has a shape similar to the pyramid cluster. A typical cluster of this type is shown in Fig.20a. The two basal layers at the foot of this cluster form a triangular prism with appropriate arrangement of vacant sites and displaced atoms (see Fig.20b for more detail), whereas the top part is a pyramid-like vacancy cluster. Approximately 20 % of vacancies form clusters of this type at 100 K. The fraction is doubled at 600 K. At low temperature and at high temperature this fraction does not depend on PKA energy, whereas at 300K it increases smoothly from 20 % to approximately 40 % as PKA energy increases from 10 keV up to 25 keV. Again, annealing is performed in order to investigate the thermal stability of clusters of the third type and, unsurprisingly, they are transformed into prismatic ones at high temperatures.

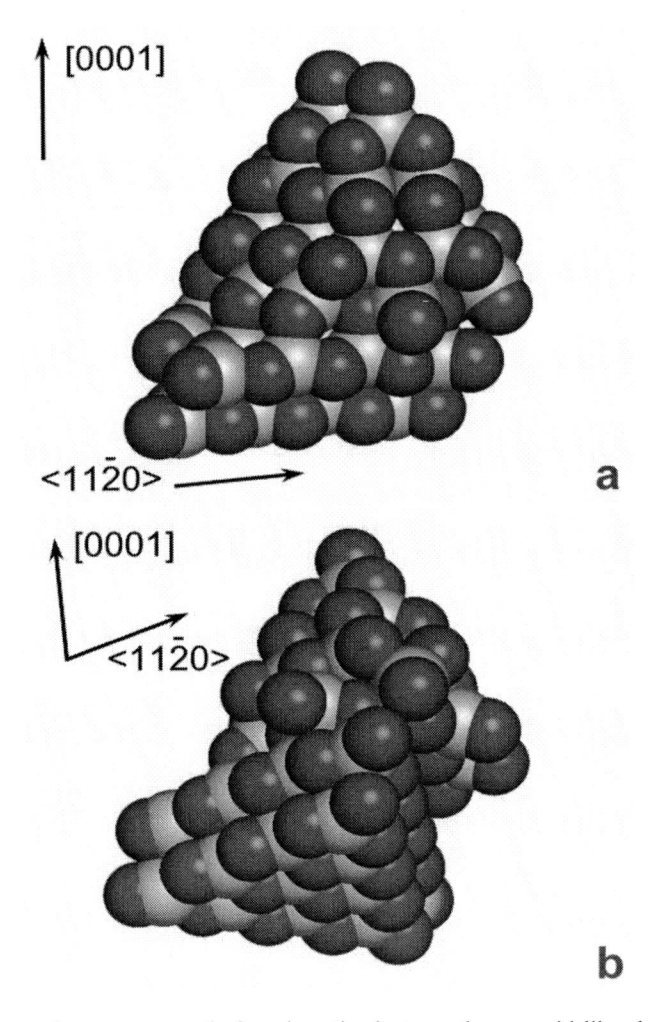

Figure 20. 32-vacancy cluster composed of a prismatic cluster and a pyramid-like cluster. (a) General view; (b) the same cluster tilted and rotated.

# 6. ATOMIC-SCALE MODELLING OF EDGE DISLOCATIONS IN α-ZIRCONIUM

The degradation of mechanical properties of structural materials exposed to fast particle irradiation is strongly influenced by the formation of point defect clusters, which are effective barriers for dislocation motion. Atomic-scale computer simulation of dislocations is an essential part of the modern multi-scale approach to predictive modelling of material service properties, particularly in relation to strength, toughness and plasticity. The results of such simulation provide detailed information about dislocation core structure, dislocation motion and interaction with point defects, their clusters and other obstacles, and so give qualitative understanding of the governing mechanisms. It also allows quantitative evaluation of the stress-strain curve, the critical resolved shear stress and the dynamic and thermal effects. Furthermore, atomic-scale dislocation modelling of special conditions (e.g. T=0) can test and justify analytical concepts of the elasticity theory of dislocations.

Using the output of a large programme of MD simulations of radiation damage in α-Zr, we simulate the interaction of gliding dislocations with point defect clusters found in displacement cascades. The importance of such interactions for localisation of plastic flow by dislocation channelling in prism and basal planes in neutron-irradiated Zr has been assessed recently by Onimus et al. [22]. Atomic-scale details of the interaction of $1/3<11\bar{2}0>$ edge dislocations with (0001) and $\{1\bar{1}00\}$ slip planes with typical vacancy and SIA clusters are investigated. To contrast the effect of different slip systems the dislocation core energy and atomic displacement in the core region have been evaluated. The information gained provides a necessary background for investigation of the interaction of the dislocations with clusters.

## 6.1. Simulation Technique and Identification of Dislocation Core

MD simulations of the dislocations are carried out using the equilibrium short-range many-body interatomic potential for Zr [8] we employed for primary damage modelling. In order to allow long-range (distance $\gg$ b) motion of the dislocation, periodic boundary conditions are applied along the direction of **b** (x-axis) as well as the dislocation line direction (y-axis). Following the method described in [23], the initial unrelaxed model with an edge dislocation is created from two half-crystals $z \geq 0$ and $z < 0$ with x dimension (N+1)b and Nb, respectively. The lower half-crystal ($z < 0$) is elastically elongated by b/2 along x-axis and the upper half is compressed by b/2: the two parts are then joined and relaxed to form a perfect dislocation with x-y slip plane. Although the two half-crystals are strained by $\pm$b/2, the overall elastic deformation of MD cell of size $\sim$200b is small ($<|0.25\%|$). Furthermore, the dislocation glide plane coincides with the neutral stress axis of the model and the dislocation core is in a region of zero stress. It has been confirmed that the effects of the boundary conditions can be neglected (see [23] for details). The model is bounded on the upper and lower x-y faces by boxes of immobile atoms. The lower box is fixed whereas the upper one can be displaced rigidly to apply either a shear strain or stress to the cell of free atoms. In this case, we simulate athermal properties (T = 0K) and so shear strain is applied (box displacement in the x-direction): the corresponding applied stress is computed from the force on the box from the inner region atoms after relaxation.

In this study, identification of the atomic structure of the defect regions is conducted using the local geometry approach described earlier, see section 2.3, where the number and position of the first neighbours are checked for each atom. 12 neighbours in HCP coordination correspond to the perfect HCP lattice, whereas 9 atoms in HCP coordination of 12 first neighbours indicate an FCC arrangement. 11 or 10 first neighbours occur at the edge of partial dislocations and lower coordination numbers can occur in the core of the $1/3<11\bar{2}0>\{1\bar{1}00\}$ dislocation. Following Frank [24] (see also [25] or [26]), three different stacking faults can arise on the basal plane. Three atomic layers with nearest neighbours in FCC coordination define the extrinsic fault, whereas one or two layers represent the $I_1$ and $I_2$ intrinsic faults, respectively.

## 6.2. Atomic Displacement and Structure of Dislocation Core

Figs. 21 and 22 show core structure of the $1/3[11\bar{2}0](0001)$ and $1/3[11\bar{2}0](1\bar{1}00)$ edge dislocations after conjugate gradients relaxation to minimise the potential energy. The former dislocation splits into two Shockley partials in the basal plane whereas the latter does not dissociate. The linear dimensions of the dislocation core can be extracted from analysis of the displacement of atoms originally in the plane $z = 0$, as shown in Figs. 23 and 24 for the basal- and prism-plane dislocations, respectively. Fig. 23 shows the distinctive features of atom displacements, $\mathbf{u}(x)$, in the core of a dissociated dislocation. (Note that only atoms of the upper half-crystal are treated here. If the displacements of atoms at the top of the lower half-crystal are subtracted from these displacements to give the displacement differences across the glide plane, they would exhibit the usual displacement discontinuity b (rather than b/2) in $u_x$.) The component $u_x(x)$ along the glide direction $\mathbf{b}$ has two regions of high gradient at the position of the partials. Their exact location is determined by the extremum of the first derivative $du_x(x)/dx$, see Fig.23, and the corresponding size of the ribbon of $I_2$ stacking fault is nearly 8b. Another characteristic feature of the extended dislocation is the non-zero displacement $u_y(x)$ along the dislocation line, corresponding to the screw component of the partials. In contrast to this, the $1/3<11\bar{2}0>\{1\bar{1}00\}$ dislocation does not create displacements along the dislocation line, see Fig.24, and the displacement parallel to $\mathbf{b}$ is accommodated in a single narrow core. The x-y planes of the crystals are bent due to the presence of both dislocations: the two models have the same dependence of $u_z$ on x outside the core region, but the behaviour in the core of the basal edge reflects the dissociated nature of this dislocation.

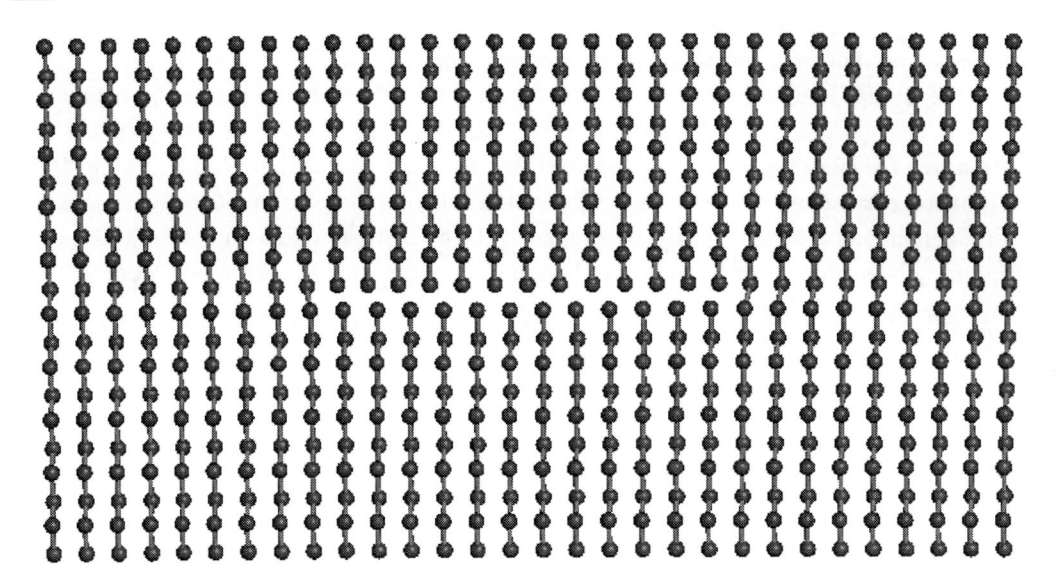

Figure 21. [1$\bar{1}$00] projection of the core structure of the 1/3[11$\bar{2}$0](0001) dislocation, which splits into partials on the basal plane. The (11$\bar{2}$0) planes perpendicular to b are shown by sticks.

Figure 22. [0001] projection of the core structure of the 1/3[11$\bar{2}$0](1$\bar{1}$00) dislocation. The (11$\bar{2}$0) planes perpendicular to b are shown by sticks.

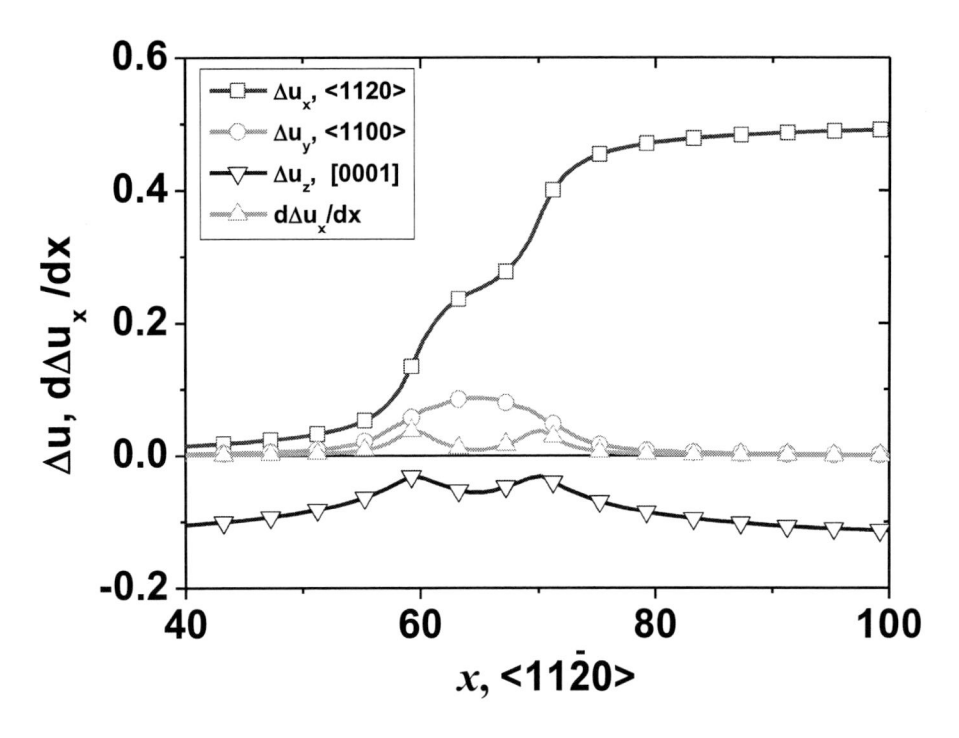

Figure 23. Displacement (units b) of atoms in the bottom plane of the upper half-crystal after relaxation to form the $1/3[11\bar{2}0](0001)$ dislocation.

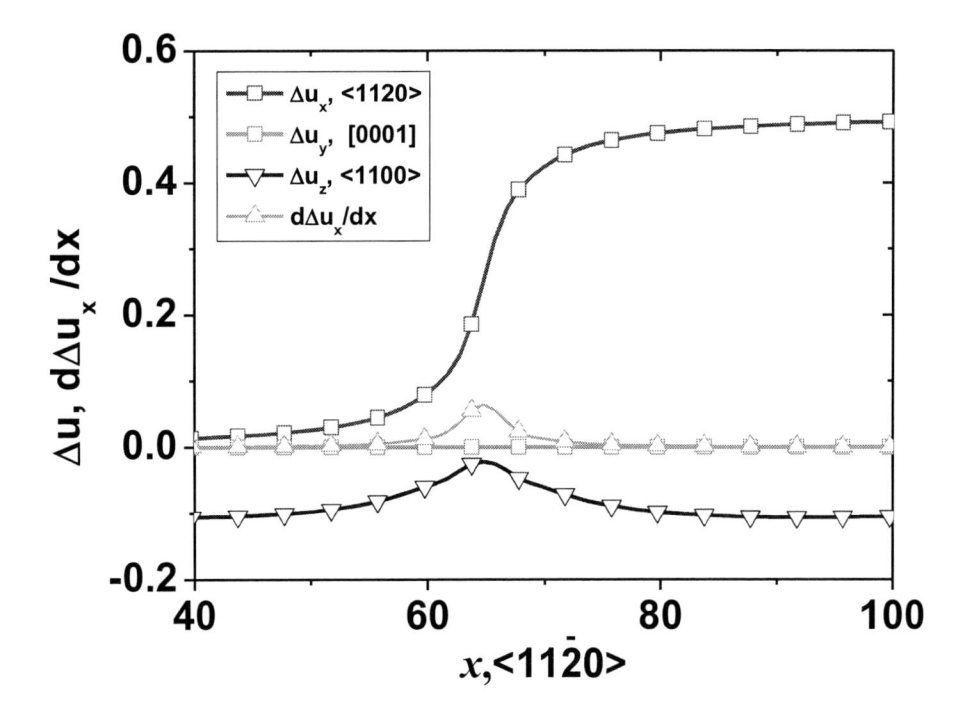

Figure 24. Displacement (units b) of atoms in the bottom plane of the upper half-crystal after relaxation to form the $1/3[11\bar{2}0](1\bar{1}00)$ dislocation.

## 6.3. Peierls Stress and Dislocation Core Energy

In order to determine the critical stress (CRSS) for glide at 0K, i.e. the Peierls stress, the following technique is applied. The initially relaxed dislocated crystal is subjected to a resolved shear strain (acting on the glide plane in the direction of **b**), which is increased in increments by shear displacement of the upper rigid box. After application of each strain increment of $10^{-4}$ the simulated crystal is relaxed using the conjugate gradients method and the corresponding applied shear stress is calculated. The critical (Peierls) stress, $\sigma_P$, is achieved when the dislocation started to glide under increasing strain without further increase in stress. The strain-stress curves for the two slip systems are plotted in Fig.25, from which it is seen that $\sigma_P$ for the basal dislocation is 1.6MPa and that for the prism-slip dislocation is 5.6MPa. The linear dependence of stress on strain before $\sigma_P$ is reached represents Hooke's law and has a gradient equal to the shear modulus for the particular shear orientation. It is 34GPa for $<11\bar{2}0>$ (0001) strain and 31GPa for $<11\bar{2}0>\{1\bar{1}00\}$ strain: these values are close to $C_{44} = $ 36GPa and $C_{66} = $ 33GPa, respectively, for a perfect crystal with the same interatomic potential [8].

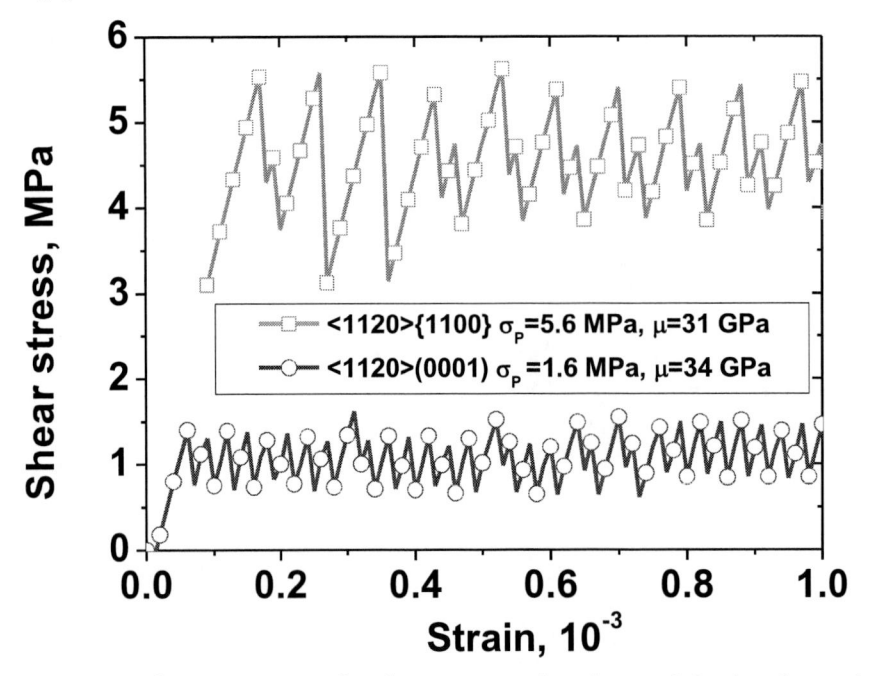

Figure 25. Strain-stress curves for the two crystals orientated for basal or prism glide. Evaluation of the core energy has been conducted using a model crystal of large size. It is relaxed and the energy of the volume inside cylinders of radius R centred on the dislocation core is calculated. The energy per unit length of dislocation line is plotted against the outer cut-off radius, R, (logarithmic scale) in Fig.26. Two regions of the curves can be distinguished. At large R the dislocation energy is proportional to the logarithm of R, as expected from the linear elasticity theory of dislocations, see e.g. [25, 26], with the difference in slope reflecting the anisotropic nature of the shear modulus. At small cut-off distances the linear continuum treatment becomes invalid. The effective radius of the extended dislocation

core is large, but the core energy is relatively low due to the dissociation, whereas the prism-plane dislocation has a small core radius and high core energy.

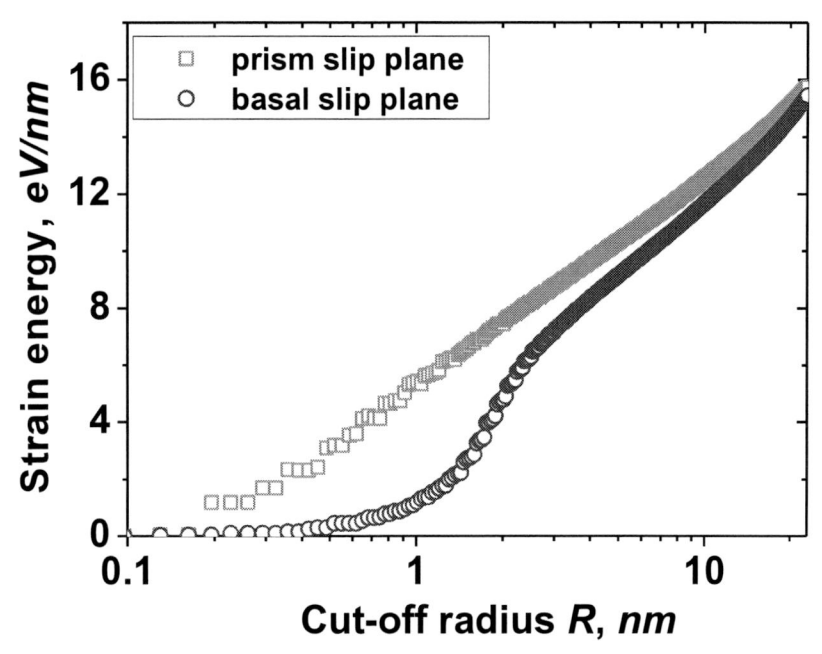

Figure 26. Energy per unit length of dislocation within cylinders of radius R centred on the $1/3[11\bar{2}0](0001)$ and $1/3[11\bar{2}0](1\bar{1}00)$ edge dislocations.

## 7. INTERACTION OF $1/3<11\bar{2}0>(0001)$ EDGE DISLOCATIONS WITH POINT DEFECT CLUSTERS CREATED IN DISPLACEMENT CASCADES IN $\alpha$-ZR

The technique described in the previous section has been adopted to conduct atomic-scale simulation of the $1/3<11\bar{2}0>(0001)$ edge dislocation in the HCP structure. With the interatomic potential used [8], this dislocation dissociates into two Shockley partial dislocations with a stacking fault on the basal plane, see Figs 21 and 23. Thermal effects are not considered here and so simulation is carried out for 0K by relaxation using the conjugate gradients method. A point defect cluster found in displacement cascade simulation is introduced in the relaxed, dislocated model and the equilibration procedure repeated again. A resolved shear strain is then applied, increasing from zero in steps of $10^{-4}$. After each step, the crystal is relaxed and the resultant applied shear stress is evaluated. A local geometry approach described above in section 2.3 is applied to visualise and identify the atomic structure of the dislocation core and defect cluster. A simulation is considered finished when the gliding dislocation overcomes the cluster and the CRSS is taken as the maximum on the stress-strain curve. The model contained more than $3x10^6$ free atoms and the distance between the periodic boundaries along the dislocation line, i.e. the spacing along the row of clusters, is 140b, where b is the magnitude of the Burgers vector **b**.

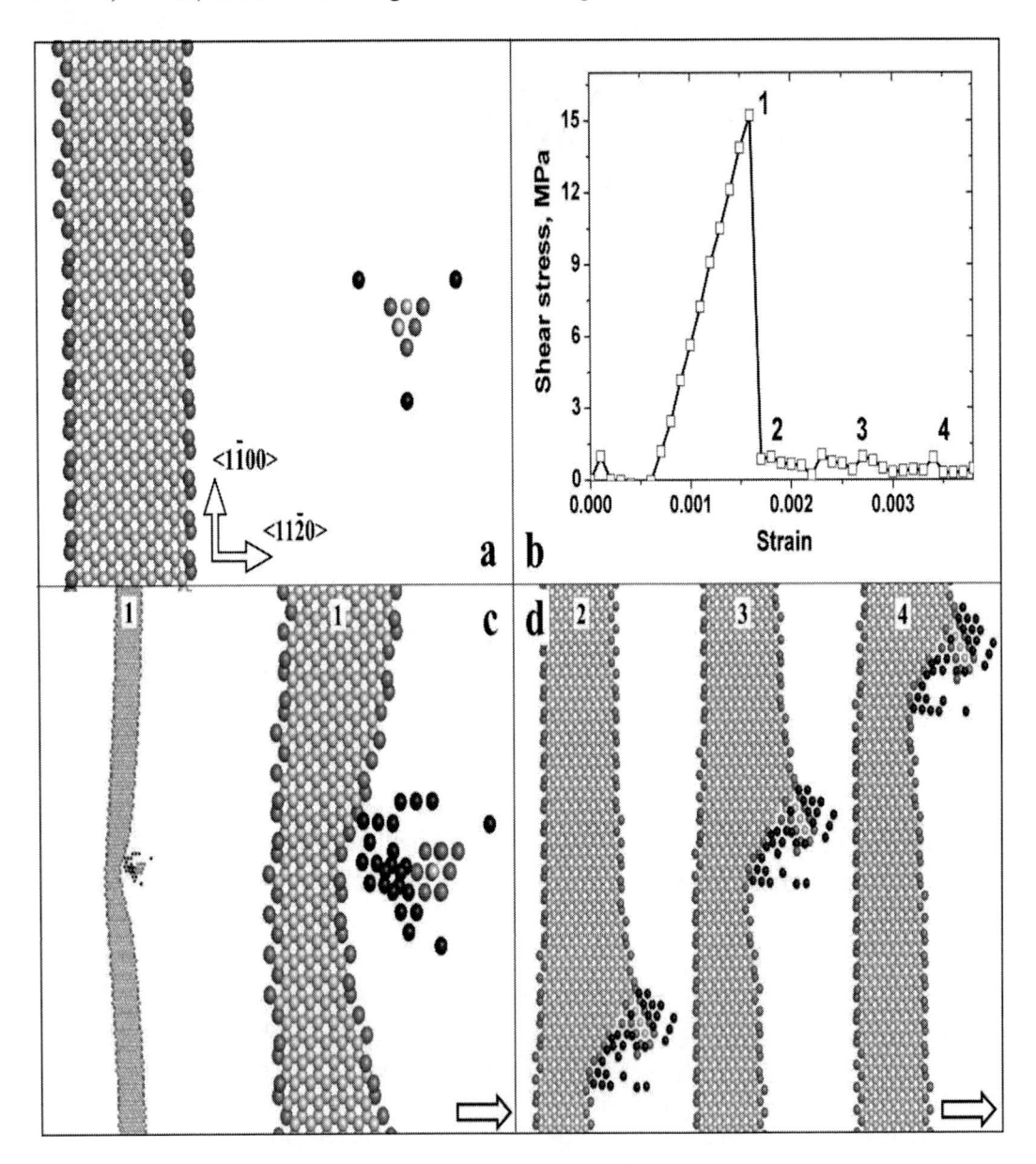

Figure 27. Interaction of the dissociated $1/3[11\bar{2}0]$ (0001) edge dislocation with 5-SIA basal triangular cluster. In this and other Figures, white, grey and dark-blue spheres correspond to 12, 10-11 and 6-8 first neighbours in FCC coordination, respectively. (a) Initial configuration; (b) strain-stress curve; (c) shape of the dislocation line (left) and fine structure at the interaction point (right) at the CRSS (point 1 on strain-stress curve); (d) drag and drift of the cluster (points 2-4 on stress-strain curve). The arrow here and in other Figures shows the direction of dislocation glide.

## 7.1. Interaction with Triangular SIA Cluster in Basal Plane

According to MD modelling of primary damage, see section 5.1 of this review, nearly 10% of all SIAs created in displacement cascades in α-Zr agglomerate in triangular clusters of 4, 5 or 6 SIAs all lying in one (0001) atomic plane. Interaction of the extended dislocation with a 5-SIA cluster (15 atoms sharing 10 lattice sites) lying in basal plane coincident with the bottom of the extra-half plane has been simulated. Fig.27 shows (a) the initial configuration, (b) the stress-strain curve, and (c, d) several snap-shots of the dislocation and cluster as strain increases. The Peierls stress of the dislocation in the absence of a cluster is 1.6MPa, according to section 6.3. Initially there is attraction of the dislocation to cluster and the shear stress falls to near zero, but further strain leads to a linear increase of the stress up to a critical value of 15MPa, see Fig.27b. The dislocation becomes bent and the faulted region is narrowed near the cluster, see Fig.27c. Beyond this, the leading partial dislocation pushes the cluster and, at the same time, the cluster drifts along the partial, see the snapshots in Fig.27d. The dislocation is no longer pinned and the applied stress falls to the Peierls stress value for a cluster-free crystal. The cluster is neither destroyed nor absorbed by the dislocation. The breakaway stress of 15MPa compares with a value of 23MPa for the prism-plane edge dislocation to overcome the same obstacles with spacing of a similar size, see section 8.1 for details.

## 7.2. Interaction with Irregular 3D SIA Cluster

It has been found, see section 5.1, that clusters of this type do not have any regular internal atomic structure and tend to be elongated in a [0001] direction. Relevant snap-shots and the stress-strain curve for interaction between the basal dislocation and such a cluster placed in the glide plane are shown in Fig.28. The cluster contains 34 SIAs and is crossed in its centre by the glide plane of the dislocation. Initially the leading partial dislocation is attracted by the cluster and the fault ribbon widens towards the cluster, see Fig.28b. The corresponding fall in the shear stress is at point 1 on the strain-stress curve, Fig.28c. Further strain increase leads to linear increase of shear stress but with a small drop (point 2 on strain-stress curve) corresponding to penetration of the first partial into the cluster, see Fig.28d. At the eventual critical stress of 24MPa the dislocation constricts and absorbs the SIA cluster completely to form a pair of super-jogs, see Fig.28f. The jogged dislocation continues gliding at a stress level of 4MPa. The breakaway stress of 24MPa compares with a value of 12MPa for the prism-plane edge dislocation for the same obstacles with spacing of a similar size; see section 8.2 of this review for details.

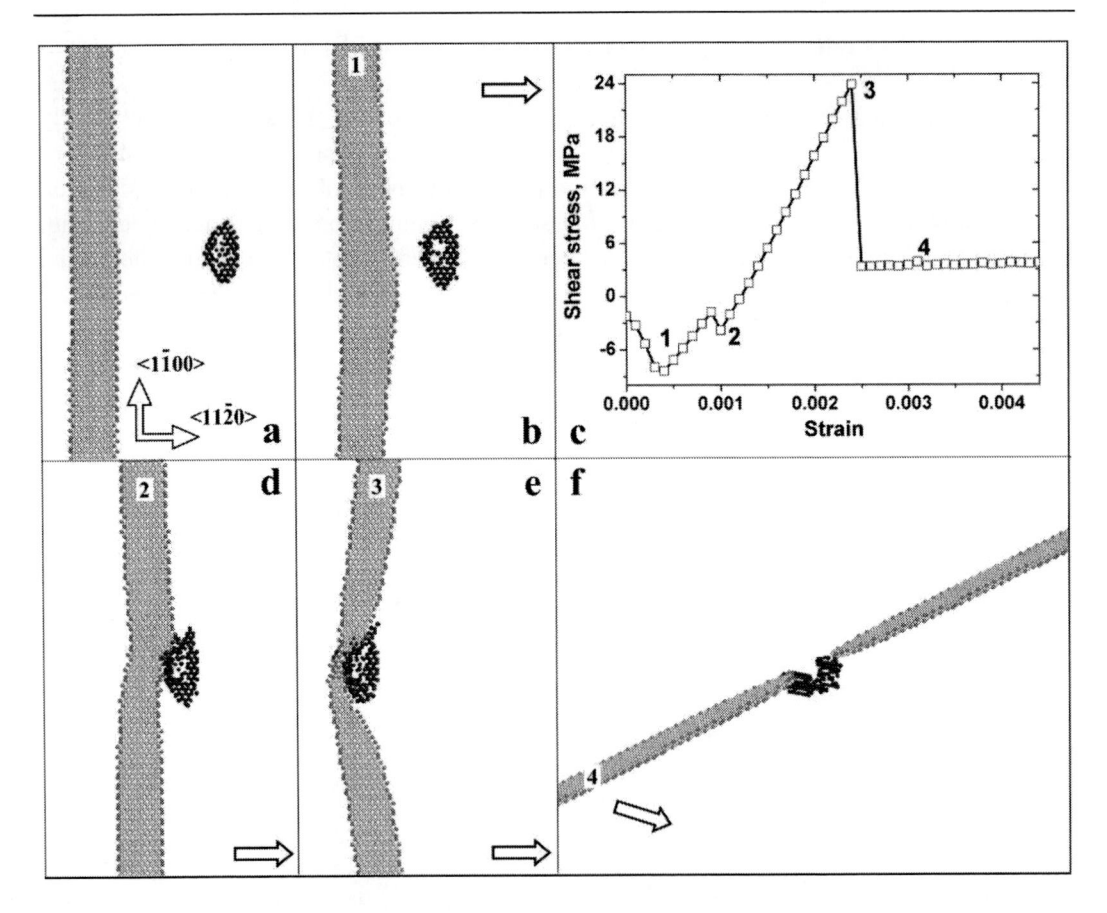

Figure 28. Interaction of the extended 1/3[11$\bar{2}$0] (0001) edge dislocation with irregular 3D SIA cluster. (a) Initial configuration; (b), (d), (e) and (f) configurations corresponding to points marked on the stress-strain curve (c).

## 7.3. Interaction with Prismatic Vacancy Loop

This vacancy cluster is formed by collapse of atoms around a vacancy platelet on a {1$\bar{1}$00} prism plane and consists of a triangular prism of displaced atoms with {11$\bar{2}$0} faces and two (0001) triangle bases, Fig.13. The initial configuration of 42 vacancies and the extended dislocation is shown in Fig.29a. The glide plane of the dislocation cuts the cluster in its centre. As in the case of the irregular SIA cluster above, the dislocation is first attracted by the cluster, but the interaction is significantly stronger, compare Figs.28c and 29c, and leads to noticeable enlargement of the faulted region and bending of the dislocation line towards the cluster. Further strain results in almost linear stress increase up to the CRSS. This vacancy cluster is a stronger obstacle to dislocation motion than the SIA clusters above; the dislocation becomes strongly curved and the CRSS is approximately two times higher, see Figs. 29c and d. The dislocation becomes constricted at the maximum stress, but, in contrast to the SIA cluster in section 7.2, only the half of the vacancy cluster above the glide plane is absorbed by the dislocation, see Figs. 29e and 29f. The jogged dislocation breaks from the remaining vacancies and continues to glide at a stress three times higher than that observed for the irregular SIA cluster, see Fig.28c and 29c.

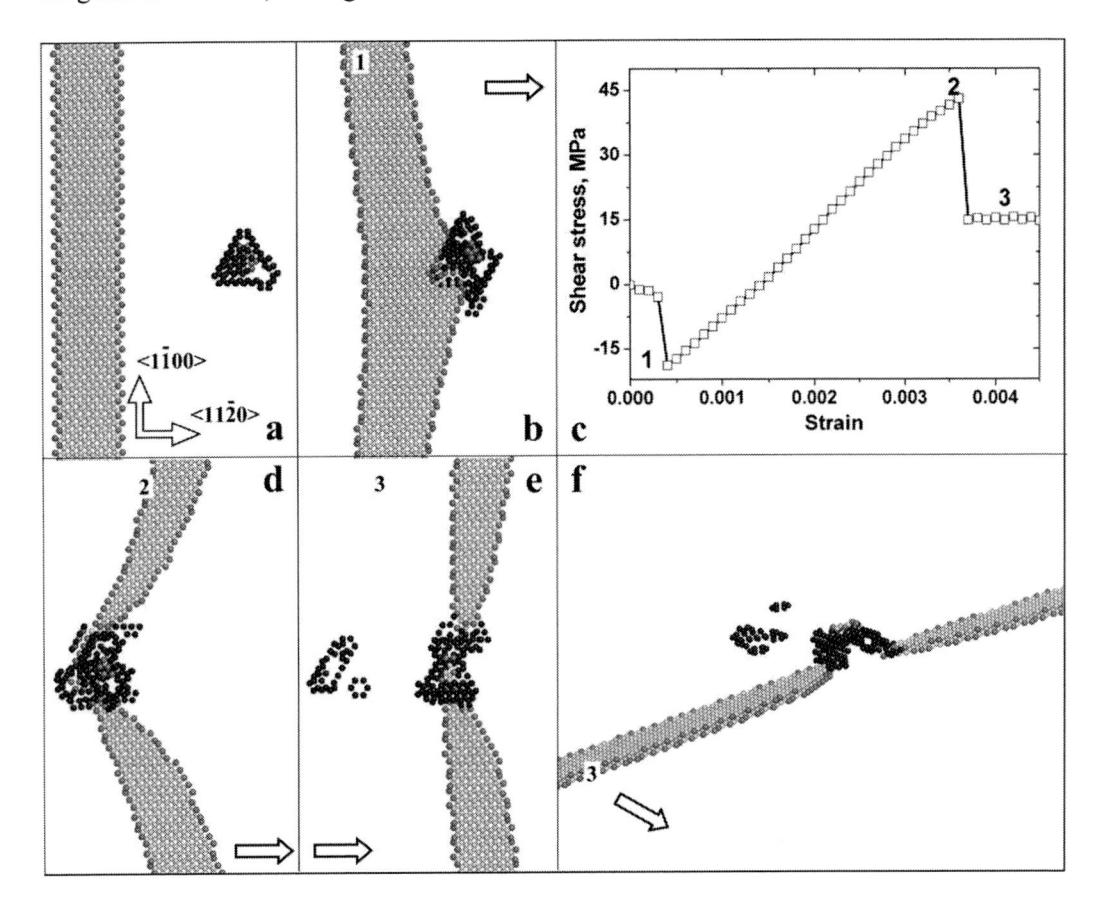

Figure 29. Interaction of the extended 1/3[11$\bar{2}$0] (0001) edge dislocation with prismatic vacancy cluster. (a) Initial configuration; (b), (d), (e) and (f) configurations corresponding to points marked on the stress-strain curve (c).

## 7.4. Interaction with Pyramid Vacancy Cluster

A pyramid vacancy cluster is an embryo of a c-type vacancy dislocation loop and is created by collapse of the lattice towards a basal platelet of vacancies, see section 5.2 and Figs. 17-18. The cluster under consideration here contains 61 vacancies and has its base three atomic planes below the dislocation glide plane. The initial configuration is shown in Fig.30a. The cluster is an obstacle to dislocation glide and shear stress increases up to a first maximum, see point 1 on Fig.30c and Fig.30b. When this first stress peak is achieved the leading partial overcomes the cluster and the trailing partial then encounters it, see Fig.30d. The maximum on the strain-stress curve (point 3) occurs just as the trailing partial breaks away, see Figs.30e and f. Note that both the vacancy cluster and the extended dislocation recover their initial form after breakaway. The obstacle strength of this pyramid vacancy cluster is nearly equal to that of irregular SIA cluster of section 7.2.

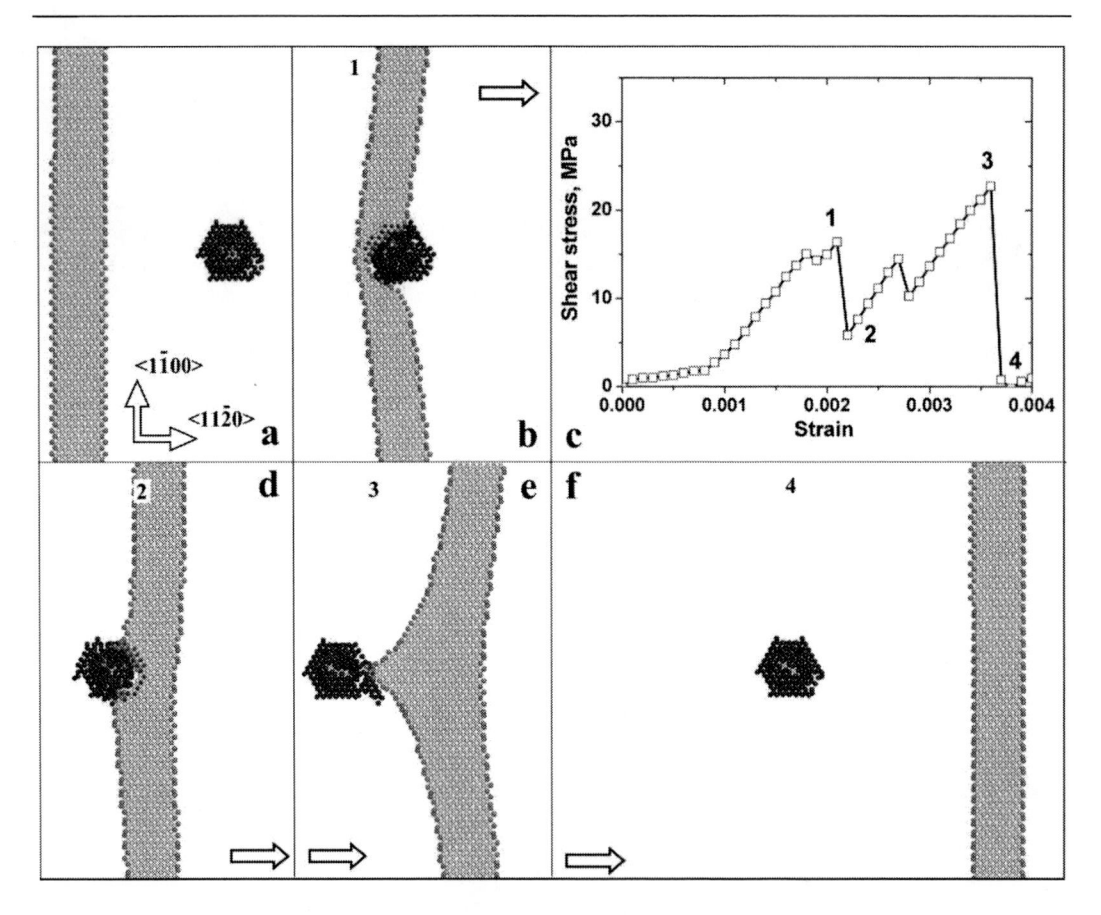

Figure 30. Interaction of the extended $1/3[11\bar{2}0](0001)$ edge dislocation with pyramid vacancy cluster. (a) Initial configuration; (b), (d), (e) and (f) configurations corresponding to points marked on the stress-strain curve (c).

# 8. INTERACTION OF $1/3<[11\bar{2}0](1\bar{1}00)\}$ EDGE DISLOCATIONS WITH POINT DEFECT CLUSTERS FOUND IN COLLISION CASCADES IN α-ZIRCONIUM

The modelling technique applied for studying interaction of $1/3<11\bar{2}0>$ (0001) edge dislocation with radiation defects has been employed for MD modelling of $1/3<11\bar{2}0>\{1\bar{1}00\}$ edge dislocation in HCP zirconium. With the interatomic potential used [8], $1/3<11\bar{2}0>\{1\bar{1}00\}$ edge dislocation does not dissociate in glide plane and according to our preliminary calculations has the Peierls stress of 5.6 MPa. Thermal effects are not considered here and relaxation is carried out using the conjugate gradients method. Following the procedure described in section 7 typical point defect clusters found in the collision cascades are placed in the glide plane of the dislocation. A shear strain is applied in steps of $10^{-4}$, the crystal is relaxed after each step and the resultant applied shear stress is evaluated. A local geometry approach, see section 2.3, we use earlier is employed to visualise and identify the atomic structure of $1/3<11\bar{2}0>\{1\bar{1}00\}$ dislocation core and a defect cluster. A simulation is accomplished when the gliding dislocation overcomes the cluster and the CRSS is taken as the maximum on the stress-strain curve. The model contains more than $3x10^6$ free atoms and the distance between the periodic boundaries along the dislocation line, i.e. the spacing along the row of clusters, is 140b, where b is the length of the Burgers vector.

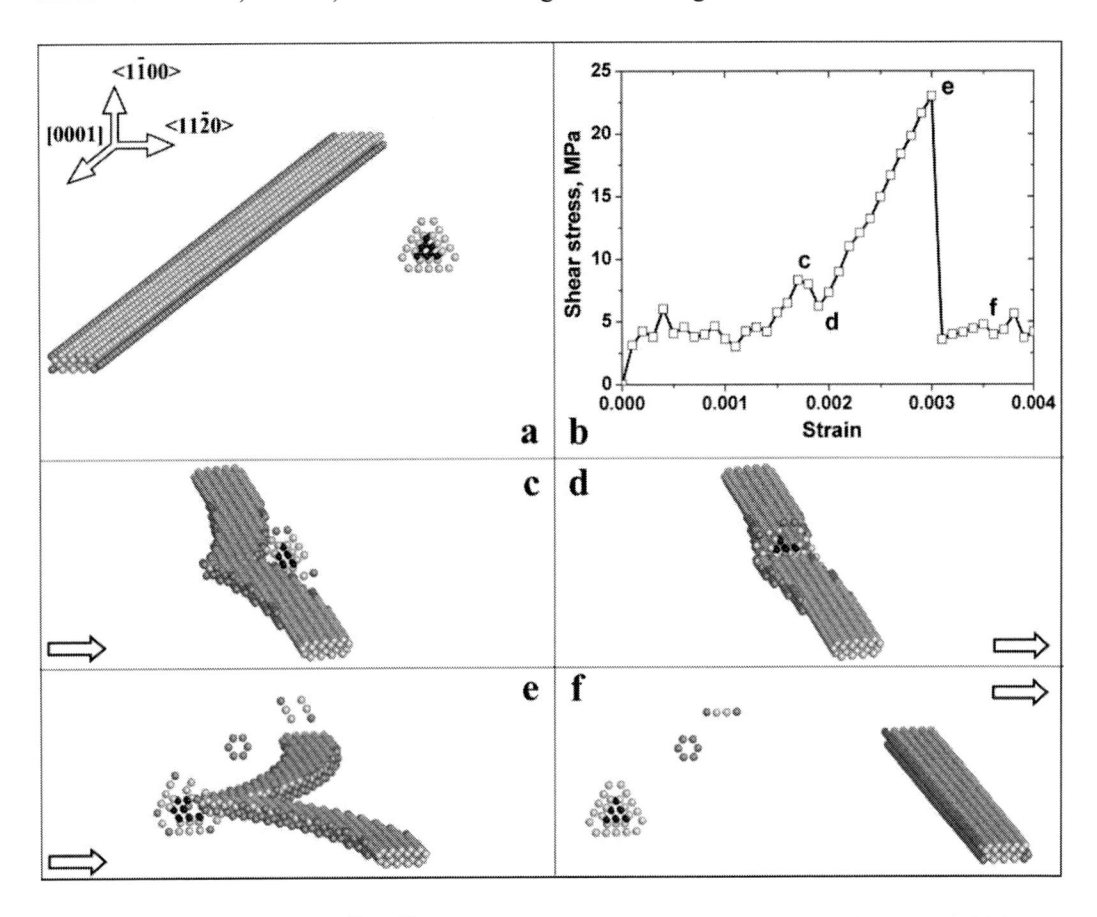

Figure 31. Interaction of $1/3[11\bar{2}0](1\bar{1}00)$ edge dislocation with 5-SIA triangular cluster. (a) Initial configuration; (c-f) typical configurations marked at strain-stress curve (b). The unmarked white arrow here and in other figure shows the direction of dislocation glide.

## 8.1. Interaction with Triangular SIA Cluster in Basal Plane

Triangular clusters of 4, 5 and 6 SIAs, see Fig.12, all lying in one (0001) atomic plane comprise nearly 10% of all SIA clusters created in displacement cascades. Interaction of the dislocation with a 5-SIA cluster (15 atoms sharing 10 lattice sites) lying in a basal plane and intersected through its centre by the dislocation glide plane has been simulated. Fig.31 shows (a) the initial configuration, (b) the stress-strain curve, and (c-f) several snap-shot visualisations taken from the computer modelling. Initially there is a slight repulsion of the dislocation by the cluster that leads to an increase of the flow stress above the Peierls-stress value. With further straining the stress increases to point c on the stress-strain plot and the dislocation then enters the cluster with a stress reduction, see points c-d in Fig.31b and Figs. 31c and 31d. The dislocation line is then pinned by the cluster and bends as further strain leads to near-linear increase of the shear stress up to the CRSS of approximately 23MPa, see Fig.31e. The dislocation passes through the cluster at this stress and both dislocation and cluster recover their initial configuration, compare Figs. 31a and 31f. In addition, the interaction creates one Frenkel pair in the crystal (visualised in Fig.1f as a regular hexagon for the vacancy and a four atom chain for the SIA). The breakaway stress of 23MPa compares with a value of 15MPa for the basal edge dislocation to overcome the same obstacles with spacing of a similar size, see section 7.1 of this review.

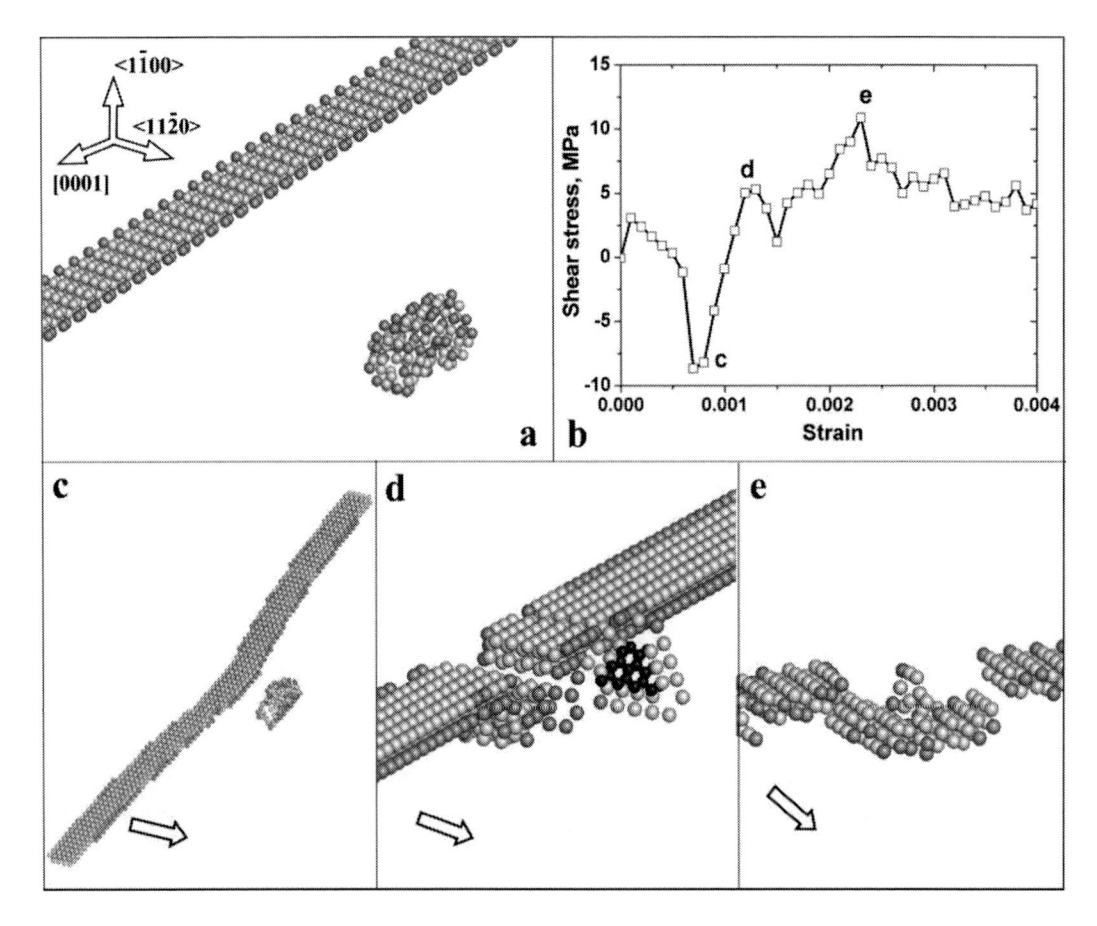

Figure 32. Interaction of $1/3[11\bar{2}0](1\bar{1}00)$ edge dislocation with irregular 3D SIA cluster. (a) Initial configuration; (c-e) typical configurations marked at strain-stress curve (b).

## 8.2. Interaction with Irregular 3D SIA Point Defect Cluster

Typical cluster of this type is shown in Fig.10. It is elongated in the [0001] direction, does not have any ordered internal structure and stable against annealing. The relevant snapshots and the stress-strain curve for interaction between the $1/3<11\bar{2}0>\{1\bar{1}00\}$ edge dislocation and such a cluster are shown in Fig.32. The cluster contains 34 SIAs and is placed in the glide plane so that the moving dislocation crosses it in its centre. The initial configuration is shown in Fig.32a. The dislocation is initially attracted by the cluster and bends towards it, see Fig.32c, with a resultant drop of the shear stress (point c on the strain-stress curve, Fig.32b). Further application of strain leads to increase of shear stress and eventual destruction of the cluster. Part of the cluster is absorbed by climb down of the dislocation and the other part forms into a large triangular SIA cluster in (0001) orientation similar to that considered above, see Fig.32d and compare with Fig.31. Because of the climb of the dislocation in this region, the glide plane of the climbed segment does not intersect the newly-formed triangular cluster. However, as soon as the dislocation core appears below the cluster, the cluster becomes absorbed by the dislocation, see Fig.32e.

The interaction of the prism edge dislocation with the irregular cluster is seen to be rather complex. It passes through several intermediate steps, but the cluster itself is a relatively weak pinning centre for moving $1/3<11\bar{2}0>\{1\bar{1}00\}$ edge dislocation, with the CRSS (point e in Fig,32b) is less than 12MPa. This compares with a value of 24MPa for the basal edge dislocation to overcome the same obstacles with spacing of a similar size, section 7.2.

## 8.3. Interaction with SIA Dislocation Loops

In this section we consider obstacles in the form of SIA dislocation loops with perfect Burgers vector. There are three possible $<11\bar{2}0>$ orientations for **b** of such glissile dislocation loops and the two treated here are those that do not coincide with that of the gliding dislocation, i.e. they have a glide prism with axis at 60° to **b** of the dislocation and adopt a habit plane between $\{11\bar{2}0\}$ (pure edge) and $\{1\bar{1}00\}$.

Fig.33 illustrates the interaction under applied strain between the dislocation and a circular dislocation loop of 117 SIAs. The positive line sense and Burgers vector of the dislocation and loop using the RH/**FS** convention [25] are shown by **L** and **b**, respectively. In this case, **b** of the loop makes an angle of 60° with that of the edge dislocation. The segments of the relaxed loop that lie in a basal plane are dissociated, see Fig.33a. The applied stress increases with increasing strain as the dislocation bows forward on each side of the loop until the maximum of 45MPa at point c on the stress-strain curve (Fig.33b): the shape of the dislocation line and SIA loop at this stress are shown in Fig.33c. The interstitial loop retains its initial form up to this stage but then its Burgers vector changes to that of the dislocation, enabling it to be absorbed by the line to form a pair of super-jogs (and two vacancies are created in the crystal), see Figs.33d and e. The jogged line continues to glide thereafter at a resolved shear stress of about 7MPa.

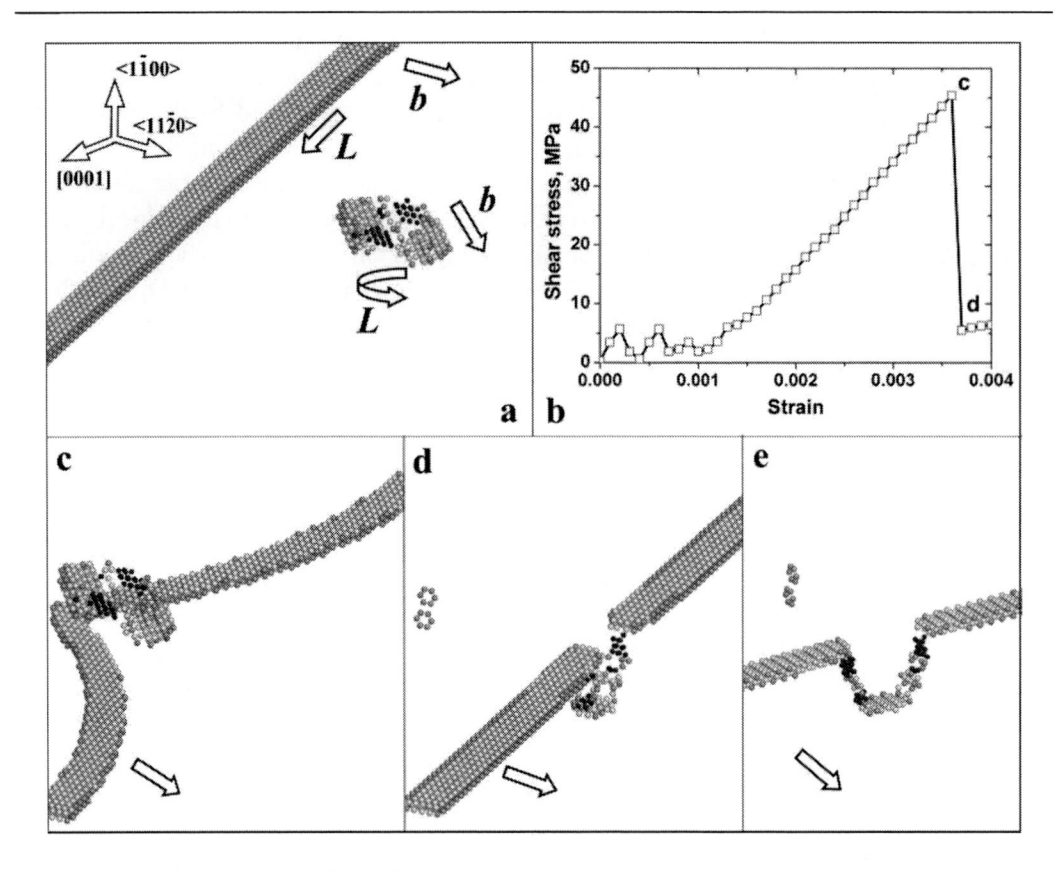

Figure 33. Interaction of $1/3[11\bar{2}0](1\bar{1}00)$ edge dislocation with SIA dislocation loop making angle of $60^0$ with the Burgers vector of dislocation. (a) Initial configuration; (c-d) typical configurations marked at strain-stress curve (b), (e) enlarged and tilted snap-shot shown in (d). The arrows b and L here and in Fig.34 show the Burgers vector and positive line sense consistent with the RH/FS convention [25].

The other loop obstacle considered has the same nature, shape and size as that in the preceding paragraph, but its Burgers vector makes angle of 120° with that of the edge dislocation, see the arrows denoting **b** and **L** in Fig.34a. As a result, the dislocation line interacts with the SIA dislocation loop in a completely different manner and the obstacle strength is 33% higher. Initially it is strongly attracted by the loop, see the strain-stress curve in Fig.34b, and bends towards it. Their contact leads to formation of a segment which, though with **b** of the form $1/3<11\bar{2}0>$, is sessile in the glide system of the dislocation and loop, see the snap-shot in Fig.34c. (The two Burgers vectors in Fig.34 are such that the two parts of line and loop that approach each other cannot form a low-energy segment of this sort.) The sessile segment firmly pins the dislocation line, see the snap-shot in Fig.34d, and the dislocation becomes strongly curved as the strain increases further. Finally, at a stress of 60MPa, the Burgers vector of the loop changes to that of the line. The loop is absorbed as super-jogs, breakaway occurs and the dislocation continues to glide at a resolved shear stress just above the Peierls stress. This dislocation reaction is accompanied by formation of several vacancies (small regular hexagons in the top left corner in Fig.34e).

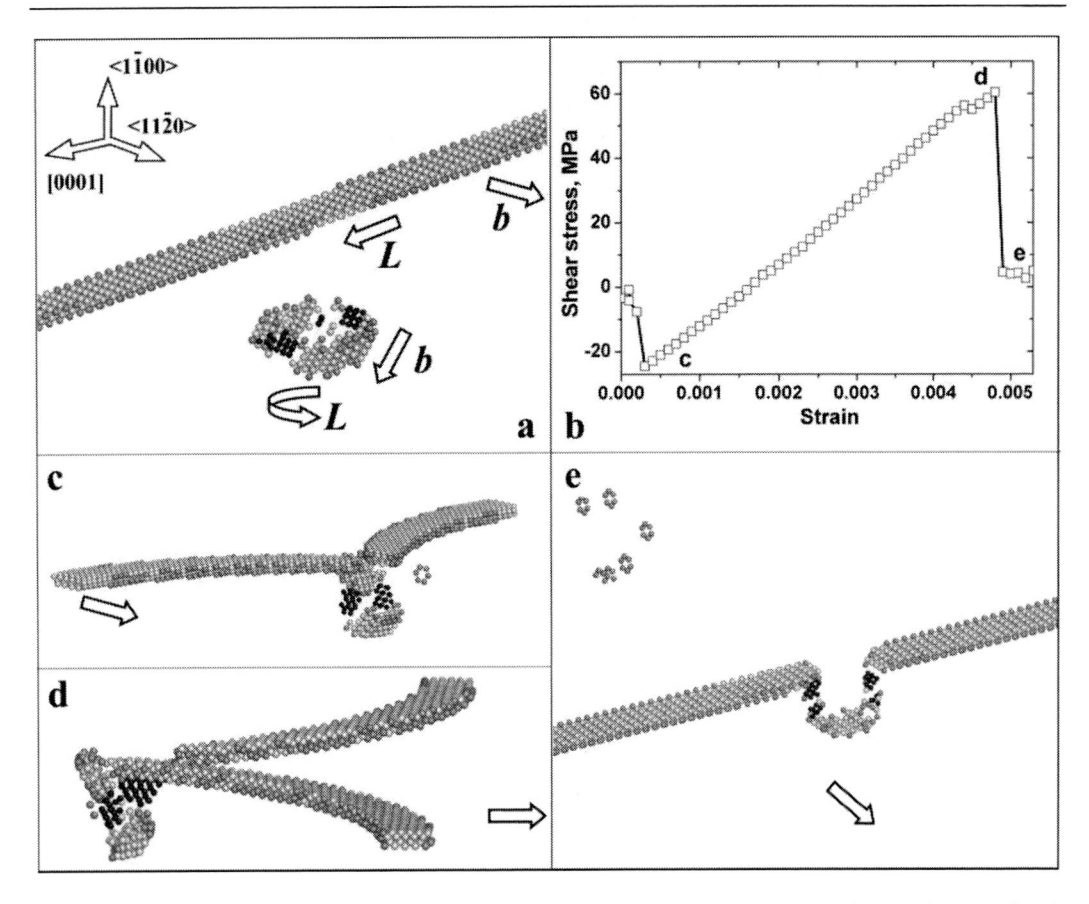

Figure 34. Interaction of 1/3[11-20](1-100) edge dislocation with SIA dislocation loop making angle of $120^0$ with the Burgers vector of dislocation. (a) Initial configuration; (c-e) typical configurations marked at strain-stress curve (b).

# 9. SUMMARY

The results described in this review are the outcome of the largest set of simulations of displacement cascades in zirconium to date. They span ranges of energy and temperature of interest for basic research into the performance of HCP metals in an irradiation environment. The defect number and configuration vary strongly from cascade to cascade, and so we have attempted to generate and analyse a sufficient number of cascades for each condition of $E_{PKA}$ and $T$ to determine data and detect trends with statistical significance. Computational techniques described in Section 2 have been adopted to increase the model size usable for a given cascade energy and to enhance the efficiency of data analysis.

The large-scale MD modelling of displacement cascades in pure zirconium is carried out for three temperatures (100, 300 and 600 K) and four PKA energies (10, 15, 20, 25 keV). At least 25 cascades for each temperature and PKA energy are simulated. The results extend and confirm those of the smaller set of simulations reported earlier in [10].

The total number of Frenkel pairs and fraction of point defects in vacancy and SIA clusters have been determined as functions of temperature and PKA energy. It is established that SIAs have higher affinity to form clusters than vacancies, i.e. the relation $\varepsilon_i / \varepsilon_v \geq 1$ is satisfied for the whole temperature and PKA energy ranges.

SIA clusters of three principal types are identified. The most frequently occurring are in the form of glissile dislocation loops with Burgers vector $1/3<11\bar{2}0>$. They appear at all temperatures and PKA energies and contain from 45 to 70 % of the SIA population in clusters. 3D irregular SIA clusters represent clusters of the second type and contain from 20 to 40 % of clustered SIAs. SIA clusters of the third type are triangular arrangements of SIAs lying in one basal plane, and they account for approximately 10% of the clustered SIAs. All SIA clusters exhibit high thermal stability and are not destroyed during annealing at temperatures up to 1200 K. The interstitial dislocation loops are mobile at all simulated temperatures and can glide rapidly in one dimension along the $<11\bar{2}0>$ direction of their Burgers vector. The 3D irregular clusters are immobile at low to intermediate temperatures and can diffuse in three dimensions at high temperatures. Triangular SIA clusters are immobile at temperatures below 600 K and exhibit planar diffusion in their basal plane at $T \geq$ 600 K.

Vacancy clusters of three general types are created in displacement cascades in pure zirconium. The dominant one is a prismatic vacancy cluster bounded by three $\{1\bar{1}00\}$ faces and capped by (0001) planes. It appears at all temperatures and PKA energies and contains from 50 to 70 % of all vacancies created in vacancy clusters. Clusters of this type are stable against temperature increase and can glide at temperatures near 600 K by conversion to prism-plane vacancy loops. Their glide direction coincides with one of the $<11\bar{2}0>$ crystallographic directions and does not change. However, mobility of these clusters is noticeably suppressed at higher temperatures. Vacancy clusters of the second type have pyramid shape with base in (0001) and four $\{1\bar{1}00\}$ pyramidal faces. They are immobile. At low temperatures these clusters can oscillate by inversion along the [0001] direction. Depending on the size and shape of the cluster, the stable extrinsic fault is formed. Further temperature increase transforms such pyramid-like clusters into prismatic vacancy clusters of the first type. Vacancy clusters of the third type are a mixture of the prismatic cluster and pyramid clusters, with shape and atom arrangement of one part similar to the former and that of the other part similar to the latter. These clusters are immobile at low temperatures and undergo transformation to prismatic vacancy clusters of the first type at high temperatures.

It is interesting to note that the variety of configurations of defect clusters formed in high-energy cascades obtained by this study is much wider than that observed earlier in modelling of high-symmetry vacancy and SIA clusters [9, 13, 27]. Moreover, the mobility properties of such clusters cover a wide range from the one-dimensional mechanism in $<11\bar{2}0>$ directions, to two-dimensional motion in the basal plane and three-dimensional diffusion. The investigation has revealed the relative population of the cascade defects that exhibit these different types of mobility. Some uncertainty in interatomic potentials still remains and more simulations should be done to compare different potentials. However, the results on cluster formation and properties presented here can help in understanding the basic phenomena occurring in anisotropic materials under irradiation in particular the radiation hardening due to pinning of mobile dislocations by residual vacancy and SIA clusters created in displacement cascades.

An auxiliary atomic-scale simulations for edge dislocations of the $<11\bar{2}0>$ (0001) and $<11\bar{2}0>\{1\bar{1}00\}$ slip systems have been conducted for a model of α-zirconium. The edge dislocation in the basal plane dissociates into two partials whereas the prism-plane dislocation remains undissociated. Dissociation results in a larger effective core size but lower core energy, and consequently the Peierls stress for glide at 0K is lower for the basal system by a factor of three, i.e. 1.6MPa compared with 5.6MPa. It should be noted that the Peierls stress is sensitive to the shape of the γ-surface and the level of energy along the displacement path **0** → **b** within the dislocation path. The interatomic potential used here gives a stable fault of relatively low energy (80 mJ/m$^2$) for $I_2$ on (0001) [6], whereas *ab initio* calculations indicate that the maximum along $<11\bar{2}0>$ is lowest on the prism plane [28].

The atomic-scale mechanisms and strengthening effects of dislocation-cluster interactions in an HCP metal are studied by using MD modelling. Static conditions (T = 0K) have been simulated to allow future comparison to be made with research on other metals and elasticity models of obstacle strengthening. In this review, interaction of the $1/3<11\bar{2}0>(0001)$ basal edge dislocation with four different but typical point defect clusters is described and four different interaction mechanisms are discussed. A triangular basal-plane cluster of 5 SIAs is pushed by the dislocation without absorption, whereas an irregular 3D SIA cluster is completely absorbed by the dislocation, which constricted and climbed down to acquire a pair of super-jogs. Only the half of the prism-plane vacancy cluster above the dislocation glide plane is absorbed by dislocation climb. In the case of a (basal) pyramid vacancy cluster, the partials of the dissociated dislocation interacted separately with the obstacle and both the dislocation and cluster remained unchanged after the interaction.

The prismatic vacancy cluster produces the highest critical stress, namely 45MPa for the obstacle size and spacing considered. This is comparable with results found for stacking-fault tetrahedra of comparable size and spacing in the FCC metal copper [29], which are known to produce substantial effects on the yield and flow stress and to produce strain localisation at high density.

Interaction of the $1/3<11\bar{2}0>\{1\bar{1}00\}$ prism-slip edge dislocation with four different but typical self-interstitial clusters has been simulated and different interaction mechanisms are revealed. A triangular basal-plane cluster of 5 SIAs is the only one that recovers its initial structure after the interaction and is not absorbed by the moving dislocation. In the basal-slip case, it is pushed by the dislocation without absorption, but at a stress of 15MPa compared with a dislocation-breakaway stress of 23MPa in this case. The irregular 3D cluster of 34 SIAs is found here to be completely absorbed by the dislocation, which climbed to acquire super-jogs. It is a relatively weak obstacle, giving a critical stress of 12MPa compared with 24MPa for the basal system with similar obstacle spacing.

Interstitial dislocation loops with perfect Burgers vector are found to be relatively strong obstacles to dislocation glide on the prism system. Their strength and interaction mechanism are determined by the orientation of **b** with respect to that of the dislocation. In both cases considered, the loop pins the dislocation before shearing to change its Burgers vector and be absorbed by the line. For the stronger of the two obstacles, a sessile segment first forms by a dislocation reaction: this pins the dislocation line until loop absorption at a critical stress of 60MPa.

# REFERENCES

[1] D.J. Bacon and T. Diaz de la Rubia, *J. Nucl. Mater.* 216 (1994) 275.

[2] D.J. Bacon, A.F. Calder, F. Gao, V.G. Kapinos and S.J. Wooding, *Nucl. Instrum. and Meth. in Phys. Res. B* 102 (1995) 37.

[3] D.J. Bacon, A.F. Calder and F. Gao, *J. Nucl. Mater.* 251 (1997) 1.

[4] D.J. Bacon, F. Gao and Yu.N. Osetsky , *J. Nucl Mater.* 276 (2000) 1.

[5] D.J. Bacon, Yu.N. Osetsky, R. Stoller and R.E. Voskoboinikov, *J. Nucl. Mater.* 323 (2003) 152.

[6] G.J. Ackland, S.J. Wooding, and D.J. Bacon, *Philos.Mag. A* 71 N3 (1995) 553.

[7] J.P. Biersack and J.F. Ziegler, *Nucl. Instrum. and Meth. in Phys. Res.* 194 (1982) 93.

[8] S.J. Wooding, *Computer Simulation of Radiation Damage in HCP metals, Ph.D.Thesis*, The University of Liverpool, 1994.

[9] S.J. Wooding, L. M. Howe, F. Gao, A. F. Calder and D. J. Bacon, *J. Nucl. Mater.* 254, N2-3 (1998) 191.

[10] F. Gao, D. J. Bacon, L. M. Howe and C. B. So, *J.Nucl.Mater* 294 N3 (2001) 288.

[11] N. de Diego and D.J. Bacon, *Philos.Mag. A* 80 (2000) 1393.

[12] Yu.N. Osetsky, N. de Diego and D.J. Bacon, *Met. and Mat. Trans. A* 33 (2002) 777.

[13] N de Diego,N., Yu.N. Osetsky and D.J.Bacon, *Met. and Mat. Trans. A* 33 (2002) 783.

[14] W.L. Wood, *Practical time-stepping schemes* , Clarendon, Oxford (1990).

[15] L.A.Marques, J.E.Rubio, M Jaraiz, L.Enriquez, J. Barbolla, *Nucl. Instrum. and Meth. in Phys. Res. B* 102 (1995) 7.

[16] D. Frenkel and B. Smit, *Understanding Molecular Simulation. From Algorithms to Applications,* Academic Press, 1996.

[17] R. E. Stoller and A. F. Calder, *J.Nucl.Mater.* 283-287 (2000) 746.

[18] F. Gao, D. J. Bacon, P. E. J. Flewitt and T. A. Lewis, *J.Nucl.Mater.* 249 N1 (1997) 77.

[19] R.E.Voskoboinikov, Yu.N.Osetsky, D.J.Bacon, *J. Nucl. Mater. 377* (2008) 385.

[20] R.E.Voskoboinikov, Yu.N.Osetsky, D.J.Bacon, *ASTM International* 2 N8 (2005) 12410.

[21] D.J. Bacon, *J.Nucl.Mater.* 159 (1988) 176.

[22] F.Omnis, I.Monnet, J.L.Bechade, C.Prioul, P.Pilvin, *J.Nucl.Mater.* 328 (2004) 165.

[23] Yu.N.Osetsky, D.J.Bacon, *Modell. Simul. Mater. Sci. Eng.* 11 (2003) 427.

[24] F.C.Frank, *Philos.Mag.* 42 (1951) 809.

[25] D.Hull and D.J.Bacon, *Introduction to Dislocations*, Butterworth-Heinemann, 2001.

[26] J.P.Hirth, J.Lothe, *Theory of Dislocations*, Krieger, 1992.

[27] V.G. Kapinos, Yu.N._Osetsky and P.A. Platonov, *J.Nucl. Mater.* 184 (1991) 211.

[28] C.Domain, R.Besson and A.Legris, *Acta Mater.* 52 (2004) 1495.

[29] Yu.N.Osetsky, R.E.Stoller, Y.Matsukawa, *J. Nucl. Mater.*, 329-333 (2004) 1228.

# INDEX

## D